Nitrogen Metabolism in Rice

Nitrogen Metabolism in Rice

P. Basuchaudhuri

M.Sc(Ag), Ph.D.
Formerly, Senior Scientist
Indian Council of Agricultural Research
Santoshpur, Kolkata, India

CRC Press is an imprint of the
Taylor & Francis Group, an **informa** business
A SCIENCE PUBLISHERS BOOK

CRC Press
Taylor & Francis Group
6000 Broken Sound Parkway NW, Suite 300
Boca Raton, FL 33487-2742

Printed on acid-free paper
Version Date: 20160125

International Standard Book Number-13: 978-1-4987-4667-0 (Hardback)

Library of Congress Cataloging-in-Publication Data

Names: Basuchaudhuri, P., author.
Title: Nitrogen metabolism in rice / author: P. Basuchaudhuri.
Description: Boca Raton, FL : CRC Press, 2016. | Includes bibliographical, references and index.
Identifiers: LCCN 2016000608 | ISBN 9781498746670 (hardcover : alk. paper)
Subjects: LCSH: Nitrogen in agriculture. | Rice-- Fertilizers. | Nitrogen-- Metabolism. | Plants-- Effect of nitrogen on.
Classification: LCC S587.5.N5 B38 2016 | DDC 631.8/4-- dc23
LC record available at http://lccn.loc.gov/2016000608

Visit the Taylor & Francis Web site at
http://www.taylorandfrancis.com

and the CRC Press Web site at
http://www.crcpress.com

Dedication

To my loving
Granddaughter
PROITI

Preface

Rice is a staple food for half of the world population. Its total production is next to wheat production. Rice is very much consumed in Asian countries. Hence, cultivation of rice is extensive in these countries. Though the land area under rice is decreasing slowly, the productivity is to increase, to feed the growing population, with the high yielding super rice varieties. Furthermore, high yielding climate resilient varieties are now also available for the service of farmers. The high yield of rice is associated with the high demand of plant nutrient fertilizer, especially nitrogenous fertilizer. The consumption of nitrogenous fertilizer has shown a gradual increase over the years. Meanwhile, The cost of fertilizer is also increasing. At this juncture it is important to increase the nitrogen use efficiency. This will also enable us to reduce the environmental pollution globally. This could only be achieved by efficient fertilizer management practices and better understanding of nitrogen utilization in plant system so that a high production and high quality harvest can be achieved. The better understanding of nitrogen utilization may help in breeding efficient rice varieties, mutants and transgenic rice. So, it is necessary to actively consider all aspects, namely management practices, understanding of biochemical and physiological aspects of nitrogen assimilation and crop improvement in rice for efficient utilization of nitrogenous fertilizer.

Rice is a model monocotyledonous plant as it is of small genome. Much of rice gene data are now available and utilized for crop improvement. However, under varied climatic situations and varying biotic–abotic stress conditions rice crop improvement is in progress. Nitrogen use efficiency in rice, i.e. nitrogenous fertilizer required per kilogram of rice grain production, is similarly important and has received much attention.

Rice grain quality is considered mainly on the basis of protein composition and the availability of essential amino acids. For human consumption, thus, proteins along with energy supplying carbohydrate in high quality rice are important. Therefore, manipulation in crop improvement is also necessary to elucidate high quality protein-rich grains as the high yield grains. Hence, nitrogen efficient high yield, high quality rice plant is a goal for coming days.

The present book discusses, in details, different facets of nitrogen metabolism in rice plant. The metabolic pathways, ontogenic changes, biotic and abiotic factor influences and genetic progress related to nitrogen metabolism of rice is discussed in a simple way. Students, researchers and teachers will find the book useful for having details of rice nitrogen metabolism with the hope that it will provide the basic as well as current concept of the subject to further research so that rice crop improvement can be achieved with high nitrogen use efficiency.

I am deeply indebted to my family members for their support and cooperation.

September 2015

P. Basuchaudhuri
Kolkata

Contents

chapter one

Introduction

Nitrogen is the essential primary nutrient for plant growth and development. Amongst the nutrients nitrogen requirement is high after carbon, hydrogen and oxygen. In air, nitrogen is the highest percent present. However, gaseous nitrogen is not generally available to plant except by micro organisms. Hence, nitrogen necessary should be made available in ionic form. Naturally, nitrogen is available in soil by decomposition of organics. But with the time and growth, demand for high growth and yield of plants and depletion of organics in soil the nitrogen availability in soil becomes insufficient. Therefore, addition of inorganic fertilizer becomes imperative as the deficiency symptom of nitrogen is predominant in plants. The deficiency of nitrogen has very much detrimental influence on growth and development. This influence of nitrogen nutrition is most vital and its maintenance requires judicious application of nitrogenous fertilizer along with other nutrient fertilizers, though organic matters deposited in the soil after decomposition provide a significant amount of nitrogen for plants.

It is well known that in nature there is a nitrogen cycle. The nitrogen cycle is the process by which nitrogen is converted in to its various chemical forms. This transformation can be carried out through both biological and physical processes. Important processes in the nitrogen cycle include fixation, ammonification, nitrification and denitrification. The majority of Earth's atmosphere (78%) is nitrogen, the largest pool of nitrogen. However, atmospheric nitrogen has limited availability for biological use, leading to a scarcity of usable nitrogen in many types of ecosystems. The nitrogen cycle is of particular interest to ecologists because nitrogen availability can affect the rate of key ecosystem processes, including primary production and decomposition. Human activities such as fossil fuel combustion, use of artificial nitrogen fertilizers and release of nitrogen in waste water have dramatically changed the global nitrogen cycle.

Most of the organic matter on earth is manufactured by plants through the assimilation of inorganic carbon and nitrogen from the environment into organic molecules, in process driven by sunlight energy. These processes, photosynthesis and nitrogen assimilation, are essential not only for plants themselves but for all other forms of life— animals, fungi, and most bacteria— which obtain their carbon and nitrogen only from organic compounds and are thus completely dependent on plants for their nutrition.

Producing more grain means more fertilizer demand. Global total nitrogen consumption has grown modestly linearly, while consumption of phosphorus and potash has been relatively stable since past decades. At a global scale, cereal yields and fertilizer nitrogen consumption have increased in a near-linear fashion during the past 40 years and are highly correlated with one another. However, large differences exist in historical trends of nitrogen fertilizer usage and nitrogen use efficiency among regions, countries and crops. The reasons for these differences must be understood to estimate future nitrogen fertilizer requirements. Global nitrogen needs will depend on (i) changes in cropped cereal area and the associated yield increases required to meet increasing cereal demand from population and income growth, and (ii) changes in nitrogen use efficiency at the farm level. Analysis indicates that the anticipated 38% increase in global cereal demand by 2025 can be met by a 30% increase in nitrogen use on cereals, provided that the steady decline in cereal harvest area is halted and the yield response to applied nitrogen can be increased by 20%. If losses of cereal cropping area continue at the rate of the past 20 years (–0.33%) per year and nitrogen use efficiency cannot be increased substantially, a 60% increase in global nitrogen use on cereals would be required to meet cereal demand.

Interventions to increase nitrogen use efficiency and reduce nitrogen losses to the environment must be accomplished at the farm or field scale through a combination of improved technologies and carefully crafted local policies that contribute to the adoption of improved nitrogen management; uniform regional or national directives are unlikely to be effective at both sustaining yield increases and improving nitrogen use efficiency. Examples from several countries show that increases in nitrogen use efficiency at the rates of 1% per year or more can be achieved if adequate investments are made in research and extension. Failure to arrest the decrease in cereal crop area and to improve nitrogen use efficiency in the world's most important agricultural systems will likely cause severe damage to environment at local, regional, and global scales due to a large increase in reactive nitrogen load in the environment (Dobermann and Cassman, 2005).

To determine the optimal amount of nitrogen fertilizer for achieving a sustainable rice production at the Taihu Lake region of China, two-year on farm field experiments were performed at four sites using various nitrogen application rates. The results showed that 22-30% of the applied nitrogen was recovered in crop and 7-31% in soils when nitrogen was applied at the rates of 100-350 kg per ha. Nitrogen losses increased with nitrogen application rates, from 44% of the applied fertilizer nitrogen at the rate of 100 kg nitrogen per ha to 69% of the nitrogen applied at 350 kg nitrogen per ha. Ammonia volatilization and apparent denitrification were the main pathways of nitrogen losses. The nitrogen application rate of 300 kg nitrogen per ha, which is commonly used by local farmers in the study region, was found to lead to a significant reduction in economic and environmental efficiency. Considering the cost for mitigating environmental pollution and the maximum net economic income an application rate of 100-150 kg nitrogen per ha would be judicious. This nitrogen application rate could greatly reduce nitrogen loss from 199 kg nitrogen per ha occurring at the nitrogen application rate of 300 kg nitrogen per ha to 80-110 kg nitrogen per ha, with the rice grain yield still reaching 7,300-8,300 kg. dry weight per ha (Deng et al. 2012).

The rice crop removes large amounts of nitrogen for its growth and grain production. The estimated amount of nitrogen removal ranges from 16 to 17 kg for the production of one ton of rough rice, including straw. Total nitrogen uptake by rice plant per hectare varies among rice varieties. Most of the rice soils of the world are deficient in nitrogen, and biological nitrogen fixation by Cyanobacteria and diazotrophic bacteria can only meet a fraction of the nitrogen requirement. Fertilizer nitrogen applications are thus necessary to meet the demand. Generally urea is the most convenient nitrogen source for rice. The efficiency of the urea-nitrogen in rice culture is very low, around 30-40%, in some cases even lower. The low nitrogen use efficiency is due to mainly ammonia volatilization, denitrification, leaching and runoff losses. However, the magnitude of nitrogen loss varies depending on environmental conditions and management practices.

About 85% of the world's rice cropped area is under wetland culture. In the wetland rice soils, rice plants take nitrogen mainly as ammonium requiring less energy to assimilate into amino acids, than nitrate. Ammonium fertilizers dissociate directly to ammonium ion while used and may decompose by catalytic hydrolysis to produce ammonium ion. Nitrates are also produced in the rice soil and taken up by the plant; however, the amount is less.

Salts of various inorganic ions are essential for plant growth and must be accumulated from the surrounding solution. They generally dissociate to ions that are freely mobile in solution, and in aquatic environment the solutions generally are stirred and nutrient ions reach the plant readily. In soil the stirring does not occur and ions move to the root by a combination of bulk flow and diffusion. Bulk flow occurs as ions are carried along water traveling to the root surface of the transpiring plants. Diffusion occurs as gradients in ion concentrations from the solution changes, when the ions accumulate or deplete next to the surface of root according to their rate of uptake compared to water. Inside the plant, the rate of ion uptake is usually independent of the rate of water uptake, but very slow water entry can decrease ion entry apparently because the ion concentration in the root xylem becomes too high. Dehydration of the soil also decreases the rate of ion uptake but the rate more or less remain in balance with the decrease in dry mass occurring at the same time. Hence, accumulated concentrations do not change much in the tissue with change in water supply.

Plants obtain the major elements that make up the plant body–carbon, oxygen, hydrogen and nitrogen–mainly as carbon dioxide, water and nitrate/ammonium. They also take up and use many other minerals and elements, in much smaller quantities. These means of nutrition, from inorganic compounds, is known as autotrophy. When organisms obtain their carbon and nitrogen only from organic compounds it known as heterotrophy. Photosynthesis and nitrogen assimilation are the two basic systems in plant which are unique in nature and very much important for the growth and development. Enzymes or block of enzymes are the key to the all metabolic processes in plant. Enzymes are formation of carbon-nitrogen skeletons through photosynthesis and nitrogen metabolism. In non-legumes nitrogen is acquired mostly by uptake of NO_3 which is reduced first to NO_2 and then NH_3. The first reduction is catalyzed by nitrate reductase, which is synthesized in presence of NO_3 and the second reduction is catalyzed by nitrite reductase, which is a constitutive enzyme in plastids. Nitrate reduction decreases during dehydration mostly because NO_3 is transported to the sites of nitrate reductase synthesis more slowly in the transpiration stream because of the decreased uptake of NO_3 by the roots. The decreased flux of NO_3 decreases the synthesis of the enzyme, and the natural degradation of the enzyme in the cell depletes the cell reductase activity. Thus, the decreased NO_3 flux is a signal from the roots to the shoots that controls this aspect of shoot metabolism. However, in rice ammonium taken up is assimilated in the ammonium assimilation process avoiding the reduction processes. Subsequently, the next major process is transamination to synthesize different amino acids with varied efficiency depending on the availability of substrates, nutrients and other metabolic processes and the environment. Amino acids polymerize to proteins depending on the demand and the prevalent conditions. Protein synthesis is also, in general, inhibited by dehydration. The inhibition is not the same for each protein. The inhibition is not caused by losses in messenger RNA or increases in the plant growth regulator abscisic acid. Decreased enzyme activities in the cell appear to result in part from this inhibition of protein synthesis followed by a decline in activity determined by the half-life of the enzyme in the cell. The regulation of nitrate reductase activities indicates that activities respond to regulator pools that often are supplied by other parts of the plant or even the soil. The changes thus depend on root/shoot signals of a specific molecular nature for each enzyme system. The control of these enzymes contains the basic components of any feedback control system that stress effects on biochemistry will be understood only in the whole plant context.

The metabolic networks responsible for the assimilation and utilization of carbon, nitrogen and other elements are highly regulated.

Regulation allows the integration and co-ordination of processes as well as rapid responses to environmental changes that directly influence plant metabolism, such as temperature, light intensity, and water availability. As the nitrogen use efficiency is the key factor for crop improvement in agriculture, the same is applicable for rice. It is necessary to view upon both genetic and management aspects. Hence, improvement of metabolic systems in nitrogen assimilation will no doubt also improve the efficiency of nitrogen use.

With the above in view an attempt has been taken up to describe systematically the facts of nitrogen metabolism in rice in a simple format in chapters with the current concepts and illustration. Rice is a model monocotyledonous plant; hence, this will serve the concepts in other monocotyledonous plants to some extend. Rice is the major food crop for half of the world's population and any attempt to improve the productivity and nitrogen use efficiency will trigger the new avenue to the goal of feeding millions.

References

Deng, M.H., X.J. Shi, Y.H. Tian, B. Yin, S.L. Zhang, Z.L. Zhu and S.L. Kimura. 2012. Optimizing nitrogen fertilizer application for rice production in the Taihu Lake region,China. *Pedosphere,* **22**: 48-57.

Dobermann, A. and K.G. Cassman. 2005. Cereal area and nitrogen use efficiency are drivers of future nitrogen fertilizer consumption. *Sci. China C Life Sci.* **48**: 745-758.

chapter two

Nitrogen Nutrition

The majority of Earth's atmosphere (78%) is nitrogen, making it the largest pool of nitrogen. However, atmospheric nitrogen has limited availability of biological use, leading to a scarcity of usable nitrogen in many types of ecosystems. Nitrogen is essential for many processes and is crucial for any life on Earth. It is a component in all amino acids, as incorporated into proteins, and is present in the bases that make up nucleic acids, such as RNA and DNA. In plants, much of nitrogen is used in chlorophyll molecules, which are essential for photosynthesis and further growth (Smil, 2000). Chemical processing or nitrogen fixation is necessary to convert gaseous nitrogen into forms usable by living organisms, which makes nitrogen a crucial component of food production.

2.1. Nitrogen in Soil

In soil nitrogenous compounds are inherently available due to decomposition of organics by microorganisms or by the fixation of gaseous nitrogen by microorganisms. As the soil types may be different depending on the ecosystem, the nitrogen content of the soil varies widely. One of the differences between acidic and neutral or alkaline soils is that in acidic soils the prominent form in which nitrogen is available to plants through the microbial mineralization of organic soil matter is, in general, the ammonium ion, and in neutral or alkaline soils it is nitrate; the reason of this is that the activity of nitrifying bacteria is reduced in acidic mineral soils.

It appears calcifuges plants grow better in NH_4-N form whereas calcicoles grow better under NO_3-N form. There was no indication of nitrogen deficiency in adversely affected plants or lack of nitrate reductase systems in the calcifuges species. There was no apparent antagonism for uptake between nitrate and phosphate ions but there were strong indications of antagonisms between ammonium and potassium ions (Gigon and Rorison, 1972).

Nutrient uptake in the soil is achieved by cation exchange, where root hairs pump hydrogen ions into the soil through proton pumps. These hydrogen ions displace cations attached to negatively charged soil particles so that the cations are available and taken up by roots. It is generally stated that the uptake of anions are more readily absorbed by plant roots when associated with monovalent cations than with di or polyvalent cations. It is well known that anion uptake is related to the expense of energy.

According to Epstein (1972) the criteria for an element to be essential for plant growth are :

(1) In the absence, the plant is unable to complete a normal life cycle or
(2) That the element is part of some essential plant constituent or metabolite.

2.2. *Nitrogen in Plants*

Nitrogen is an essential component of all proteins. Nitrogen deficiency most often results in stunted growth, slow growth, and chlorosis. Nitrogen deficient plants will also exhibit a purple appearance on the stems, petioles and underside of leaves from an accumulation of anthocyanin pigments. Most of the nitrogen taken up by plants is from the soil in the forms of NO_3, although in acid environments such as boreal forests where nitrification is less likely to occur, ammonium NH_4^+ is more likely to be the dominating source of nitrogen. Amino acids and proteins can only be built from NH_4^+ so NO_3^- must be reduced. Under many agricultural settings, nitrogen is the limiting nutrient of high growth. Some plants require more nitrogen than others, such as corn. Because nitrogen is mobile, the older leaves exhibit chlorosis and necrosis earlier than the younger leaves. Soluble forms of nitrogen are transported as amines and amides (Huner and Hopkins, 2008).

Nitrogen increases plant height, panicle number, leaf size, spikelet number, and number of filled spikelets, which largely determine the yield capacity of a rice plant. Panicle number is largely influenced by the number of tillers that develop during vegetative stage. Spikelet number and number of filled spikelets are likely determined in the reproductive stage. Farmers use split applications for nitrogen. The number and rate of application can be varied. Ability to adjust number and rate allow the synchronization to real time demand by the crop. The initial symptom of nitrogen deficiency in rice is a general light green to yellow color of the plant. It is first expressed in the older leaves because nitrogen is translocated within the plant from the older leaves to the younger ones. Prolonged nitrogen deficiency causes plant stunting, reduced tillering and yield reduction.

Models and experimental studies of the rhizosphere of rice plants growing in anaerobic soil show that two major processes lead to considerable acidification (1-2 pH units) of the rhizosphere over a wide range of root and soil conditions. One is generation of H^+ ion in the oxidation of ferrous ion by O_2 released from the roots. The other is release of H^+ from roots to balance excess intake of cations over anions, N being taken up chiefly as NH_4^+. CO_2 exchange between the roots and soil has a much smaller effect. The zone of root influence extends a few mm from the root surface. There are substantial differences along the root length and time. The acidification and oxidation cause increased sorption of NH_4^+ ions on soil solids, thereby impeding the movement of N to absorbing root surfaces. But they also cause solubilization and enhance uptake of soil phosphate (Kirk, 1993). Nitrogen fixing bacteria in the rhizosphere of the rice plant exhibit diurnal cycles that mimic plant behavior, and tend to supply more fixed nitrogen during growth stages when the plant exhibits a high demand for nitrogen (Sims and Dunigan, 1984). Rice roots require the presence of oxygen in the rhizosphere, which protects them from the toxic effect of reduced compounds like H_2S, CH_4 etc. Oxidation by roots of surrounding soil in which reducing processes predominate is an adaptive reaction characteristic of rice. Free iron and manganese, which can be present in the soils of rice field in abundance play the same role. The presence of their layer of ironoxide covering rice roots has an effect on their oxidative processes (Bai et al. 2008).

2.3. *Ammonium or Nitrate*

Nitrogen is one of the essential macronutrients for rice growth and one of the main factors to be considered for developing a high yielding rice cultivar. In a paddy field, ammonium (NH_4^+) rather than nitrate (NO_3^-) tends to be considered the main source of nitrogen for rice (Wang et al. 1993). However, in recent years, researchers have paid more and more attention to the partial NO_3^- nutrition (PNN) of rice crops, and their results have shown that lowland rice was exceptionally efficient in absorbing NO_3^- formed by nitrification in the rhizosphere (Kirk and Kronzucker, 2005; Duan et al. 2006). Rice roots can aerate the rhizosphere by excreting oxygen (O_2). Kirk (2001) reported that substantial quantities of NO_3^- were produced in the rhizosphere of rice plants through nitrification, and microbial nitrification was partially responsible for maximum overall rate of microbial O_2 consumption. Most recently, using model calculations and experiments, Kirk and Kronzucker (2005) and Kronzucker et al. (1999, 2000) concluded that NO_3^- uptake by low land rice might be far more important than was previously thought; its uptake rate could be comparable with that of NH_4^+ and it could amount to one third of the total N absorbed by rice plants. Therefore, although the predominant species of mineral nitrogen in bulk soil for paddy rice fields is likely to be NH_4^+, rice roots are actually exposed to a mixed nitrogen supply in the rhizosphere (Briones et al. 2003; Li et al. 2003).

When rice plants in solution culture were fed with a mixture of NH_4^+ and NO_3^- compared with either of the nitrogen sources applied alone at the same concentration, yield increase of 40-70% were observed (Heberer and Below, 1989; Qian et al. 2004). The growth and the nitrogen acquisition of rice were significantly improved by the addition of NO_3^- to nutrition solution with NH_4^+ alone (Cox and Reisenauer, 1972; Duan et al. 2006). The increased nitrogen acquisition could be attributed to the increased influx of NH_4^+ by NO_3^- (Kronzuker et al. 1999), NH_4^+ is taken up by plant roots through ammonium transporters (AMTS).

Rice is known as ammonium (NH_4^+) tolerant species. Nevertheless, rice can suffer NH_4^+ toxicity, and excessive use of nitrogen (N) fertilizer has increased NH_4^+ in many paddy soils to levels that reduce vegetative biomass and yield. Ammonium (NH_4^+), one of the two inorganic nitrogen sources used by plants (NH_4^+ and NO_3^-), is beneficial for plant growth under many circumstances and indeed serves as a ubiquitous intermediate in plant metabolism (Glass et al. 1992). Its assimilation is simple and has been shown to act as an inducer of resist further more entails lower energy costs compared to NO_3^- (Mehrer and Mohr, 1989). Additionally, studies have shown that NH_4^+ can improve the capacity to tolerate water stress in rice in combination with NO_3^- (Guo et al. 2007). Nevertheless, NH_4^+ frequently reaches levels in soils that affect plant growth negatively. These negative affects manifest in stunted root growth (Table 2.1), yield depression and chlorosis of leaves (Britto and Kronzucker, 2002; Balkos et al. 2010; Li et al. 2010). However higher plants display widely differing responses to NH_4^+ nutrition (Marchner, 1995) and, accordingly, can be divided in to tolerant and sensitive species (Britto and Kronzucker, 2006). Despite its reputation as NH_4^+ tolerant species, rice can be affected negatively by high NH_4^+, particularly at low K^+ (Balkos et al. 2010) which in turn may be relieved by elevated K^+ similar to conclusions reached in Arabidopsis. Several studies have shown declines in K^+ bearing clay minerals over extended cultivation periods in many rice growing areas of China.

TABLE 2.1 Root morphology of W23 and GD supplied with normal and high (15 molm^{-3}) NH_4^+

Cultivar	NH_4^+ level (N)	Total length (cm)	Volume (cm^3)	Surface area (cm^2)	Average diameter (cm)	Tip number
W23	CK	229±35.8	2.37±0.08	82.5±7.62	1.16±0.07	170±14.6
	15 molm^{-3}	220±19.2	2.59±0.31	84.3±3.63	1.23±0.12	172±49.9
GD	CK	362±44.8	2.75±0.26	112±8.73	0.98±0.06	185±30.6
	15 molm^{-3}	282±13.9	2.68±0.31	94.0±3.92	1.08±0.07	189±15.0

2.4. *Nitrogen Uptake*

Following the broadcasting of fertilizer on the rice field flood water, the concentration of N in the flood water and soil solution near it are initially so large that rates of uptake are not limited by root properties. However, after the N in the flood water has been exhausted, whether by uptake or gaseous loss, the crop relies on N in the soil, and there the concentration in solution is much smaller because NH_4^+ cation – the main form of plant available N – is absorbed on soil clays and organic matter. Therefore, rates of N transport to and absorption by root surfaces may limit N acquisition. In addition, morphological and physiological adaptations to anoxic soil conditions may affect the root's N absorption capacity. An important difference is that the concentration of N as NH_4^+ in the soil solution of a flooded soil is likely to be one to two orders of magnitude smaller than that of NO_3^- in a dry soil with an equal quantity of N (Kirk, 2001).

Basuchaudhuri and Dasgupta (1983), working with two high yielding indica rice varieties viz. Sashyashree and Jaya on utilization of major nutrients, noted that while considering the nutrient concentrations in different organs of the rice cultivars at various stages of growth, it is evident that nitrogen concentrations in different plant parts, decrease with ageing. However a sharp decline in nutrient concentrations in leaves and stem was noted during ripening. Reduction in concentration of nutrients in plant parts, with time may possibly be attributed to slow rate of uptake along with dilution effect caused by gradient movement of nitrogen to the developing grains (Ishizuka, 1965).

TABLE 2.2 Concentration of nitrogen in rice varieties

Variety	Plant part	Stage of growth		
		Panicle initiation	Flowering	Maturity
Sashyashree	Culm & Sheath	2.40	2.20	0.65
	Leaf blade	3.15	3.00	0.97
	Panicle & Grain	—	1.35	1.30
Jaya	Culm & Sheath	2.25	2.05	0.60
	Leaf blade	3.00	2.80	0.80
	Panicle & Grain	—	1.10	1.25

Amongst plant parts, the leaf blades exhibited maximum concentration of nitrogen at panicle initiation stage (Table 2.2). It may be argued that the leaf blades are the active centers of physiological activity; the nitrogen is initially translocated to the leaves, where assimilation, photosynthesis and other interrelated metabolic processes take place. On the otherhand, at maturity, the concentration of nitrogen was more in panicles and grains. During ripening, the panicle being active site for physiological activities, movement of current and reserve protosynthates take place rapidly and thus, the nutrient concentration is relatively higher in panicles than in leaves and stem. Nutrient accumulation or uptake in different plant parts, as well as in the whole plant, during different phases of growth, clearly indicate higher nitrogen accumulation in stems and leaves during vegetative phase. Hence, nitrogen accumulation in different organs and in the whole plant follows a parabolic pattern during ontogenic development of the plant.

Takahashi et al. (1955) in their studies on the composition of the leaf blade and stem of rice plants at the elongation stage under different levels of nitrogen supply observed the gradual increase in the concentration of nitrogen and protein with the increasing levels of applied nitrogenous fertilizer in different plant parts(Table 2.3). But total sugar and starch concentration decreased gradually. Between the leaf blade and stem the nitrogen concentration was higher in leaf blades. It was noted to be due to more chlorophyll synthesis and protein synthesis in the leaf blades with the increasing level of nitrogen.

TABLE 2.3 Changes in the level of composition in plant parts of rice with increasing level of nitrogenous fertilizer

Component	Plant Part							
	Leaf blade (%)				Stem (Sheath & Culm) (%)			
	N_0	N_1	N_2	N_3	N_0	N_1	N_2	N_3
Total N	1.93	2.38	3.15	3.61	0.91	0.88	1.49	1.74
Protein	1.68	2.09	2.74	3.14	0.73	0.68	1.21	1.34
Total sugar	6.05	5.24	5.30	4.61	3.79	4.04	3.15	1.52
Starch	5.16	2.34	4.69	3.78	12.10	12.90	7.50	3.60

N_0 = No nitrogen; N_1 – 37.5 kgN ; N_2 – 113 kgN ; N_3 – 188 kgN

2.4. *Critical Concentration*

The critical nitrogen concentration of a plant can be defined as the minimum nitrogen concentration required for maximum growth rate at anytime. It has been suggested that the relationship between the critical nitrogen concentration and dry matter per unit ground area for a wide range of crops is the same and is independent of climatic zone. However, the critical concentration of nitrogen varied with stages of growth. Results support the concept of a critical nitrogen content dilution curve for yield of rice, which may be independent of climatic zone. The similarity between the nitrogen dilution curves for temperate and tropical environments indicates that there is no intrinsic difference in the ratio of carbon-nitrogen capture in those environments even though final aboveground biomasses differed. Both the rate and duration of resource capture are probably limiting yields in tropical environments (Sheehy et al. 1998).

The critical limit (0.92%) has been identified for total nitrogen of rice grain in relation to the optimum grain yield at different combinations of water and nitrogen levels (Prasad et al. 2002) (Table 2.4).

TABLE 2.4 Effect of fertilizer nitrogen and water level on percent grain nitrogen of rice

Nitrogen (kgha^{-1})	Water level (cm)	Grain yield (kgha^{-1})	Grain nitrogen (%)
0	0	2863	0.77
40	0	3433	0.80
80	0	4867	0.84
120	0	5167	0.90
0	4	327	0.77
40	4	4730	0.87
80	4	6120	1.00
120	4	5740	1.13
0	8	4730	0.85
40	8	5673	0.91
80	8	6597	1.02
120	8	7197	1.14
0	12	3853	0.83
40	12	4767	0.93
80	12	6220	1.10
120	12	6270	1.18

2.6. *Nitrogen Response*

The plant root system is an important organ which supplies water and nutrients to growing plants. Information is limited on influence of nitrogen fertilization on upland rice root growth. A green house experiment was conducted to evaluate influence of nitrogen (N) fertilization on growth of root system of 20 upland rice genotypes. The nitrogen rate used was 0 mg kg^{-1} (low) and 300 mg kg^{-1} (high) of soil. Nitrogen × genotype interactions for root growth (length) and root dry weight were highly significant ($P < 0.01$), indicating that differences among genotypes were not consistent at two nitrogen rates. Overall, greater root length, root-dry weight and tops-roots ratio were obtained at a nitrogen fertilization rate of 300 mg kg^{-1} compared with the 0 mg kg^{-1} soil. However, genotypes differ significantly in root length, root dry weight and top-root ratio. Nitrogen fertilization produced fine roots and more root hairs compared with absence of nitrogen fertilizer. Based on root dry weight efficiency index (RDWEI) for nitrogen use efficiency, 70% genotypes are classified as efficient, 15% as moderately efficient and 15% were classified as inefficient. Root dry weight efficiency index trait can be incorporated in upland rice for improving water and nutrient efficiency in favor of high yields (Fageria, 2007).

The limited understanding of the mechanisms that govern the partitioning of captured resources (carbohydrates, mineral nutrients) between different plant parts and organs is considered to be the main factor restricting the development of process-based modeling of whole plant growth (Dewar, 1993; Carmell and Dewar, 1994). For biomass partitioning between shoot and roots, Thornley (1972) has proposed a simple model which is widely used. In the model, growth is dependent on the supply of carbon from the shoot to roots (pholem transport) and that of nitrogen from roots to shoot (xylem transport). The fluxes are dependent on the concentration gradients of carbon and nitrogen between the two compartments, shoot and roots. According to Thornley's model, conditions which lead to an increase in carbon concentration should, therefore, lead to all increase in biomass partitioning towards the roots, where as an increase in nitrogen concentration should favor biomass partitioning towards the shoot. In principle, the model is also considered suitable to take into account the effects of various environmental factors including mineral nutrients on the shoot, root ratio (Wilson,1988).

The well documented increases in both carbon allocation to roots and in the root-shoot dry weight rated under conditions of nitrogen limitation are consistent with the Thornley concept, despite nitrogen cycling from shoot to roots (Cooper and Clarkson, 1989) and the key role played by root-borne phytohormones, particularly cytokinins, in the effect of nitrogen supply on the root shoot ratio (Fetune and Beck, 1993). However, the effect of mineral nutrition at status on shoot-root partitioning of photoassimilates and shoot-root dry weight ratio is markedly element specific.

According to Fageria (2007) the root growth of upland rice under 0 to 300 mg nitrogen per kg soil with two sources of nitrogen viz. urea and ammonium sulphate indicated that the root growth in terms of root dry weight per plant increased exponentially when urea is used. On the other hand root dry weight per plant increased linearly under ammonium sulphate fertilizer registering maximum root dry weight accumulation under 300 mg per kg of soil indicating that ammonium sulphate influenced root growth gradually and continuously than urea.

A pot experiment was conducted to study the effects of combined application of organic and inorganic fertilizers on the nitrogen uptake by rice and the nitrogen supply by soil in a wheat-rice rotation system, and approach the mechanisms for the increased fertilizer nitrogen use efficiency of rice under the combined application of organic and inorganic fertilizers from the view point of microbiology. Comparing the applying inorganic fertilizers combined application of inorganic and organic fertilizers decreased the soil microbial biomass carbon and nitrogen and soil mineral nitrogen contents before tillering stage, but increased them significantly from heading to grain filling stage. Under the combined fertilization the dynamics of soil nitrogen supply matched best the dynamics of rice nitrogen uptake and utilization which promoted the nitrogen accumulation in rice plant and the increase of rice yield and biomass, and increased the fertilizer nitrogen use efficiency of rice significantly. Combined application of inorganic and organic fertilizers also promoted the propagation of soil microbes, and consequently more mineral nitrogen in soil was immobilized by the microbes at rice early growth stage and immobilized nitrogen was gradually released at the mid and late growth stages of rice being able to better satisfy the nitrogen demand of rice in its various growth and development stages (Liu et al. 2012).

Nitrogen accumulation in the apical spikelet of the primary branch (superior spikelet) and the second spikelet of the lowest secondary branch (inferior spikelet) of the panicle on the main stem of the rice plant (cv. Sasanishiki) was characterized during grain filling. In the superior spikelet the accumulation of dry matter and nitrogen, which started immediately after flowering, proceeded rapidly, and reached the maturation level at 20 days after flowering. In the inferior spikelet, however, the amount of dry matter and nitrogen accumulation was minimal immediately after flowering. It increased when grain filling of the superior spikelet was almost completed. ^{15}N-leveled ammonia was administered to the plants at different stages of ripening and the amount of incorporation in the spikelets was analyzed at harvest. The labeled nitrogen administered at the early stages of ripening was the main source of labeled nitrogen incorporated in the superior spikelet. However, the labeled nitrogen incorporated in the inferior spikelets largely consisted of the labeled nitrogen administered at the late stages of ripening. When all the spikelets except for the five inferior spikelets were removed from the panicle at various stages of ripening, the amount of dry matter and nitrogen accumulation increased immediately irrespective of the stage of ripening (Wasaki et al. 1992).

Nitrogen deficiency is one of the most important yield limiting nutrients in lowland and plant tissue analysis is an important criteria for diagnosis of nutritional disorders in crop plants. A field experiment was conducted during three consecutive years in central part of Brazil on a Haplaquepts. Nitrogen rates used were 0, 30, 60, 90, 120, 150, 180 and 210 kg nitrogen per hectare. Nitrogen concentration in the shoot at different growth stages was significantly ($P < 0.01$) affected by nitrogen fertilization. Optimum nitrogen concentration for maximum dry matter yield was 43.4 gkg^{-1} at initiation of tillering, 12.7 gkg^{-1} at initiation of panicle, 12.8 gkg^{-1} at booting, 11.0 gkg^{-1} at flowering and 6.5 gkg^{-1} at physiological maturity stage. In the grain, the optimum nitrogen concentration was 10.9 gkg^{-1}. Nitrogen uptake varied from 16 to 185 $kgha^{-1}$ in the shoot from initiation to flowering. At physiological maturing, nitrogen uptake was 71 $kgha^{-1}$ in the shoot and 76 $kgha^{-1}$ in the grain. Accumulated nitrogen at harvest produced 9545 $kgha^{-1}$ straw and 6450 $kgha^{-1}$ grain yield. Shoot dry weight increased with the increase in shoot nitrogen uptake up to flowering. At harvest nitrogen uptake in the shoot decreased due to translocation to the grain. Rice needs nitrogen during its whole growth cycle; however, relatively initiation of panicle, flowering and physiological maturity were the most critical growth stages for nitrogen tissue analysis to determine optimum concentration or nitrogen uptake for maximum shoot and grain yield (Fageria, 2003).

A field experiment conducted for two years (1977 and 1978) at IARI, New Delhi, showed that yield and nitrogen uptake by rice was more in the case of medium duration (135 days) variety, Improved Sabarmati, than in the case of short duration (105 days) variety Pusa-33. Highest yield and nitrogen uptake by rice was recorded when it was transplanted and lowest when rice was direct seeded (drilled in moist soil). Broadcasting sprouted seeds on a puddle seedbed gave a yield and nitrogen uptake level in between transplanting seeds and direct seeding, and fairly acceptable method of planting. Rice responded well to nitrogen and the economic optimum dose was found to be 160-170 kg nitrogen per hectare. It is noted that urea briquettes give the highest yield and nitrogen uptake by rice and was superior to sulphur coated urea or neem-cake coated urea with respect to nitrogen uptake. All these nitrogen fertilizers were better than urea (Prasad and Prasad, 1980).

Considering the dynamics of nitrogen in rice seedlings the following were observed :

1. The nitrogen source to make-up new leaves comes not only from the root (medium nutrient) but also from old leaves.
2. Nitrogen balance in a mature leaf is controlled by the influx and efflux of nitrogen.

3. Amino acids move out from a leaf partly before incorporating into peptides-proteins and their subsequent breakdown.
4. Retransferred nitrogen is transported mainly to young leaves and roots with very little going to old organs (Yoneyama and Sano, 1978).

2.7. *Source of Nitrogen*

The nitrogen source had little effect on growth (Table 2.6), gas exchange, Chl-a fluorescence parameters and photosynthetic electron allocation in rice plants, except that NH_4^+ grown plants had a higher O_2 independent alternative electron flux than NO_3^- grown plants (Table 2.7). NO_3^- reduction activity was rarely detected in leaves of NH_4^+ grown cucumber plants, but was high in NH_4^+ grown rice plants (Zhou et al. 2011).

TABLE 2.6 Effects of different nitrogen forms on dry mass of shoot, root and total plant

Dry mass (g per plant)			
N form	Shoot	Root	Total plant
NO_3^-	11.94±0.69	4.88±0.55	16.99±0.32
NO_4^+	14.19±2.46	4.70±0.79	18.89±1.77

CO_2 assimilation rate, stomatal conductance, intercellular CO_2 and transpiration rate were also noted slightly high in NH_4^+ plants (Zhou et al. 2011).

TABLE 2.7 Electron flux pattern under different forms of nitrogen nutrition in rice leaves

	NO_3^-	NO_4^+
JPSII (mmoles per m^2.s)	111.9±2.9	107.4±2.4
Jc (μmoles per m^2.s)	80.1±2.1	80.0±1.9
Jo (μmoles per m^2.s)	18.9±1.3	17.5±0.5
Ja (O_2-dependent) (μmoles per m^2.s)	2.8±0.6	4.6±0.8
Ja (O_2-independent) (μmoles per m^2.s)	10.2±1.3	5.3±0.8

2.8. *Rate of Nitrogen*

Maintenance respiration rate R_M, irrespective of growth stages, increased with increase in the nitrogen supply. The R_M increased almost in proportion to net photosynthetic rate. Biomass production increased significantly during early growth stages, while it declined after anthesis. Significant positive correlation was observed between biomass production and P_N at all growth stages except tillering. Though R_M was positively correlated with biomass production during early growth stages, it was negatively correlated with the rate of increase in shoot biomass after flowering, which could indicate a possibility of identify certain cultivars endowed with low maintenance expenses despite building up biomass (Swain et al. 2000) (Table 2.8).

TABLE 2.8 Net photosynthetic rate (P_N) and maintenance respiration rate (R_M) (μmoles CO_2 m^{-2} s^{-1}) of two rice cultivars at various growth stages under different nitrogen supply (kgha-1)

Cultivar	N-supply	Tillering		Primodial Initiation		Flowering		15 DAF	
		P_N	R_M	P_N	R_M	P_N	R_M	P_N	R_M
Ratna	30	19.1	0.086	20.5	0.111	21.0	0.126	17.8	0.124
	60	21.0	0.096	22.0	0.126	22.6	0.145	18.7	0.135
	90	21.9	0.118	22.2	0.137	23.0	0.170	19.8	0.165
	120	22.3	0.144	22.7	0.163	23.6	0.189	19.1	0.193
Swarna	30	21.5	0.080	22.3	0.093	23.2	0.134	20.6	0.120
Prabha	60	22.2	0.089	23.2	0.107	24.2	0.160	20.9	0.125
	90	24.1	0.110	25.6	0.126	26.1	0.184	22.9	0.151
	120	25.0	0.131	26.9	0.146	27.3	0.224	23.9	0.183
	Fc	<1	50.57**	93.75**	<1	21.87**	6.61*	622.63**	805.00**
	F_A	<1	411.89**	29.97**	1.11NS	64.98**	44.84**	76.11**	7854.18**
	$F_{C\text{-}A}$	<1	1.69NS	6.52**	1.02NS	6.72**	2.51*	16.10**	39.50**

** Significant at 1% level NS - not significant

Two green house experiments were conducted using ammonium sulphate and urea as nitrogen sources for upland rice grown on a Brasilian Oxisol. The nitrogen rates used were 0, 50, 100, 150, 300 and 400 mg nitrogen kg^{-1} of soil. Yield and yield components were significantly increased in a quadratic fashion with increasing nitrogen rate. Ammonium sulphte × Urea interaction was significant for grain yield, shoot dry weight, panicle number, plant height and root dry weight, indicating a different response magnitude of these plant parameters, to two sources of nitrogen.

Based on regression equation maximum grain yield was achieved with the application of 380 mg Nkg^{-1} by ammonium sulfate and 271 mg Nkg^{-1} by urea. Grain yield and yield components were reduced at higher rates of urea (>300 mgNkg^{-1}) but these plant parameters responses to ammonium sulfate at higher rates was constant. In the intermediate nitrogen rate range (125 to 275 mgkg^{-1}) urea was slightly better compared to ammonium sulfate for grain yield. Grain yield was significantly related with plant height, shoot dry weight, panicle numbers, grain harvest index and root dry weight (Table 2.9). Hence, improving these plant characteristics by using appropriate soil and plant management practices can improve upland rice yield (Fageria et al. 2011).

TABLE 2.9 Influence of N sources and rates on grain harvest index, grain sterility and 1000 grain weight of upland rice

Nitrogen (mgkg^{-1} Soil)	Grain harvest index	Grain sterility (%)	1000 grain weight (g)
0	0.48	13.22	28.58
50	0.51	14.55	28.93
100	0.53	16.58	19.49
150	0.54	18.84	29.70
300	0.53	13.27	28.31
400	0.51	20.14	27.38
Average	0.52	16.10	28.73
F-test			
N sources (S)	NS	NS	NS
N level (L)	NS	NS	**
SXL	NS	NS	NS

Yoseftabar (2013) studied on nitrogen management on panicle structure and yield in rice showed that panicle number, panicle length, panicle dry matter, number of primary branches, total number of grains and grain yield were highest by the application of 300 kg/ha of nitrogen than 100 or 200 kg/ha of nitrogen. Effect of different split application of nitrogen fertilization increased significantly with increase of split application up to three splits. According to the study panicle structure such as number of panicles (heads), spikelet density, panicle length, panicle curvature and the number of grains per panicle are determined by the nitrogen application (Table 2.10).

TABLE 2.10 Effect of rate of nitrogen fertilization on morphological parameters of hybrid rice at different stages

Nitrogen ($kgha^{-1}$)	Panicle number (No)	Panicle length (cm)		Panicle dry matter (g per plant)	Panicle dry matter (%)	Number of primary branches per panicle	Total grain per panicle	Grain yield (kg per ha)
		Flowering stage	Harvesting stage					
100	10.66	22.8	27.71	892.94	46.09	10.65	190.31	6989.0
200	11.22	21.5	28.2	817.75	48.07	11.05	199.35	7690.0
300	12.36	23.7	28.64	954.93	50.94	11.17	209.85	8611.0

In lowland rice losses of applied nitrogen take place through (a) ammonia volatilization, (b) denitrification, (c) leaching and (d) run off. The recovery of fertilizer nitrogen applied to rice seldom exceeds 30-40%. Fertilizer nitrogen use efficiency in lowland rice may be maximized through a better timing of application to coincide with the stages of peak requirement of the crop and placement of nitrogen fertilizer in the soil. Other possibilities are the use of controlled release nitrogen fertilizer and exploitation of varieties suitable to nitrogen efficiency utilization.

In the anaerobic environment of lowland rice soils, the only stable mineral form of nitrogen is NH_4^+; nitrate NO_3^- forms of nitrogen, if applied, will enter the anaerobic zone and be subjected to heavy denitrification losses. At planting time, the base dressing of nitrogen should never be supplied as nitrate. For topdressing the growing plants,however, NH_4^+ and NO_3^- forms may be used with almost equal efficiency. Fully established rice can rapidly take up applied NO_3^- before it is leached down to the anaerobic soil layer and can become denitrified.

The early nitrogen application (65 to 100 percent of total nitrogen) should be applied as an ammonium nitrogen (NH_4^+) source onto dry soil, immediately prior to flooding at around the 4 to 5 leaf growth stage. There is not an exact time to apply early nitrogen, but actually a window of a couple of weeks that the early nitrogen can be applied. Once the early nitrogen is applied, flooding should be completed as quickly as possible, preferably within five days of nitrogen application. The flood incorporates the nitrogen fertilizer into the soil where it is protected against losses via, ammonia volatilization and/or nitrification/denitrification as long as a flood is maintained. The flood should be maintained for at least three weeks to achieve maximum uptake of the early applied nitrogen (Tables 2.11 and 2.12).

TABLE 2.11 Effect of preflood nitrogen application timing and soil moisture on rice grain yield

Time before flood (days)	Soil moisture	Uptake of applied nitrogen (kg per ha)	Nitrogen use efficiency (%)	Grain yield (kg per ha)
10	Dry	95	71	6255
10	Mud	52	42	5145
5	Dry	112	82	6507
5	Mud	80	59	5296
0	Dry	120	83	6659
0	Mud	76	64	5599
0	Flooded	41	31	3783

Nitrogen applied @ 130 kg/ha.

TABLE 2.12 Percent nitrogen uptake by rice crop at different times after nitrogen application

Nitrogen application timing	Days after application	% nitrogen plant uptake
Preflood	7	11
Urea applied on a dry soil	14	21
Surface and flood immediately	21	63
	28	65
Mid season	3	70
Urea applied into the flood	7	67
	10	76

The upland rice cultivar Zhongham 3 (Japonica) and the paddy rice cultivar Yangjing 9538 (japonica) were field grown under moist cultivation (Mc) and dry cultivation (Dc) with three levels of nitrogen viz. 100 kgha^{-1} (LN), 200 kgha^{-1} (NN) and 300 kgha^{-1} (HN) compared with 200 kgha^{-1} of nitrogen, 300 kgha^{-1} of nitrogen reduced grain yield for both upland and paddy rice cultivars under dry cultivation and for the paddy rice cultivar under moist condition, whereas it increased the yield of upland rice under moist cultivation. With an increase in nitrogen level, both upland and paddy rice showed higher productive tillers, more or fewer spikelets per panicle, and lower percentage of ripened grains under moist cultivation and dry condition. However, the seed setting rate reduced to a greater extent in paddy rice than in upland rice. There was no significant difference in 1000 grain weight for the upland rice among the three nitrogen levels, whereas the 1000 grain weight reduced with the increase in nitrogen level in the paddy rice. Compared with moist cultivation, dry cultivation had no significant influence on grain weight of upland rice; however, a significantly negative effect was observed in paddy rice, dry cultivation increased the seed setting rate in both cultivars, with more increase in upland rice than paddy rice. The upland rice had less number of adventitious roots, lower nitrogen absorption ability, lower productive tillering ability, fewer panicles, fewer spikelets per panicle and lower grain yield than paddy rice. However, upland rice showed more rapid increase in adventitious roots and a slower decline in leaf nitrogen content from jointing to heading and a faster decline in leaf chlorophyll content (SPAD value) after anthesis. In addition, upland rice had a weak negative response to water stress and a strong positive response to nitrogen level. The responses of cultivation methods and nitrogen level varied largely between upland and lowland rice (Zhang et al. 2008)

Through variation and frequency analysis, the genotypic differences in nitrogen utilization efficiency (NUE), nitrogen absorption efficiency (NAE) and nitrogen utilization efficiency response (NUER) of 88 rice germplasms were studied. NUE, NAU and NUER of the tested rice at seedling stage had notable genotypic discrepancy, and were influenced by nitrogen application. NUE differed significantly under low, medium and high nitrogen and dropped with the increase of nitrogen application. NAE was not significantly different under low and medium nitrogen, but was notably smaller than that under high nitrogen. NUER between low and high nitrogen was more notably higher than that between low and medium nitrogen or between medium and high nitrogen, and the latter two had not significant difference. The coefficients of variation among NUE, NAE and NUER of rice germplasms at seedling stage were greatly different (above 20%) and the order of them was NUER>NAE>NUE. The phenotypic distributions of NUE, NAE and NUER of rice germplasms at seedling basically resembled a normal curve, the order of good of fitness was NUE>NAE>NUER (Cheng et al. 2005).

It is noted that total N uptake, physiological nitrogen use efficiency (PUNE), apparent nitrogen use efficiency (ANRE) and agronomic nitrogen use efficiency varied in different cultivars significantly. Total nitrogen uptake, physiological nitrogen use efficiency, agronomic nitrogen use efficiency varied significantly with the increment of nitrogen applied. Total nitrogen uptake increased with increase in nitrogen fertilizing contents but physiological nitrogen use efficiency ANUE decreased. There were significant differences in the effects of applying nitrogen fertilizer on nitrogen use efficiency and characteristics of nitrogen uptake (Tayefe et al. 2011) (Table 2.13).

TABLE 2.13 Nitrogen use efficiency indices for rice varieties under different nitrogen level

Nitrogen rate ($kgha^{-1}$)	NHI		ANRE		ANUE		PNUE	
	2007	2008	2007	2008	2007	2008	2007	2008
0	63	76	0.47	0.51				
30	67	77	0.47	0.51	20.01	19.96	58.53	48.47
60	71	77	0.33	0.47	18.47	20.28	55.57	49.16
90	69	75	0.35	0.40	14.25	18.36	36.73	45.76

NHI = Nitrogen Harvest Index.

Nitrogen use efficiency (NUE), defined as the ratio of grain yield to supplied nitrogen, is a key parameter for evaluating a crop cultivar and it is composed of nitrogen uptake efficiency and nitrogen physiological use efficiency (De Macle and Velk, 2004).

Nitrogen uptake efficiency is the nitrogen accumulation relative to its supply, while nitrogen physiological used efficiency represents grain yield relative to nitrogen accumulation (Moll et al. 1982). While the amount of nitrogen available from soil and fertilizer is difficult to measure, grain yield can be used for evaluating the NUE, and high NUE cultivars can be defined by their ability to produce higher grain yield than others under the same experimental conditions. As PNN can be attributed to improve nitrogen uptake, cultivar with a higher NUE has a more positive response to PNN than with a low NUE, suggesting that there might be a relationship between PNN and NUE (Table 2.14).

TABLE 2.14 Effect of partial NO_3^- nutrition on grain yield and physiological nitrogen use efficiency

Cultivar	NO_4^+ / NO_3^-	Grain yield ($gpot^{-1}$)	Physiological nitrogen use efficiency (%)
Nanguang	100/0 75/25	14.7 ± 1.07 17.7 ± 0.82	18.5 ± 0.69 18.8 ± 0.56
Elio	100/0 75/25	9.8 ± 0.76 10.5 ± 0.88	12.6 ± 1.02 13.3 ± 0.87

2.9. Method and Timing of Nitrogen Application

Experiments regarding timing of nitrogen application had shown the clear superiority of pre-plant nitrogen application (Table 2.15).

TABLE 2.15 Yield response to time of nitrogen application

Method/Timing[1]	Grain yield (kg/ha)[2]
Pre-plant	7,548
Pre-flood	6,124
Post-flood	5,778

[1] 100 kg/ha of nitrogen
[2] Yield of 3 yrs. Average

The least nitrogen losses due to leaching or volatilization took place when the only application was closer to the flooding time (Table 2.16).

TABLE 2.16 Yield response to time of nitrogen application in dry seeded rice

Timing	Yield (kg/ha)	Yield loss (%)
9 days pre-flood	7,415	8
6 days pre-flood	7,525	7
3 days pre-flood	7,559	6
0 days pre-flood	8,117	
3 days post-flood	6,484	21
6 days post-flood	6,140	24

Urea is generally the nitrogen fertilizer of choice. Most of the nitrogen fertilizer should be applied pre-flood and pre-plant in water seeded rice if the soil is not allowed to dry during the growing season. Nitrogen fertilizer should be placed either on dry soil and flooded immediately or shallow incorporated and flooded within 3-5 days. If several days elapse between the period of nitrogen application in ammonical form and flooding, much of the nitrogen will convert to nitrate. When the soil is flooded nitrate is broken down by bacteria and released to the atmosphere as a gas, a denitrification process.

Denitrification losses can be avoided by flooding the soils within 3-5 days after nitrogen application. These losses are greatest when nitrogen is applied into water on young rice. When most of the nitrogen is applied pre-plant, rice field should not be drained or drained only temporarily. In this situation, if the field must be drained during the growing season, the field should not be allowed to dry up before re- flooding. The field should be maintained in a saturated condition to protect the pre-plant nitrogen.

From internode elongation (green ring) through the beginning of head formation nitrogen must be available in sufficient quantity to promote the maximum number of grains. Nitrogen deficiency at this time reduces the number of potential grains and limits yield potential. Sufficient nitrogen should be applied pre-plant or pre-flood to assure that the rice plant needs no additional nitrogen until the panicle initiation (green ring) or the panicle differentiation stage. When additional nitrogen is required, it should be top dressed at either of these plant stages or whenever nitrogen deficiency symptoms appear. Usually only 20-50 kg per ha are required if the earlier nitrogen application was sufficient, if nitrogen deficiencies are observed prior to these growth stages, apply nitrogen top dressing immediately (Table 2.17). Early nitrogen deficiency may greatly reduce yields.

TABLE 2.17 General guidelines for efficient nitrogen management

Situation	Strategy
Upland (dry land)	Broadcast and mix basal dressing in top 5cm of surface soil. Incorporate top-dressed fertilizer by hoeing in between plant rows and then apply light irrigation, if available.
Rainfed deep water	Apply full amount as basal dressing
Lowland (submerged)	Use non-nitrate sources for basal dressing
Soil very poor in N	Give relatively more nitrogen at planting
Assured water supply	Can top dress every three weeks upto panicle initiations drain field before top dressing and reflood two days later.
Permeable soils	Emphasis on increasing number of split applications
Short duration varieties	More basal N and early top dressing preferred
Long duration varieties	Increased number of top dressing
Colder growing season	Less basal nitrogen and more as top dressing
Overaged seedlings used	More nitrogen at planting
High pH soil	Nitrogen may be applied as ammonium sulphate.

While considering the nitrogen nutrition in rice and wheat, some of the aspects, which appear important are :

1. Uptake of nitrogen is usually high in rice than wheat.

Crop	N applied (kg N ha^{-1})	Nuptake (kg N ha^{-1})
Wheat	0	39.7 ± 3.5
	100	92.3 ± 4.3
	250	161.6 ± 11.6

Rice	0	57.1 ± 0.8
	100	99.1 ± 4.1
	250	180.1 ± 10.8

2. Wheat had consistently higher nitrogen recovery than rice. This is because of the fact that in rice non-fertilizer nitrogen fixation and subsequent absorption during the growing season is appreciably high; hence apparent reduction of recovery.
3. In both wheat and rice, however, nitrogen absorption varies at different growth stages, of which nitrogen absorption reaches peak at jointing anthesis stage.
4. The amount and properties of Rubisco and the CO_2 diffusion resistance differ greatly between rice and wheat. Rice allocates 25-30% of leaf nitrogen to Rubisco with higher affinity for CO_2 (20% lower km for CO_2) than wheat, whereas wheat allocates 20 to 25% of leaf nitrogen to Rubisco with greater Km (50% higher Vmax for carboxylation) than rice (Makino et al. 1988). The CO_2/O_2 specificity for carboxylation and oxygenation does not differ between the two species.

References

Bai, Z.G., D.L. Dent, L. Olsson and M.E. Schaepman. 2008. Proxy global assessment of land degradation. *Soil use and Manag.* **24:** 223-234.

Balkos, K.D., D.T. Britto and H.T.Kronzucker. 2010. Optimization of ammonium acquisition and metabolism by potassium in rice (*Oryza sativa* L). *Plant Cell Environ,* **33:** 23-34.

Basuchaudhuri, P. and D.K. Dasgupta. 1983. Utilization of major nutrients by improved rice varieties. *Ind. J.Plant Nutr.* **2:** 5-11.

Briones, A.M., S. Jr. Okabe, Yumemiya, N.B. Ramsing, W. Reichardt and H. Okuyama. 2003. Ammonia-Oxidising bacteria on root biofilms and their possible contribution to N use efficiency of different rice cultivars. *Plant and soil,* **250:** 335-348.

Britto, D.T. and H.J.Kronzucker. 2002. NH_4^+ toxicity in higher plants: a critical review. *J. Plant Physiol* **159:** 567-584.

Britto, D.T. and H.J. Kronzuker.2006. Futide cycling at the plasma membrance: a hallmark of low-affinity nutrient transport. *Trends Plant Sci.,* **11:** 529-534.

Carmell, M.G.R. and R.C. Dewar. 1994. Carbon allocation in trees: a review of concepts for modeling. *Advances in Ecological Research,* **25:** 60-102.

Cheng, J.F., T.B. Dai, W.X. CaO and D. Jiang. 2005. Genotypic differences on nitrogen nutrition characteristics of rice germplasms at seedling stage. *Chinese J.Rice Sci.* **19:** 533-538.

Cooper, H.D. and D.T. Clarkson. 1989. Cycling of amino nitrogen and other nutrients between shoots and roots in cereals: a possible mechanism integrating shoot and root in regulation of nutrient uptake. *J.Exp. Bot.* **40:** 753-762.

Cox, W.J. and H.M. Reisenauer. 1972. Growth and ion uptake by wheat supplied nitrogen as nitrate or ammonium or both. *Plant and Soil,* **38:** 363-380.

De Macle, M.A.R. and P.L.G. Velk. 2004. The role of Azolla cover in improving the nitrogen use efficiency of lowland rice. *Plant and Soil*, **263:** 311-321.

Dewar, R.C.1993. A root shoot partitioning model based on carbon–nitrogen–water interactions and Munch phloem flow. *Functional Ecology*, **7:** 356-368.

Duan, Y.H., Y.L. Zang, Q.R. Shen and S.W. Wang. 2006. Nitrate effect on rice growth and nitrogen absorption and assimilation at different growth stages. *Podosphere*, **16:** 707-717.

Epstein,E. 1972. In Mineral Nutrition of Plants: Principles and Perspectives. John Wiley and Sons, Inc. New York.

Fagaria,N.K. 2003. Plant tissue test for determination of optimum concentration and uptake of nitrogen at different growth stages in lowland rice. *Soil Sci. & Plant Analysis*, **34:** 259-270.

Fagaria, N.K. 2007. Root growth of upland rice genotypes as influenced by nitrogen fertilization. *J.Plant Nutr.*, **30:** 843-879.

Fagaria, N.K., A. Moreira and A.M. Coelho. 2011. Yield and yield components of upland rice as influenced by nitrogen sources., *J.Plant Nutr.* **34:** 361-370.

Fetune, M. and E. Beck.1993. Reversal of the direction of photosynthate allocation in *Urtica dioica* plants by increasing cytokinin import into the shoot. *Botanica Acta*, **106:** 235-240.

Glass, A.D.M., J.E. Shaff and V. Kochian. 1992. Studies of the uptake of nitrate in barley 4. Ecophysiology. *Plant Physiol.* **9:** 456-463.

Gigon, A. and I.H. Rorison.1972. The response of some ecologically distinct plant species to nitrate and ammonium nitrogen. *J. Ecology.*, **60:** 93-102.

Guo, S.W., G.Chen, Y. Zhou and Q.R. Shen. 2007. Ammonium nutrition increases photosynthesis rate under water stress at early development stage of rice (*Oryzasativa* L). *Plant and Soil*, **296:** 115-124.

Heberer, J.A. and E.E. Below. 1989. Mixed nitrogen nutrition and productivity of wheat grown in hydroponics. *Ann Bot.*, **63:** 643-649.

Huner, N.P. and W. Hopkins. 2008. Introduction to plant physiology, 4th Edn John Wiley & Sons. Inc.

Ishizuka, Y. 1965. Nutrient uptake at different stages of growth. *In*: The Mineral Nutrition of the Rice Plant. The Johns Hopkins Press, Baltimore, Marryland .

Kirk, G.J.D. 1993. Root Ventilation, rhizosphere modification and nutrient uptake. *In*: Systems, approaches for agricultural development. Academic Publishers, Dordrecht.

Kirk, G.J.D. 2001. Plant mediated processes to acquire nutrients: nitrogen uptake by rice plants. *Plant and Soil*, **232:** 129-134.

Kirk, G.J.D. and H.J. Kronzucker. 2005. The potential for nitrification and nitrate uptake in the rhizosphere of wetland plants: a modeling study. *Ann Bot.*, **96:** 639-646.

Kronzucker, H.J., M.Y. Siddiqi, A.D.M. Glass and G.J.D. Kirk. 1999. Nitrate-ammonium synergism in rice: a subcellular flux analysis. *Plant Physiol*, **119:** 1041-1045.

Kronzucker, H.J., A.D.M. Glass, M.Y. Siddiqi and G.J.D. Kirk. 2000. Comparative kinetic analysis of ammonium and nitrate acquisition by tropical lowland rice: implication of rice cultivation and yield potential. *New Phytologist*, **14:** 471-476.

Li, P., B .Velde and D.C. Li. 2003. Loss of K bearing clay minerals in flood irrigated, rice growing soil in Jiangxi province. *China Clayminer*, **51:** 75-82.

Li, Q., B.H. Li, H.J. Kronzucker and W.M. Shi. 2010. Root growth inhibition by NH_4^+ in *Aabidopsis* is mediated by the root tip and is linked to NH_4^+ efflux and GMPase activity. *Plant Cell Environ*, **33:** 1529-1542.

Liu, Y.R., X. Li, J. Yu, Q.R. Shen and Y.C. Xu. 2012. Mechanisms for the increased fertilizer nitrogen use efficiency of rice in wheat rice rotation system under combined application of inorganic and organic fertilizers. *Ying Yong Sheng Tai XUe Bao,* **23:** 81-86.

Makino, A, T. Mae and K. Ohira. 1988. Differences between wheat and rice in the enzyme properties of ribulose-1, 5-bisphosphate carboxylase/oxygenase in the relationship to photosynthetic gas exchange. *Planta,* **174:** 30-38.

Marchner, H. 1995. Mineral nutrition of higher plants. Academic, London.

Mehrer, I. and H. Mohr. 1989. Ammonium toxicity: description of the syndrome in Synapis alba and the search for its causation. *Physiol Plant,* **77:** 545-554.

Moll, R.H., E.J. Kamprath and W.A. Jakson. 1982. Analysis and interpretation of factors which contribute to efficiency to nitrogen utilization. *Agron. J.* **74:** 562-564.

Prasad, L.K., B. Saha, A. Haris, K. Rajan and S.R. Singh. 2002. Critical grain nitrogen content for optimizing nitrogen and water in rice *(Oryza sativa L.). J. Biol Sci.,* **2:** 746-747.

Prasad, M. and R. Prasad. 1980. Yield and nitrogen uptake by rice as affected by variety, method of planting and new nitrogen fertilizers. *Fert. Res.,* **1:** 207-213.

Peuke, A.D., W. Hartung and W.D. Jeschke. 1994. The uptake and hoe of C, N and ions between roots and shoots in *Ricinus communis* II grown with low and high nitrate supply. *J. Expt. Bot.,* **45:** 733-740.

Qian, X.Q., Q.R. Shen, G.H. Xu, J.J. Wang and M.Y. Zhor. 2004. Nitrogen form effects on yield and nitrogen uptake of rice crop grown in aerobic soil. *J. Plant Nutr.,* **27:** 1061-1076.

Sheehy, J.E., M.J.A. Dionora, P.L. Mitchell, Speng, K.G. Cassman, G. Lamaire and R.L. Williams. 1998. Critical nitrogen concentrations: implecations for high yielding rice *(Oryza sativaL)* cultivars in the tropics. *Field Crops Res,* **59:** 31-41.

Sims, G.K. and E.P. Dunigan. 1984. Diurnal and seasonal variations in nitrogenase activity (C_2H_2 reduction) of rice roots. *Soil Biol. Biochem.***16:** 15-18.

Smil, V. 2000. Cycles of Life. Scientific American Library, New York.

Swain, P., M.J. Baig and S.K. Nayak. 2000. Maintenance respiration of *Oryza sativa* leaves at different growth stages as influenced by nitrogen supply. *Biol Plant,* **43:** 587-590.

Takahashi, J., M. Yanagisawa, M. Kono, F. Yazawa and T .Yoshida. 1955. Studies on nutrient absorption by crops (in Japanese, English summary). *Bill Watt Inst. Agree Sci,* **B-4:** 1-83.

Tayefe, M., A. Gerayzade, E. Amiri and A.N. Zade. 2011. Effect of nitrogen fertilizer on nitrogen uptake, nitrogen use efficiency of rice. IPCBEE, 24.

Thornley, J.H.M. 1972. Mathematical Models in Agriculture. CABI Publishing Wang, M.Y., M.Y. siddiqi and A.D.M. Glass. 1993. Ammonium uptake by rice roots.1. Fluxes and subcellular distribution of 13 NH_4^+. *Plant Physiol,* **103:** 1249-1258.

Wasaki, T., A. Mae, K. Makino and O. Nihiko. 1992. Nitrogen accumulation in the inferior spikelet of rice ear during ripening. *Soil Sci & Pl. Nutr.,* **38:** 517-525.

Wilson, J.B. 1988. A review of evidence on the control of shoot root ratio, in relation to models. *Ann. Bot.,* **61:** 433-449.

Yoneyawa, T. and C. Sano. 1978. Nitrogen nutrition and growth of the rice plant II. Correlations concerning the dynamics of nitrogen in rice seedlings. *Soil Sci Pl. Nutr.,* **24:** 191-198.

Yoseftabar, S. 2013. Effect of nitrogen management on panicle structure yield in rice (*Oryza sativa* L). *IJACS,* **5:** 1659-1662.

Zhang, YJ., Y.R. Zhou, B.Du and Y.C. Young. 2008. Effects of nitrogen nutrition on grain yield of upland and paddy rice under different cultivation methods. *Acta Agronomica Sinica,* **34:** 1005-1013.

Zhou, Y.H., Y.L. Zhang, X.M. Wang, J.X. Cui, X.J. Xia and J.Q. Yu. 2011. Effects of nitrogen form on growth, CO_2assimilation, chlorophyll fluroescence and photosynthetic electron allocation in cucumber and rice plants. *J. Zhejiiang Univ. Sci. B.,* **12:** 126-134.

chapter three

Nitrate Reduction

Nitrogen is one of the important nutrients for plant growth and development. In natural conditions, the form of nitrogen uptake is mainly determined by its abundance and accessibility, which makes nitrate (NO_3^-) and ammonium (NH_4^+) become the major sources of inorganic nitrogen taken up by higher plants. In well aerated agriculture soils, nitrate is the main form of available inorganic nitrogen utilized by higher plants (Sasakawa and Yamamoto, 1978). Nitrate is easy to be lost through drainage water and denitrification and low efficiency in fertilization (Huang et al. 2000).

3.1. Nitrate in Plant

Plants have already developed an intrinsic regulation mechanism to adopt themselves to uneven nitrogen distribution environments and utilize mineral nutrition efficiently. Plant root system is plastic to uneven distribution of nitrogen by increased proliferation of the lateral roots preferentially in a nitrogen-rich zone. It has been demonstrated that a local supply of NO_3^- on root could stimulate the development and growth of lateral roots (Hackett, 1972; Zhang and Forde, 2000). Besides the increased length and number of lateral roots in the NO_3^- supplied side, it was also found that root metabolic activity and partitioning of assimilates to the NO_3^- supplied root were increased (Granato and Raper, 1989). In addition, a significant increase of endogenous IAA content in the zone of nitrate supply was also observed (Sattelmacher et al. 1993). It has been also reported, the NO_3^- could serve as a signal to stimulate lateral root elongation, and a putative model was suggested that root branching be modulated by opposing signals from plants, internal nitrogen status and the external supply of nitrate (Zhang et al. 1999). However, the physiological and molecular processes leading to the root morphological reaction induced by nitrate have not been well understood.

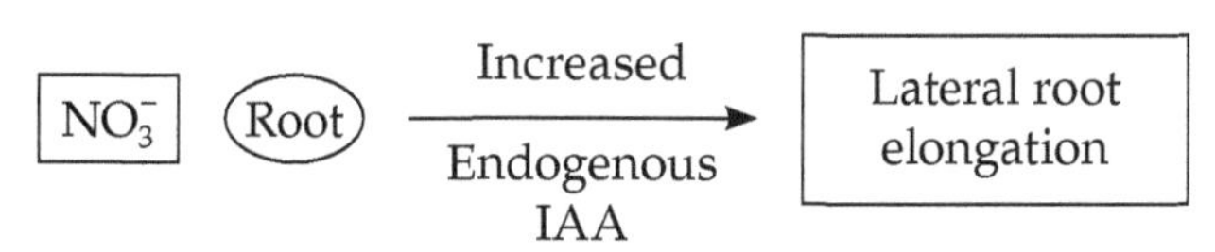

Nitrate regulates a lot of genes involved in primary nitrogen metabolism, including transport system genes NRT1 and NRT2, nitrite reductase

(NII) and the enzymes for incorporating ammonium into amino acids such as glutamate synthetase (GOGAT) (Stitt, 1999). Many genes involved in organic acid metabolism, redox metabolism and starch synthesis are also changed at transcription level in response to nitrate supply (Stitt, 1999). Genomic analysis identified that genes for water channels, root phosphorus and potassium transporters, transcriptional regulation, stress response genes and ribosomal proteins were all induced by nitrate, implying the collaboration and interaction of nutrient absorption among nitrate and other nutrients, like phosphorus and potassium (Wang et al. 2001).

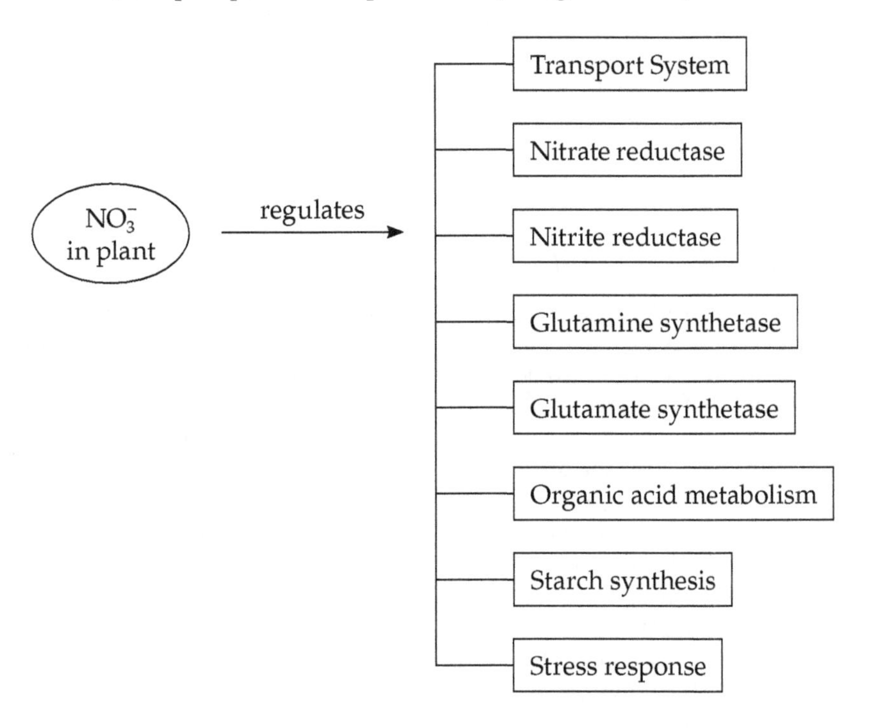

3.2. *Nitrogen Utilization*

Crop productivity during the last 50 years is highly correlated with fertilizer nitrogen input. Maximized efficiency of use of soil and fertilizer nitrogen is one of the main concerns of fertilizer input technology and research imperatives directed towards this goal have often been emphasized.

Nitrogen utilization efficiency at the plant level involves a number of steps, such as uptake of nitrogen from the soil, reduction of nitrate and/or biosynthesis of amino acids in either root or shoot, transport of either reduced or unreduced forms of nitrogen, biosynthesis of enzymes involved in the process and their regulation, compartmentation and remobilization of storage nitrogen. It is thus obvious that the problem is a complex one and involves the integration of all these aspects. Nitrate reductase (NR) (EC 1.6.6.1) is the first enzyme in the assimilation path of nitrate — the predominant form of nitrogen available for crop plants in the field. Significant positive correlation exists between this enzyme and the nitrogen status of some higher plant systems, and that growth, yield and protein content are sometimes correlated with the enzyme levels in seeds and leaves (Hagemann, 1979; Srivastava, 1980).

Recently, Arima and Kumazawa (1975) have shown by the 15N tracer method that glutamine is a primary product of ammonium assimilation and is synthesized from glutamic acid and newly absorbed ammonium by the catalytic activity of the enzyme glutamine synthase (GS) (EC 6.3.1.2). The synthesis of 14C glutamine from 14C glutamate was differentially affected by the source of nitrogen (Iyer et al. 1981) and GS activity is reversibly repressed with increase in the external ammonium concentration (Rhodes et al. 1975; Arima et al. 1976). On the otherhand universal existence of glutamate dehydrogenase (GDH) (EC 1.4.1:4) in plants is well known, but the physiological significance of the enzyme is still not clear, especially as regards its high km to ammonia and inhibition by ATP. The assimilatory nitrate reduction pathway is a vital biological process in higher plants, algae and fungi as it is the principal route by which inorganic nitrogen is incorporated into organic compounds. It involves energy-dependent uptake of nitrate through the roots and its reduction to nitrite and further to ammonium in a two step process catalyzed by the highly regulated enzymes nitrate reductase (NR) and nitrite reductase (NiR) in cytosol and chloroplast respectively. The reduced nitrogen ammonium is converted into organic nitrogen forms, glutamine (Gln) and glutamate (Glu) by combined action of GS and GOGAT in a cyclic manner (Sivasankar and Oaks, 1996). These amino acids subsequently serve as the nitrogen donor for all other amino acids.

3.3. *Nitrate Assimilation*

Nitrate assimilation is a highly regulated process because of its dependence on photosynthesis for energy and reductants as well as the toxicity of the metabolites of this pathway, nitrite and ammonium. Accordingly gene expression and enzyme activity of the various proteins involved in this pathway are regulated by both internal and external stimuli such as nitrate itself, carbon and nitrogen metabolites, growth regulators, light, temperature and carbon dioxide concentration (Aslam et al. 1997). However, the strongest responses are induction by nitrate and repression by ammonium or its derivatives in a combined action of feed forward and feedback regulation mechanism (Crawford, 1995). The activity of nitrate reductase is controlled according to nitrogen status of the plants by hierarchy of transcriptional, post-transcriptional and post-transcriptional regulation (Crawford 1995; Daniel–Vedele et al. 1998). Recently Lea et al.

(2006) have shown that the post-transcriptional regulation for setting the levels of amino acids, ammonium, and nitrate in *N. plumbaginifolia* control of NR gene transcription facilitates long-term responses to the nitrate signal (hours to days), whereas post-transcriptional regulation allows rapid changes in NR activity (mins to hours). Nitrate is never allowed to accumulate in the cells because of its toxicity. In maize it has been shown to inhibit NR transcription level and activity (Raghuram and Soproy, 1999). Ammonium and glutamine have been shown to have varied responses (no effect, inhibition or stimulation) depending on the species, genotype, tissue and experimental conditions (Aslam et al. 1979; Shankar and Srivastava, 1998). On the otherhand the roles of glutamate and 2-oxoglutarate in regulating NR and NiR are not available.

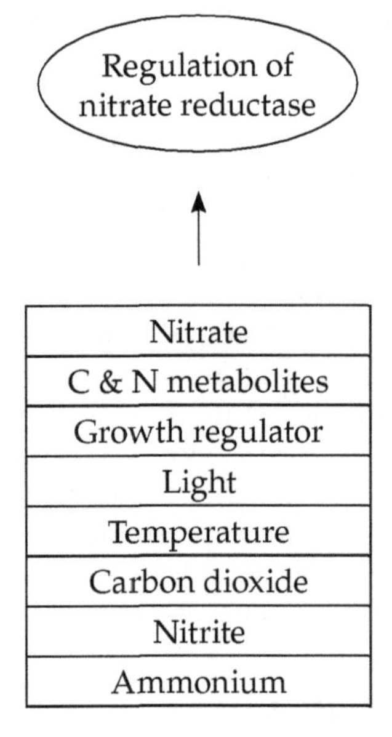

Reduction of nitrate to nitrite is the first step of reduction and is associated with a transfer of two electrons. Again, the reduction is from plant system where NAD(P) or NADH is oxidized. NAD(P)H and NADH are obtained from plant metabolic cycles.

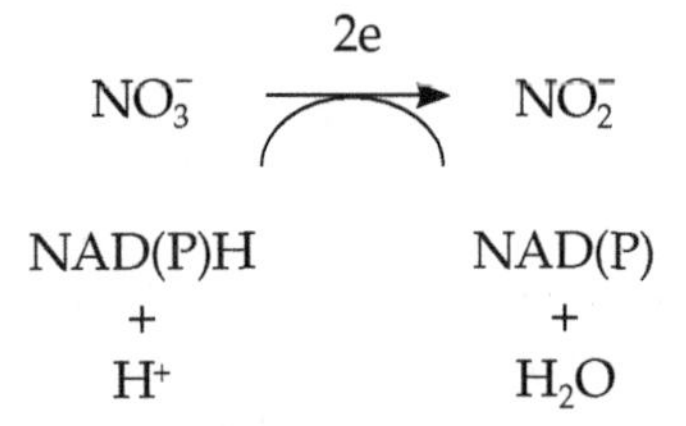

In nitrate reductase (NR) the electron transfer may be given in a better way as, NAD (P)H $\rightarrow$ (FAD $\rightarrow$ $Cytb_{557}$ $\rightarrow$ Mo) $\rightarrow$ NO_3^-

3.4. *Nitrate Reductase*

Nitrate reductase is a homodimer with two identical sub units of 100-114 KDa. The enzyme contains Apo Enzyme of protein nature synthesized from amino acids pool of the plant system in association with mRNA. The other part is molybdenum cofactor.

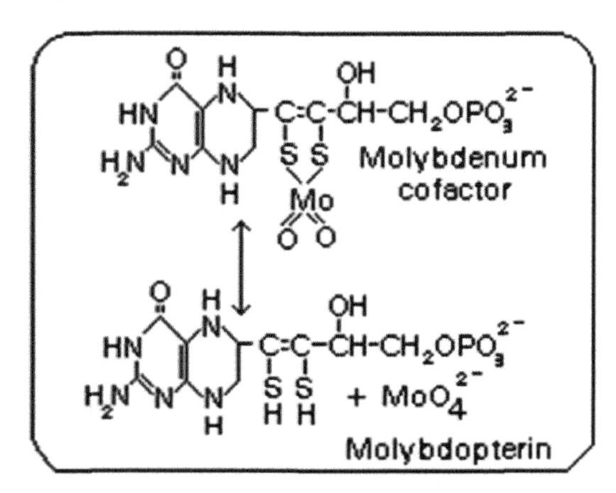

FIGURE 3.1 Molybdenum cofactor.

The reaction involves the oxidation of the sulphur atoms and not the molybdenum as previously suggested. The mechanism involves a molybdenum and sulphur based redoxchemitry instead of redox chemistry based only on the Mo ion. Molybdopterins are synthesized from guanosine triphosphate.

According to Rajagopalan (1989) the enzyme system dimer and the reduction process may be as follows:

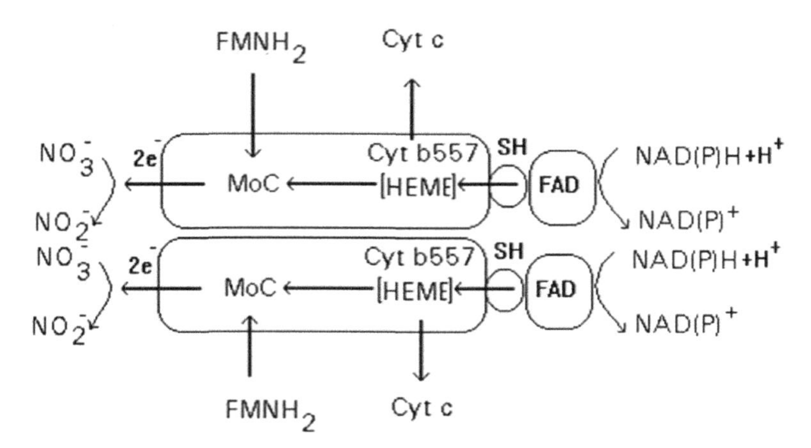

FIGURE 3.2 Nitrate reductase dimer system.

Nitrate reductase (NR: EC 1.6.6.1-3) catalyzes NAD(P)H reduction of nitrate to nitrite. NR serves plants, algae and fungi as a central point for integration of metabolism by governing flux of reduced nitrogen by several regulatory mechanisms. The NR monomer is composed of a ~100 kDa polypeptide and one each of FAD, heme iron and molybdenum–molybdo/pterin (Mo-MPT). NR has eight sequence segments.

a. N-terminal 'Acidic' region
b. Mo-MPT domain with nitrate reducing active site
c. Interface domain
d. Hinge 1 containing serine phosphorylated in reversible activity regulation with inhibition by 14-3-3 binding protein
e. Cytochromeb domain
f. Hinge 2
g. FAD domain and
h. NADPH domain.

The cytochrome b reductase fragment contains the active site where NAD(P) H transfers electrons to FAD (Campbell, 1999).

Nitrate reductase is a water soluble enzyme consisting of a polypeptide bound to the Mo-MPT, heme iron, and FAD. It is known that dimerization of the enzyme is necessary for its activity, and thus the native enzyme is homodimer having a tendency to further dimerization to a homotetramer (dimer of dimer). The enzyme has two active centres carrying out the internal electron transfer from FAD onto the heme iron and then to Mo-MPT. In the first active centre electrons are transferred from FAD via NaDH (or NADPH) onto enzyme. In the second active centre two electrons are transferred from reduced Mo(iv) to nitrate and reduce the latter to nitrite and hydroxide. Eukaryotic NRase exhibits its activity with such electron acceptors as ferricyanide and cytochrome by obtaining electrons directly from FAD and heme iron, respectively (Morozkina and Zuyagi-Iskaya, 2007).

3.5. *Induction of Enzyme*

Nitrate simultaneously induced NADH and NADH nitrate reductase activity in rice seedlings. Chloramphenicol, other organic nitro compounds such as o-nitro aniline and 2,4 dinitrophenol and nitrate also induce nitrate reductase in rice seedlings. The nitrate and nitrite induced nitrate reductase also accepted electrons more efficiently from NADH than NADPH. However, when this enzyme was induced by organic nitro compounds it could accept electrons more effectively from NADPH than NADH (Shen, 1972). Nitrate reductase is induced by nitrate in excised embryos and germinating intact seedlings of rice. The enzyme is induced

24 hours after imbibitions. The rate of enzyme formation increases with the age of seedlings. There is a lag period of 30 to 40 minutes between the addition of substrate and the formation of nitrate reductase. Formation of the enzyme is promoted by the presence of ammonium, chloramphenicol, actinonmycin D and cycloheximide effectively inhibit the formation of nitrate reductase.

Rice seedlings can assimilate nitrate from the beginning of the germination. However, the utilization of nitrate is completely suppressed by the presence of ammonium. As soon as ammonium is depleted from the medium, nitrate utilization is resumed. Ammonium inhibits the first step of nitrate reduction i.e. NO_3^- to NO_2^- but does not inhibit the assimilation of nitrite. Ammonium formed has a feedback inhibition (Shen, 1969).

Barley nitrate reductase cDNA clone $bNRp^{10}$ was used as a hybridization probe to screen a genomic DNA library of rice (*Oryza sativa* L) cultivar M201. Two different lambda clones were isolated, subcloned to plasmids, and partially characterized. The subclone $pHBH_1$ was tentatively identified as encoding a $NADH^-$ nitrate reductase. Southern and dot blot analysis suggest that, in rice, nitrate reductase is encoded by a small gene family. Regulation of $NADH^-$ nitrate reductase was investigated in rice cultivars Labell and M201 representing the subspices indica and japonica, respectively. In the absence of nitrate, only trace levels of nitrate reductase activity and mRNA were detected in seedling leaves. Upon addition of nitrate to seedling roots, nitrate reductase activity and mRNA increased rapidly in leaves. Nitrate reductase activity continued to increase over 24 hours period, but the mRNA accumulation peaked at about 6hour and then declined. Western blot analysis with a barley NADH nitrate reductase antiserum showed the presence of two bands of approximately 115 and 105 kDa. These protein bands were not detected in the extracts of tissue grown in absence of nitrate (Hamat et al. 1989).

The induction and regulation of NR and NiR by various nitrogen metabolites in excised leaves of rice (cv. Panvel I) seedlings grown hydroponically (nutrient starved) for 10-12 days and adapted for two days in darkness was examined. Nitrate induced the activity of both the enzymes reaching an optimum at 40mM in six hours. Nitrite and ammonium inhibited in light in a concentration dependent manner. Glutamine, which had little effect of its own on NR in light and no effect on NiR in both light and dark, strongly inhibited nitrate induced NR in dark. When the activities of these enzymes were measured from leaves treated with glutamate and 2-oxoglutarate, a similar pattern of induction was observed for NR and NiR. The transcript levels of NR and NiR increased in a similar extent in the presence of nitrate (Table 3.1). However, light did not cause any significant change in transcript levels. These results indicate that both the enzymes are under light regulation by nitrogen metabolites and light, and are co-regulated under certain conditions (Ali et al. 2007).

TABLE 3.1 Nitrogen metabolites and light reactions on nitrate reductase

Nitrogen metabolite	Light	Dark
Nitrite	Inhibit depending on concentration	–
Ammonium	"	–
Glutamine	Little effect	Strongly inhibit
Glutamate	Little effect	Strongly inhibit
2-oxoglutarate	Little effect	Strongly inhibit

Studies also revealed that

1. In rice seedlings synthesis of methyl viologen–nitrate reductase was stimulated by light, as was that of NADH–nitrate oxidoreductase. A small residual effect of light on the synthesis of enzymes persisted in the dark for a short time.
2. In etiolated seedlings exposed to light and nitrate, a lag period of three hours was necessary before enzyme synthesis commenced, whereas in green seedlings kept in dark for 36 hours, synthesis of both the enzymes started as soon as light and nitrate were provided.
3. Experiments with cycloheximide suggested that fresh protein synthesis in light was necessary for formation of active enzymes. Mere activation by light of inactive enzymes or their precursors was not involved.
4. In green seedlings, synthesis of nitrite reductase was more sensitive to chloramphenicol than that of nitrate reductase. In chloramphenicol treated etiolated seedlings, however, synthesis of both enzymes was inhibited to the same extent on subsequent light treatment.
5. A close correlation was observed between inhibition of the Hill reaction by 3-(3, 4 dichlorophenyl)–1, 1–dimethyl urea and simazin (2-chloro–4, 6–bis (ethylamino)–S–triazine) (at high concentration) and the inhibition of enzyme synthesis. At lower concentrations, however, simazin stimulated nitrate reductase.
6. In a single leaf, synthesis of enzymes was observed only in portions exposed to light, whereas little activity was present in the dark covered part.
7. CO_2 deprivation severely inhibited the synthesis of enzymes in the light. Sucrose could not reverse this effect.
8. In excised embryos cultured in synthetic media containing sucrose, light was also essential for enzyme formation.
9. It is suggested that redox changes taking place in the green tissues as a result of Hill reaction create conditions favourable for the inducd synthesis of nitrate reductase and nitrite reductase (Sawhney and Naik, 1972).

3.6. Nitrate Reductase and Stress

Unfavourable temperature, CO_2 levels, water availability and salinity reduce the activity of the enzyme (Beevers and Hageman, 1969; Morilla et al. 1973) largely because of an inhibition of protein synthesis. Desiccation of leaves resulted in markedly lower (60 to 70%) levels of nitrate reductase activity at leaf water potentials of –6 to –8 bars (Morilla et al. 1973). Photosynthesis had declined only 10 to 20 percent at these water potential, however (Boyer, 1970).

In addition to the importance of nitrate reductase as an indicator of protein synthesis by the crop, the decline in nitrate reductase activity represents another equally important effect during dry times. Nitrate reductase is the first enzyme involved in enzymic reduction of nitrate and its eventual incorporation into proteins as the amino form. Because the enzyme may have low enzyme activities to control the flux of reduced nitrogen for the plant, a lower activity of the enzyme may mean a lower flux of reduced nitrogen and consequently decreased capability for protein production (Beevers and Hageman, 1969).

Plant leaves are a sink for nitrogen during the vegetative stage and afterwards, this nitrogen is remobilized for later use. The method used in determining total nitrogen is time consuming and may be hazardous as it involves the use of concentrated sulphuric acid. Therefore, leaf tissue and sap nitrate are used as indicators of plants nitrogen status, and many farmers use this as a measure in making decisions regarding fertilizer application rates. These measurements of leaf tissue nitrate allocated to the leaves is temporarily stored in vacuoles and then remobilized especially when nitrogen supply is insufficient to meet demand (Bartenbach et al. 2010) or during senescence (Table 3.2).

TABLE 3.2 Nitrate content of shoot and root of T(N) 1 and Ponni grown in nitrate for a duration of five days

N-concentration	T(N) 1		Ponni	
(mg/ml)	Shoot	Root	Shoot	Root
21	3.73	183.3	172.0	212.2
42	276.80	402.7	167.7	487.2
84	526.20	483.2	128.2	259.3

Nitrate in mg/100 mg dry weight

Nitrogen is transported mainly via amino acids. In rice and wheat upto 80% of grain nitrogen contents are derived from leaves (Kant et al. 2011). Since most plants perform nitrate vacuole storage and tolerate high ion concentrations, it is reasonable to assume that nitrate has an important function as an osmotic agent (Wicker et al. 2007). During the growth and development of plants, nitrogen is moved into and out of proteins in different organs and transported between organs in a limited number of transport compounds. Some of the organic N is moved between compounds via the activity of transaminases and glutamine amide transferases, but a significant portion is released as NH_3 and reassembled via glutamine synthase (GS) (Miflin and Habash, 2002). There are two GS isoenzymes (Cytosolic GS1 and plastid GS2) in most plant species, having a different subcellular compartmental location (Tobin and Yamaya, 2001; Cao et al. 2008).

Nitrate nitrogen is used in various processes including absorption, vacuole storage, xylem transport, reduction and incorporation into organic forms (Wicker et al. 2007). Primary nitrate assimilation takes predominantly in the roots of the plant, being strongly dependent on the age and limitation of space for root growth (Marquez et al. 2007). Nitrate taken up by plants is reduced to nitrite by nitrate reductase (Kuoadio et al. 2007; Cao et al. 2008; Rosales et al. 2011). The enzyme catalyzes the reduction of nitrate to nitrite with pyridine nucleotide in nitrogen assimilation in higher plants (Ahmad and Abdin, 1999). Since nitrite is highly reactive, plant cells immediately transport the nitrite from the cytosol into chloroplast in leaves and plastid in roots. In this organelles, nitrite is further reduced to NH_4^+ by nitrite reductase (Rosales et al. 2011).

Nitrate In Cytosol	— NO_2^-	Nitrate Reductase	Transported	Chloroplast in leaves Plastids in root	Transported

In higher plants, cytosolic NAD(P)H–nitrate reductase is rapidly modulated by environmental conditions such as light, CO_2 or oxygen availability in leaves. Nitrate reductase is activated by photosynthesis, reaching an activation state of 60-80%. In the dark or after stomatal closure, leaf nitrate reductase is inactivated down to 20 or 40% of its maximum activity. In roots hypoxia and anoxia activate nitrate reductase, whereas high oxygen supply inactivates nitrate reductase (Kaiser et al. 1999).

Malonate inhibited completely anaerobic dark reduction of nitrate in rice leaves and the inhibition was reversed by fumarate. Under dark anaerobic conditions $14CO_2$ evolution from endogenously labeled substrates was dependent on nitrate reduction. When nitrate reduction was inhibited by tungstate and cyanide $14CO_2$ evolution was inhibited and conversely and these results suggest a mitochondria origin of NADH for nitrate reduction and this suggestion was confirmed by following the evolution of $14CO_2$ from endogenous substrates, most organic acids, labeled in darkness. In comparison, the reduction of nitrite was very slow and commenced only after a time lag of 20 min. This could be a detoxification mechanism for the removal of nitrite which does not normally accumulate in leaves under natural conditions. NADH acted as a physiological reductant for dark aerobic nitrite reduction (Ramarao et al. 1981).

3.7. Genotypic Variations

To determine the activities of these enzymes in rice, 53 cultivars from six varietal groups based on isoenzyme variations were randomly selected. Seedlings were analyzed in vitro for nitrate reductase activity (NRA) and nitrate content. Significant differences in NADH-NRA, NADPH-NRA and nitrate content among rice cultivars and six varietal groups were observed. Differences in NRA among cultivars within a group were significant for some groups but not so in nitrate content. Among rice genotypes, therefore, differences in NADH–NRA were found. Based on NADH-NRA, rice genotypes could be classified into three NRA categories — high, moderate and low. NRA expresses the ability of the plants to use nitrate. Results suggest that certain cultivars and varietal groups were good nitrate assimilators. Cultivars with high NRA were identified. Of the six varietal groups, Groups II, V and I (Indica) had higher NRA than Japonica, floating and deep water rices. In all cultivars, activity of NADH-NR was higher than that of NADPH-NR (Table 3.3). In separate experiments, 14 cultivars with varying NRA levels were evaluated under irrigated lowland conditions (Table 3.4). Significant and positive correlations were observed between NRA and grain yield and between NRA and biomass (Barlaan and Ichii, 1996).

TABLE 3.3 Mean values of nitrate reductase activities and nitrate content of rice cultivars from six varietal groups

Varietal group	Cultivar	Conventional group	NADPH – NRA (nmol NO_2^- min^{-1} gfw^{-1})	NADH-RNA	NO_3^- content (μmol NO_3^- gfw^{-1})
I	Bhasamanik	Indica	8.7	334.6	54.0
	BPI 76	"	9.4	467.7	61.0
	BR1	"	4.6	286.8	57.6
	British Guiana 79	"	12.8	605.0	57.0
	Kwang Lu Az 4				
	Mudgo	"	5.9	386.8	56.6
	LEB Mue Nahng	"	1.6	176.7	58.8
	III Padi Bali				
	PTB 10	"	3.2	336.2	54.3
		"	6.6	209.6	54.6
	Suweon 287	Javanica	10.1	389.5	57.4
		Indica			
		"	6.0	376.7	52.1

II	Boro 1	Indica	8.4	504.2	56.3
	Chinsurah Boro 1	"	7.3	451.4	56.5
	Dharial	"	9.8	527.3	55.5
	Jhona 349	"	12.8	494.0	54.6
	Kamal Kati (aus)	"	8.9	302.8	56.3
	Kalu Balawee	"	5.5	272.9	52.9
	Kele	"	14.9	551.6	55.7
	M 142	"	18.3	412.1	53.1
	N 22	"	5.7	379.4	62.1
	Surjamukhi	"	4.4	371.4	52.5
III	Aswania	Indica	5.7	279.1	68.2
	Bamoia 341	"	5.9	180.6	68.1
	Bhadoia 233	"	10.3	399.8	63.3
	Goai	"	7.8	238.9	64.9
	Laki	"	15.1	363.6	69.6
	Taothabi	"	9.2	341.0	66.8
IV	Rayada 16.02	Indica	5.7	329.1	67.2
	Rayada 16.03	"	11.7	340.4	64.0
	Rayada 16.04	"	9.6	254.0	66.8
	Rayada 16.08	"	14.4	299.6	68.6
	Rayada 16.10	"	24.7	410.8	70.5
	Rayada 16.13	"	15.6	249.4	63.5
V	Atte	Indica	12.4	319.8	59.0
	Basmati	"	9.4	437.3	62.8
	Doma Siyah	"	14.4	461.5	61.9
	Doma Zard	"	10.5	510.8	58.8
	Firooze	"	10.5	526.6	58.5
	Hansraj	"	10.8	345.6	56.9
	Kataribogh	"	7.6	398.8	57.8
	Mehr	"	12.8	464.7	63.4
	Radhum pagal	"	7.6	325.5	62.3
	T 3	"	11.4	371.9	66.9
VI	Azucena	Indica	5.0	294.8	56.5
	Chikushi	Japonica	4.8	221.7	57.3
	E 425	Indica	17.4	385.6	58.0
	Ratio	Intermediate	4.8	385.2	59.2
	Kotobouki Mochi	Japonica	4.8	212.3	56.3
	Naukong 31	"	6.0	231.8	58.0
	Sungliao 2	"	10.0	292.8	65.4
	Tg (9)	Indica	9.6	493.3	63.0
	Taichung 65	Japonica	8.2	288.2	56.5
	Takao Mochi	"	8.9	229.0	60.8

TABLE 3.4 Means of leaf NR activity at 15 days after transplanting, grain yield and biomass of rice cultivars evaluated under irrigated lowland.

Cultivar/ Line	NADH-NRA (n mol NO_2^- mm^{-1} gfw^{-1})	NADPH-NRA	Grain Yield (gm^{-2})	Biomass (gm^{-2})
British Guiana 79	282.7	5.8	666.2	1880.3
Firooze	239.6	6.0	598.6	1514.9
Mudgo	245.0	7.0	682.2	2396.1
Dharial	126.6	6.2	621.8	1475.8
Norin 8	177.2	5.7	605.5	1445.9
M 819	3.0	5.7	441.2	1336.0
Nipponbare	181.4	5.3	604.2	1463.5
NR 827	89.2	4.8	323.2	642.2
NR 676	87.8	5.2	279.7	559.1
IR 30	247.1	6.0	523.7	1193.2
C 25	7.5	5.0	339.8	1175.4
C 27	25.8	5.3	235.0	1004.9
C 32	38.9	5.0	301.0	1182.1
C 33	31.8	5.8	139.4	987.1
Mean	127.4	5.7	454.4	1304.0
LSDO5	44.9	2.4	138.3	291.7
P (F test)	**	Ns	**	**
CV (%)	21.0	24.9	18.1	10.6

** significant at 1% level.

Nitrate reductase in leaves of Taichung 65 and Ai-chaw-Wu-chien showed higher levels of activities during their growth regardless of crop seasons. The first peak occurred at the beginning of tillering, the nitrate reductase activity in roots was higher than that in leaves. In the second crop season the nitrate reductase activity in leaves at tillering had the same high level as at flowering; however, nitrate reductase activity in roots is very low (Yuan and Shien, 1980).

3.8. Factors Affecting Nitrate Reductase

Shieh and Liao (1987) reported that nitrate reductse activities were enhanced by nitrogen concentration. No detectable activity of nitrate reductase was found in rice roots grown at nitrogen levels below 20 ppm nitrogen.

Barlaan et al. (1998) studied seven cultivars and seven NR mutants grown under irrigated low land conditions bio assayed *in vitro* for NRA at different growth stages and evaluated for grain yield, biomass and other agronomic characters. NADH-NRA was present at minimal levels in all cultivars before transplanting. Fifteen days after transplanting (DAT), NADH-NRA in all cultivars reached optimum levels and gradually declined until maturity. A similar pattern was observed in most of the NR mutants but with lower level of activity: NADH-NRA in cultivars and mutants gradually increased from 15 to 45 DAT and then fluctuated between 5 to 10 n mol NO_2^- min^{-1} gfw^{-1} until maturity. Nitrate availability in the soil may significantly influence NRA and nitrate accumulation in rice in irrigated lowlands. Effects of low or deficient levels of NR on traits evaluated varied among mutants. A decrease of 27 to 73% in grain yield was observed in mutants relative to original cultivars.

When five N-nitrate (N-NO_3^-) : N–ammonium (N–NH_4^+) ratios (100:00, 80:20, 60:40, 50:50 and 40:60) in water culture solution was maintained and analysis carried at three different growth stages, it was observed that N supply in the exclusive form of nitrate, or ammonium at higher proportions than nitrate, decreased dry matter, especially during panicle exsertion, affecting the yield. The maximum dry matter production of rice cultivars occurred at nitrate rates between 58 and 68%. The maximum grain yield was obtained at nitrate ratios between 75 and 78%. The excessive accumulation of nitrate in plant tissues due to low activity of nitrate reductase in the initial growth phase, and excess of ammonium were the main causes of decline in rice growth and yield, when nitrate was the only N form or when ammonium was used at higher proportions than nitrate in the nutrient solution.

3.9. Grain Yield and Nitrate Reductase Activity

Considerable variation in foliar nitrate reductase activity (NRA) was observed in 24 varieties of *Oryza sativa*. NRA was positively correlated with grain weight. Dry weight of 10 days old seedlings was positively correlated both with grain weight and with foliar NRA (Yang and Sung, 1980).

Rice genotypes have shown wide variability and considerable potential for nitrate–N assimilation. This study relates weed competitiveness to the ability of the rice plant to assimilate nitrate at early seedling stage corresponding to high nitrate reductase activity at early seedling stage. The genotypic differences in rice cultivars representing traditional and improved glaberima, japonica and indica groups were studied by growing the rice in culture solutions comprising low nitrate, high nitrate, ammonium nitrate and ammonium. Leaf nitrate reductase activity was measured at 7, 14 and 21 days. Preliminary results showed large differences in nitrate reductase activity among genotypes. Leaf nitrate reductase activity was positively correlated with the presence of ammonium in japonica and indica types but was not in traditional lowland types. Results suggest that there is a genotypic nitrate accumulation threshold level that triggers nitrate reductase activity (Ouko et al. 2002).

3.10. Influence of Abiotic Stress

When NR activity was analyzed at 28° and 6°C it showed that approximately 40% of the enzyme activity remains in plants exposed to 6°C (0.17 μ moles NO_2^- h^{-1} g^{-1} MF), in comparison with the activity observed in control plants (28°C) (0.45 μ moles NO_2^- h^{-1} g^{-1} MF) (Borges et al. 2004). Nitrate reductase activity was analyzed and revealed the involvement of this enzyme in chlorate action, even at low temperature (Table 3.1).

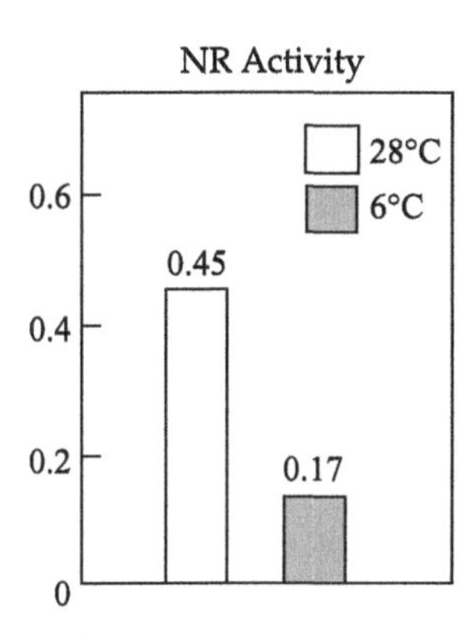

FIGURE 3.3 Cold influence on nitrate reductase activity in rice.

Nitrate reductase activity showed a decline under water stress conditions. In light conditions NR activity was greater than in the dark ones in all the treatments including unstressed seedlings (Pandey and Agarwal, 1998). An NR inactivating factor was found in extracts of leaf blades, leaf sheaths and roots of rice seedlings. The factor was nondialyzable, precipitable with $(NH_4)_2SO_4$ and heat labile. The factor from rice roots inactivated NADH nitrate reductase, $FMNH_2$ nitrate reductase and NADH cytochronie c reductase from rice shoots, but had no effect on the activities of NADH diaphorase and nitrate reductase. The factors from rice shoots, and maize roots inactivated NADH nitrate reductase prepared from cultured rice cells. The factor from cultured rice cells also inactivated rice shoot NADH nitrate reductase (Yamaya and Ohira, 1978). Nitrate reductase inhibitor is usually found in the roots of rice plants (CV. MRT) but it was also produced in the shoots of aging plants. The inhibitor was inducible in the shoot of the rice seedlings by dark, minus nitrate or plus ammonium treatment. There appears to a general involvement of the inhibitor in the control of nitrate assimilation in the plant (Leong and Shen, 1982).

Activities of enzymes nitrate reductase (NR) and nitrite reductase (NiR) were determined in rice seedlings differing in salt tolerance raised under increasing levels of NaCl salinity. Salinity caused marked increase in *in vivo* NR activity in roots and shoots of salt tolerant CVs CSR-1 and CSR-3 whereas in salt sensitive CVs Ratna and Jaya a marked inhibition in *In vivo* NR activity was observed under salinization. Under both controls as well as salt treatments in all cultivars roots always maintained higher level of *In vivo* NR activity than shoots. *In vitro* NR activity increased in both roots and shoots of all cultivars during early days of growth with maximum at 10-15 days and decreased thereafter. In salt tolerant cultivars salinity caused an increase in *in vitro* NR activity in shoots but not in roots whereas in salt sensitive activity of the enzyme was always more in salt stressed seedlings compared to controls. Salinity increased NR activity in seedlings of sensitive cultivars whereas in tolerants suppression in root NiR activity was observed due to salinity(Table 3.5). Like NR the activity of NiR was also higher in roots than shoots. 1M NaCl in the enzyme assay medium suppressed *in vivo* NR activity in roots of 15 days old nonsalinized seedlings with more suppression in sensitive cultivars than tolerant. Results suggest possible different behaviours of nitrogen assimilatory enzymes in rice cultivars differing in salt tolerance and that salt tolerance ability is associated with high *in vivo* NR activity in seedlings and its further activation under salinization (Katiyar and Dubey, 1992).

TABLE 3.5 Effect of salinity stress on NRA *in vivo* and *in vitro* in salt tolerant and sensitive cultivars of rice

Enzyme	Salt tolerant	Salt sensitive
In vivo	Increase in NRA in roots and shoots 1M NaCl supress NRA	Inhibition of NRA 1M NaCl supress more NRA.
In vitro	Increase of NRA in shoot but not in root	Increase of NRA

Application of 2, 4-D foliar spray or when used as an herbicides showed that the nitrate content is increased (significant at 2.5 to 5% level). Similarly, the nitrate reductase activity, estimated *in vivo,* exhibited an increase which was at 0.5% level of significance after seven days of 2, 4-D supply to the plants. Nevertheless, the activity did not show significant increase when estimated *in vitro.* NR activities showed a consistent decrease in activity, the decrease doubled after three days of herbicide supply and remained almost same after seven days. The results (Table 3.6) indicate that the herbicide is probably not suppressing the uptake of nitrate from the soil but enhancing the process of uptake and as a result there is an increase in nitrate content and nitrate reductase from the shoot (Borges et al. 2004).

TABLE 3.6 Effect of 2,4-D on the nitrate reductase and nitrate content from rice shoot

Days after 2,4-D	Nitrate content (μmol NO_3^- gfw^{-1})		Nitrate reductase (μmol NO_2^- 30 min^{-1} gfw^{-1})	
Addition	Control	Experiment	Control	Experiment
0	5.10±0.35	–	0.90±0.12	–
1	5.30±0.29	5.7±0.3	0.85±0.25	1.8±0.21
3	5.50±0.15	6.5±0.7	1.25±0.45	1.8±0.21
7	5.80±0.48	7.1±0.2	1.40±0.20	2.3±0.30

However, Huffaker and Peterson (1974) have reported chemical interference with nitrogen metabolism and decreased protein content of crops herbicidal phytotoxicity. Hence, the existence of the interdependence between NO_3^-, NRA and protein production which has been established may be influenced by herbicides.

Although butachlor has not been reported to inhibit photosynthesis, in experiments with rice, crude protein, amino acids and nitrate reductase activity differed suggesting that butachlor may act as an inhibitor of protein synthesis (Chen et al. 1981).

Nitrate, a major source of nitrogen for higher plants, is taken up from the soil by energy dependent nitrate transporters present in roots where it may be reduced to ammonium, stored in the vacuole or transported to leaf. The majority site of nitrate assimilation is leaf for most of the crop plants and root for woody plants. Once inside the cell, nitrate is converted to ammonium (reduced nitrogen) in two successive steps catalyzed by nitrate reductase (NR) and nitrite reductase (NiR) in cytosol and chloroplast, respectively. The first step is assumed to be the rate limiting step. The enzymes of nitrate assimilation are regulated at both transcriptional and post transcriptional levels by various endogenous and exogenous facts like nitrate, CO_2, light, hormones, temperature, carbon and nitrogen metabolites:

1. Nitrate and light play major roles in the regulation of nitrate assimilation pathway. Although the exact mechanism is not known, the involvement of several signaling intermediates have been shown to mediate this regulation. For instance, the involvement of a G-protein in the regulation of NR has been reported in maize leaves.

2. Calcium is an important second messenger which mediates the responses of various external and internal signals.
3. Presence of calcium-dependent protein kinases and their roles in regulating NR have been identified from various plants. Similarly, phosphoinositides have been shown to play a role in NR regulation. In spite of all the developments there have been very few reports on the regulation of these enzymes in rice, which is an ammonium utilizing crop plant.

3.11. *Mechanism of Regulation of Nitrate Reductase Activity*

Ali et al.(2007b) in a study was to determine the effects of activators and inhibitors of signalizing pathways on nitrate induced NR activity and transcript levels in rice leaves. Rice plants (*Oryza sativa* L) were grown hydropornically (without nutrients) on germination paper under continuous white light for 10-12 days. Leaves, adapted for darkness for two days, were treated with signaling agents both in light and dark. The results show that Okadaic acid (protein phosphatases/and 2A inhibitor) and lithium (inhibitor of IP_3 pathway) specifically inhibit activity and mRNA levels of NR under light conditions with no effect on NR activity. Both NR and NiR activities are inhibited and stimulated in the dark by PMA and Bisindolyl maleimide (BIM) (PKC activator and inhibitor), respectively. Cholera toxin specifically enhances NR activity and steady state levels of mRNA in the dark showing the involvement of G proteins. Calcium has a stimulatory effect on both the enzymes, with an increase in NR mRNA

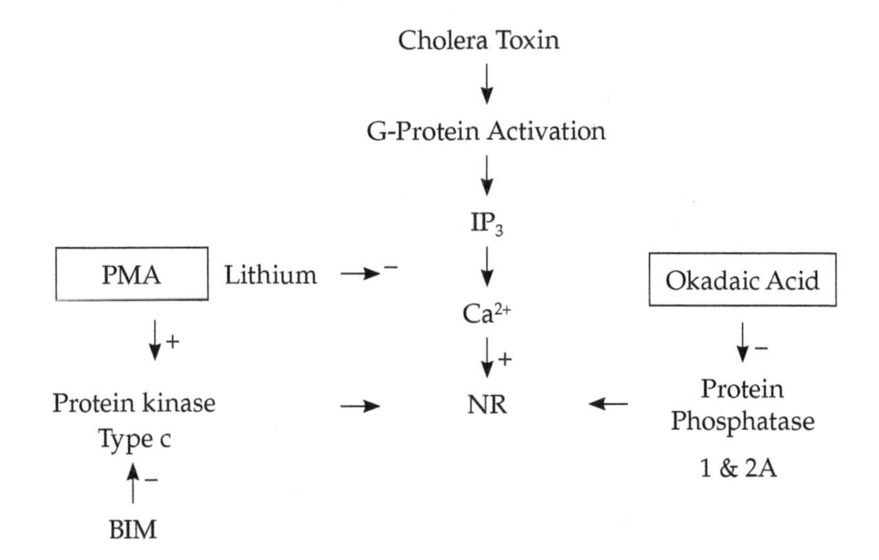

levels albeit to a lesser content. These results suggest that the activities of nitrate assimilatory enzymes in rice are regulated independently by G protein and IP_3 mediated pathways and co-regulated by PKC and calcium.

References

Ahmad, A. and M.Z. Abdin. 1999. NADH: Nitrate reductase and NAD(P)H: Nitrate reductase activities in mustard seedlings. *Plant Sci.,* **14:** 1-8.

Ali, A., S. Sivakami and N. Raghuram. 2007a. Effect of nitrate, nitrite, ammonium, glutamate, glutamine and 2-oxoglutarate on RNA levels and enzyme activities of nitrate reductase and nitrite reductase in rice. *Physiol. Mol. Biol. Plants,* **13:** 17-25

Ali, A., S. Sivakami and N. Raghuram. 2007b. Regulation of activity and transcript levels of NR in rice (*Oryza sativa*): Roles of protein kinase and G-Proteins. *Plant Sci.,* **172:** 406-413.

Aslam, M., R.C. Huffaker, D.W. Rains and K.P. Rao. 1979. Influence of light and ambience carbon dioxide concentration on nitrate assimilation by intact barley seedlings. *Plant Physiol.,* **63:** 1205-1209.

Aslam, M., R.L. Travis, D.W. Rains and R.C. Huffaker. 1997. Differential effect of ammonium on the induction of nitrate and nitrite reductase activities in root of barely (*Hordeum vulgare*) seedlings. *Physiol. Plant.,* **101:** 612-619.

Arima, Y. and K. Kumazawa. 1975. A kinetic study of amide and amino acid synthesis in rice seedling roots fed with 15 N labeled ammonium (Part 2): Physiological significance of glutamine on nitrogen absorption and assimilation in plants. *J. Sci. Soil and Manure, Jpn.,* **46:** 355-361.

Arima, Y., T. Horinouchi and K. Kumazawa. 1976. Variation and regulation of glutamine synthetase activity in rice seedlings fed with ammonium and nitrate (Part 4): Physiological significance of glutamine on nitrogen absorption and assimilation in plants. *J. Sci. Soil and Manure, Jpn.,* **47:** 198-203.

Barlaan, E.A. and M. Ichii. 1996. Genotypic variability in nitrate assimilation in rice *In:* Rice Genetics (ed) G.S.Khush., IRRI, Philippines pp. 434-440.

Barlaan, E.A., H. Sato and M. Ichii. 1998. Nitrate reductase activities in rice genotypes in irrigated lowlands. *Crop Sci.,* **38:** 728-734.

Bartenbach, J.V.D.F., M. Bogner, M. Dynowski and U. Ludewig. 2010. CLC-b-mediated NO_3^-/H^+ exchange across the tonoplast of Arabidopsis vacuoles. *Plant Cell Physiol,* **51:** 960-968.

Beevers, I. and R.H. Hageman.1969. Nitrate reductase in plants. *Annu. Rev. Plant Physiol,* **20:** 495-522.

Borges, R., E.C. Miguel, J.M.R. Dias, M. da Cunha, R.E. Bressan-Smith, J.G. de Oliveira and G.A. de Sowa Filho. 2004. Ultrastructural, physiological and biochemical analysis of Chlorate toxicity on rice seedlings. *Plant Sci.,* **166:** 1057-1062.

Boyer, J.S. 1970. Differing sensitivity of photosynthesis in low leaf water potentials in corn and soybean. *Plant Physiol,* **46:** 236-239.

Campbell, W.H. 1999. Nitrate reductase structure, function and regulation: Bridging the gap between biochemistry and physiology. *Annu. Rev, Plant Physiol, Plant Mol. Blol.,* **50:** 277-303.

Cao, Y., X.R. Fan, S.B. Sun, G.H. Xu, J. Hu and Q.R. Shen. 2008. Effect of nitrate on activities and transcript levels of nitrate reductase and glutamine synthetase in rice. *Pedosphere,* **18:** 664-673.

Chen, Y.L., C.C. Lo and Y.S. Wang. 1981. Effects of butachlor on carbon and nitrogen assimilation in rice and barnyard grass. *Weed Res (Japan).*

Crawford, N.M. 1995. Nitrate: Nutrient and signal for plant growth. *Plant Cell,* **4:** 859-868.

Daniel-Vedele, F., S. Filleur and M. Caboche. 1998. Nitrate transport: A key step in nitrate assimilation. *Curr. Opin. Plant Biol.,* **1:** 235-239.

Granato, T.C. and C.D. Raper. 1989. Proliferation of maize roots in response to localized supply of nitrate. *J. Exp. Bot.,* **40:** 263-275.

Hackett, C. 1972. A method of applying nutrients locally to roots under controlled conditions and some morphological effects of locally applied nitrate on branching of wheat roots. *Aust. J. Biol. Sci.,* **25:** 1169-1180.

Hageman, R.H. 1979. Integration of nitrogen assimilation in relation to yield. pp. 591-611. *In:* Nitrogen Assimilation of plants. (ed.) E. J. Hewitt and C.V. Cutting, Academic Press, London.

H. M., Hamat., A. Kleinhofs and R.L.Warner. 1989. Nitrate reductase induction and molecular characterization in rice (*Oryza sativaL*). *Mol. Gen. Genet,* **218:** 93-98.

Huang, Y.Z., Z.W. Feng and F.Z. Zhang. 2000. Study on loss of nitrogen fertilizer from agriculture fields and counter measure. *J. Graduate School Acad, Sin.,* **17:** 43-58.

Huffaker, R.C. and L.W. Peterson. 1974. Protein turn over in plants and possible means of its regulation. *Annu, Rev. Plant Physiol.,* **25:** 363-392.

Iyer, R.K., R. Tuli and J. Thomas. 1981. Glatamine synthetases from rice: Purification and preliminary characterization of two forms in leaves and one form in roots. *Arch. Biochem. Biophys.,* **209:** 628-635.

Kaiser, N.M., H. Weiner and S.C. Huber. 1999. Nitrate reductase in higher plants: A case study for transduction of environmental stimuli into control of catalytic activity. *Physiol. Plant.,* **105:** 381-389.

Kant, S., Y.M. Bi and S.J. Rothstein. 2011. Understanding plant response to nitrogen limitation for the improvement of crop nitrogen use efficiency. *J. Exp. Bot,* **62:** 1499-1509.

Katiyar, S. and R.S. Dubey. 1992. Influence of NaCl salinity on behaviors of nitrate reductase and nitrite reductase in rice seedings differing in salt tolerance. *J. Agron. Crop Sci.,* **169:** 289-297.

Kouadiao, J.Y., H.T. Konakon, M. Kone, M. Zouzou and P.A. Anno. 2007. Optimum condition for cotton nitrate reductase extraction and activity measurement. *Afr. J. Biotechnol.,* **6:** 923-928.

Lea, U.S., M.T. Leydecker, I. Quillere, C. Meyer and M. Lillo. 2006. Post transitional regulation of nitrate reductase strongly affects the levels of free amino acids and nitrate, whereas transcriptional regulation has only minor influence. *Plant Physiol.,* **140:** 1085-1094.

Leong, C.C. and T.C. Shen. 1982. Occurrence of nitrate reductase inhibitor in rice plants. *Plant Physiol.,* **70:** 1762-1763.

Marquez, A.J., M. Betti, M. Garcia-Calderon, A. Credai, P. Diaz and J. Monza. 2007. Primary and secondary nitrogen assimilation in Lotus Japonicus and the relationship with drought stress. *Lotus Newsl.,* **37:** 71-73.

Miflin, B.J. and D.Z. Habash. 2002. The role of glutamine synthetase and glutamate dehydrogenase in nitrogen assimilation and possibilities for improvement in the nitrogen utilization of crops. *J. Exp. Bot.,* **53:** 979-987.

Morilla, C.A., J.S. Boyer and R.H. Hageman. 1973. Nitrate reductase activity and polyribosomal content of corn (*Zea mays* L) having low leaf water potentials. *Plant Physiol.,* **51:** 817-824.

Morozkina, E.V. and R.A. Zuyagi iskaya. 2007. Nitrate reductases: Structure, functions and effect of stress factors. *Biochemistry (Moscow)*, **72:** 1151-1160.
Okuo, M., F. Asch and M. Becker. 2002. Screening of rice genotypes for early leaf nitrate reductase activity. *In:* Challenges to organic farming and sustainable land use in tropics and subtropics. Wotzenhausen.
Pandey, R. and R.M. Agarwal. 1998. Water stress induced changes in proline contents and nitrate reductase activity in rice under light and dark conditions. *Physiol. Mol. Biol, Plants*, **4:** 53-57.
Raghuram, N. and S.K. Soproy.1999. Role of nitrate, nitrite and ammonium ion in phytochrome regulation of nitrate reductase gene expression in maize. *Biochem Mol. Biol. Intl.*, **47:** 139-249.
Rajagopalan, K.V. 1989. Chemistry and biology of molybdenum cofactor. pp. 212-227. *In:* Molecular and genetic aspects of nitrate assimilation. Oxford Science Publications, Oxford.
Ramarao, C.S., Srinivasan and M.S. Naik. 1981. Origin of reductant for reduction of nitrate and nitrite in rice and wheat leaves *in vitro*. *New Phytol*, **87:** 517-523.
Rhodes, D., G.A. Rendon and G.R.Stewart. 1975. The control of glutamine synthetase level in *Lemna minor* L. *Planta*, **125:** 201-211.
Sasakawa, H. and Y. Yamamoto. 1978. Comparison of the uptake of nitrate and ammonium by rice seedlings — Influences of light, temperature, oxygen concentration, exogenous sucrose and metabolic inhibitor. *Plant Physiol.*, **62:** 665-669.
Sattelmacher, B., J. Gerendar, K. Thomas, H. Bruck and N.H. Bagdady. 1993. Interaction between root growth and mineral nutrition. *Environ. Exp. Bot.*, **33:** 63-73.
Sawhney, S.K. and M.S. Naik. 1972. Role of light in the synthesis of nitrate reductase and nitrite reductase in rice seedling. *Biochem. J.*, **130:** 475-485.
Shankar, N. and H.S. Srivastava. 1998. Effect of glutamine supply on nitrate reductase isoforms in maize seedling. *Phytochem.*, **47:** 701-706.
Shen, T.C. 1969. The induction of nitrate reductase and the preferential assimilation of ammonium in germinating rice seedlings. *Plant Physiol.*, **44:** 1650-1655.
Shen, T.C. 1972. Nitrate reductase of rice seedlings and its induction by organic nitrocom pounds. *Plant Physiol.*, **49:** 548-579.
Shieh, Y.J. and W.Y. Liao. 1987. Influence of growth temperature and nitrogen nutrition on photosynthesis and nitrogen metabolism in the rice plant (*Oryza sativa*L). *Bot. Bull. Academia Sinica*, **28:** 151-167.
Sivashankar, S. and A. Oaks. 1996. Nitrate assimilation in higher plants: The effect of metabolites and light. *Plant Physiol Biochem.*, **34:** 609-620.
Srivastava, H.S. 1980. Regulation of nitrate reductase in higher plants. *Phytochem.*, **19:** 725-733.
Stitt, M. 1999. Nitrate regulation of metabolism and growth. *Curr.Opin.Cell Biol.*, **2:** 178-186.
Tobin, A.K. and T. Yamaya. 2001. Cellular compartmentation and ammonium assimilation in rice and barley. *J. Exp. Bot.*, **52:** 591-604.
Wang, Y.H., D.F. Garvin and L.V. Kochian. 2001. Nitrate induced genes in tomato roots: Array analysis reveals novel genes that may play a role in nitrogen nutrition. *Plant Physiol.*, **127:** 345-359.
Wicker, T., F. Sabot, A. Hua-van, J.L. Bennetzen, P. Capy, B. Chalhoub A. Flavell, P. Leroy, M. Morgante, O. Panaud, E. Paux, P. SauMiguel and A.H. Schulman. 2007. A unified classification system for eukaryotic transposable elements. *Nat. Rev. Genet.* **8:** 973-982.

Yamaya, T. and K. Ohira. 1978. Nitrate reductase inactivating factor from rice seedlings. *Plant Cell Physiol.*, **19:** 211-220.
Yang, C.M. and J.M. Sung. 1980. Relations between nitrate reductase activity and growth of rice seedlings. *J. Agric. Assoc. China.* **111:** 15-23.
Yuan, H.F. and Y.S. Shien. 1980. Seasonal variations of nitrate reductase, Glutamate dehydrogenase and the soluble nitrogenous compounds during rice growth. *Bot. Bult Academia siniica,* **21:** 35-52.
Zhang, H., A. Jennings, P.W. Barlowt and B.G. Forde. 1999. Dual pathways for regulation of root branching by nitrate. *Proc. Natl. Acad, sci. USA,* **96:** 6529-6534.
Zhang, H. and B.G. Forde. 2000. Regulation of *Arabidopsis* root development by nitrate availability. *J. Exp. Bot.,* **51:** 51-59.

chapter four

Nitrite Reduction

4.1. Nitrite Reductase

Ferredoxin nitrate oxidoreductase referred to as nitrite reductase catalyzes the six electron reduction of nitrite to ammonia during an early step in the nitrogen assimilation pathway of oxygenic photosynthetic organisms, using reduced ferredoxin as the physiological electron donor (Hase et al. 2006). These ferredoxin-dependent assimilatory nitrite reductases are soluble enzymes with molecular masses of approximately 65 kDa and are located in the stromal space of chloroplasts in photosynthetic eukaryotes (Hase et al. 2006). Ferredoxin-dependent nitrite reductases are characterized by a unique active site that contains a siroheme coupled to a [4Fe–4S] cluster via a bridging sulphur atom from a cystenine residue (Swamy et al. 2005; Hase et al. 2006). Electron paramagnetic resonance (EPR) and resonance Raman studies have elucidated the electronic state of the coupled prosthetic group arrangement found in oxidized nitrite reductases (Krueger and Siegel, 1982a, b). This 'as isolated' form of nitrite reductases contains an oxidized, EPR-silent cluster in the [4Fe–4S] state and a siroheme that contains a high spin, six coordinate Fe^{4+} with water or some other weak ligand in the sixth axial position (Kuznetsova et al. 2004a, 2004b) with the fifth axial position being occupied by the cysteinyl sulphur that provides the bridge to the iron-sulphur cluster (Swamy et al. 2005). Although the [4Fe–4S] cluster and the siroheme are magnetically coupled, they behave as independent one-electron couples during electrochemical titrations, with E values of −290 and −370 mV for the siroheme and $([4Fe{-}4S]^{2+})^{+}$ cluster respectively.

Nitrite + 6 reduced ferredoxin + $7H^+$ = NH_3 + $2H_2O$ + 6 oxidised ferredoxin.

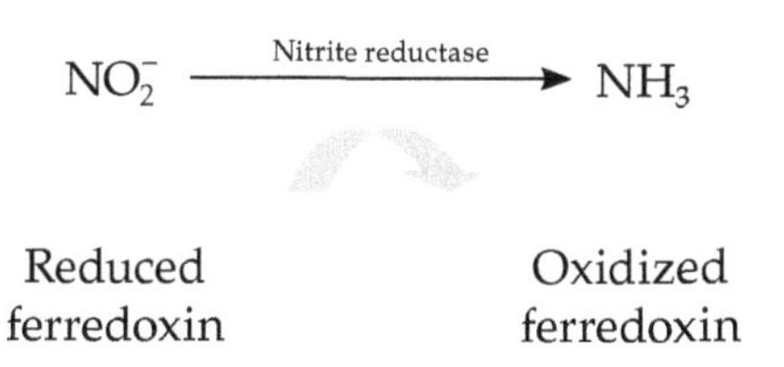

4.2. Siroheme

Siroheme is synthesized by the tetrapyrrole pathway, begins with the synthesis of 5-aminolevulinate (ALA) either by the condensation of the succinyl CoA and glycine, as in animals, yeast, and the a-group of the proteobacteria or by the transformation of glutamate, as in higher and lower plants and most other bacteria. Eight molecules of ALA are converted into Uroporphyrinogen III by the concerted action of three enzymes, namely ALA-dehydrotase, porphobiligen deaminase, and Urogen III synthase, Urogen III serves as the central template from which all biologically functional modified tetrapyrroles are derived by a number of enzyme-mediated modifications that include peripheral alterations to the acetate and propionate side chains, methylations, oxidation state of the macrocycle, ring size and the central metal ion.

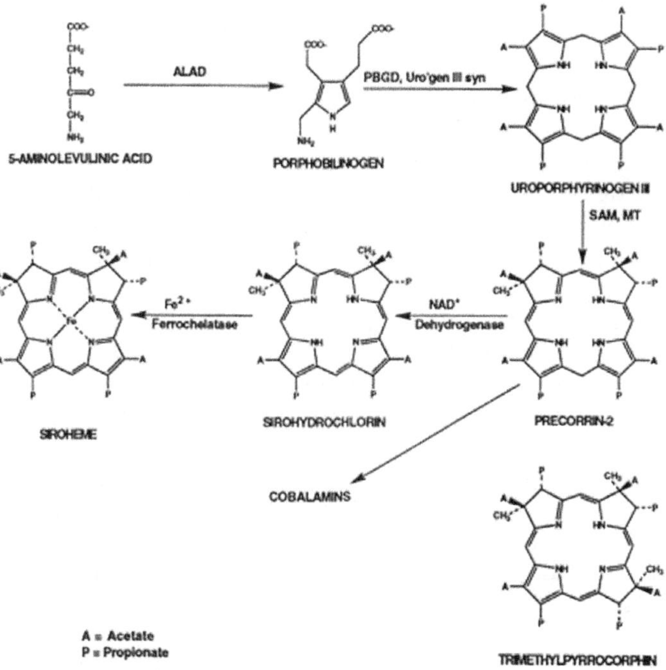

4.3. Nitrogen Assimilation

Nitrate assimilation by plants is one of the two primary pathways in the biological conversion of inorganic nitrogen into reduced nitrogenous compounds in the global nitrogen cycle. Nitrate assimilation occurs mostly in leaf mesophyll cells, and the resulting reduced nitrogen species are then used to synthesize all required amino acids and amino acid derived compounds.

Reactions throughout the course of the nitrate assimilation pathway occur in separate cells and cellular compartments; nitrate is first transferred from root epidermal cells to the cytosol of leaf cells via vascular tissues, from chloroplasts to sink organs (Forde, 2000). The nitrogenous metabolites involved in this pathway are mostly ionic at physiological pH and need specific transporters to allow them to penetrate the biological membranes.

Nitrite reductase uses siroheme, an iron containing mesobacteriochlorin, along side a [4Fe – 4S] cluster to perform the six-electron reduction of nitrate to ammonia. X-ray analysis of the structure provides in sight into the role of the siroheme in the powerful redox reaction, both as an anchor for the acid / base chemistry that directs substrate formation and as an electronically-flexible cofactor that derives the electron transfer reaction.

4.4. Induction of Nitrite Reductase

Changes in ferredoxin-nitrite reductase in etiolated rice seedlings were followed during induction by nitrate and light. Etiolated seedlings showed maximal induction of the enzyme activity in etiolated seedlings receiving nitrate in darkness increased half as much as that in nitrate-treated greening plants. The increase in nitrite reductase activity during induction coincided with an increase in the content of proteins immunoprecipitated by antibodies raised against spinach nitrite reductase. Light had no effect on the induction of the extractable nitrite reductase in the presence of nitrate. Poly (A)+RNA extracted from nitrate-treated greening shoots directed the synthesis in a rabbit reticulocytelysate of polypeptides immunoprecipitated by spinach nitrate reductase antibodies. One major polypeptide larger than the native enzyme was found among the translation products, suggesting that nitrite reductases in greening rice shoots are synthesized as a precursor form. Analysis of two dimensional electrophoretograms indicated the existence of isoforms of nitrite reductase in rice seedlings which had been immunoprecipitated with spinach nitrite reductase antibodies (Orgawa and Ida, 1987).

Nitrite reductase, which catalyzes the next step in assimilation of nitrate to ammonia, has been reported to be dependent for its activity on reductants generated in the light reaction of photosynthesis, but the role of light, if any, in the synthesis of the enzyme itself is not known (Ramirez et al. 1966). Nitrate, ammonium and glutamine had no effect on nitrite reductase in both light and dark. Glutamate and 2-oxoglutarate also showed a similar pattern of induction. The transcript levels of nitrite reductase increased to a similar extent in presence of nitrate (Ali et al. 2007). It is also noted that glutamine had no effect on nitrite reductase in both light and dark. Glutamate and 2–oxoglutarate had shown a similar pattern of induction. The enzyme is tightly regulated by nitrogen metabolites and light and are coregulated under certain conditions.

Nitrite is a natural intermediate of nitrate assimilation in higher plants. However, because of the concerted action of nitrate reductase and nitrite reductase normally it does not appear in high concentration in the cytoplasm. An internal accumulation of nitrite (caused by an external increase of herbicides) is toxic to plants, leading to extensive morphological and metabolic changes. In the experiments, the effects of nitrite and chlorate (converted *in vivo* to chloride, a slowly metabolizable, indirect model compound for nitrite) on the growth parameters and on the *in vivo* measured nitrate reductase and *in vitro* measured nitrite reductase activities were compared in 14 days old seedlings of rice (Cv. Oryzella) to confirm the use of the chlorate model in nitrite stress studies. The growth parameters (fresh weight, root and shoot lengths) and nitrate reductase and nitrite reductase activities were tested in hydroponic cultures, chlorate and nitrite being applied through the roots, in a whole plant system. The effect of chlorate proved to be very similar to nitrite, but the higher mobility of chlorate in the xylem caused a more pronounced inhibition in the shoots. Unlike nitrite toxicity the chlorate/chlorite effect is hardly alleviated by nitrite reductase. Both compounds decreased the nitrate reductase activity, but in the case of chlorate no induction was observed in low concentration range. Nitrite reductase was insensitive to both ions (Pecsvaradi and Zsoldos, 1996).

It has been known for sometimes, however, that the ultraviolet radiation in sunlight photochemically reduces nitrate in aqueous solution to nitrite. It had also been reported that ultraviolet light had a much stronger effect on nitrate assimilation in wheat leaves than light of longer wave lengths. Significant accumulation of nitrite in rice leaves exposed in sunlight, also been observed, even when nitrate reductase activity was negligible (Naik et al. 1976)

4.5. *Regulation of Nitrite Reductase*

The enzymes of nitrate assimilation are regulated at both transcriptional and post-transcriptional levels by various endogenous and exogenous factors like nitrate, CO_2, light, hormones, temperature, carbon and nitrogen metabolites. Even though nitrite reductase is an important enzyme of nitrate assimilation pathway, its regulation has received less attention. However, in a study it had been noted that calcium caused a significant stimulation of the activities of nitrite reductase at 5 and 10mM. This effect was observed only in dark with no effect in light.

4.6. *Variation in Genotypes*

Significant difference of nitrite reductase activity was observed among the rice genotypes (Barlaan and Ichii, 1996). However, no significant relationship between nitrite reductase and grain yield was observed (Table 4.1).

TABLE 4.1 Mean values of nitrite reductase activities and nitrate content of rice cultivars

Cultivar	Nitrite reductase activity (n mol NO_2^- min^{-1} g fw^{-1})	NO_3^- content (μ mol NO_3^- gfw^{-1})	Cultivar	Nitrite reductase activity (n mol NO_2^- min^{-1}gfw^{-1})	NO_3^- content (μ mol NO_3^- gfw^{-1})
Bhasamanik	99.6	54.0	Aswaina	72.3	68.2
BPI76	97.6	61.0	Bamoia 341	65.6	68.1
BR1	105.9	57.6	Bhadola 133	96.3	63.3
British Guiana 79	89.6	57.0	Goai	73.6	64.9
Kwang LuAi 4	89.3	56.6	Laki	92.6	69.6
Mudgo	82.0	58.8	Taothabi	83.6	66.8
LEB MueNahng III	114.9	54.3	Rayada 16.02	114.6	67.2
Padi Bali	71.3	54.6	Rayada 16.03	101.6	64.0
PRB10	93.3	57.4	Rayada 16.04	88.0	66.8
Suweon 287	122.6	52.1	Rayada 16.08	97.0	68.6
Boro 1	99.6	56.3	Rayada 16.10	100.0	70.5
Chinsurah Boro 1	100.9	56.5	Rayada 16.13	99.3	63.5
Dharial	59.0	55.5	Atte	69.0	59.0
Jhona 349	110.9	54.6	Basmati	101.3	62.8
Kamal Kati (ans)	97.0	56.3	Doma Siyah	122.9	61.9
Kalu Balawee	121.9	52.9	Firooze	93.3	58.5
Kele	90.6	55.7	Hansraj	77.0	56.9
M142	91.0	53.1	Katari Bhog	88.6	57.8
N22	55.3	62.1	Mehr	90.3	63.4
Surjamukhi	105.6	52.5	Radhunipagal	82.3	62.3
			T3	99.3	66.9

after Barlaan and Ichii (1996)

Rice plants grown under ammonium and nitrate nitrogen showed significant differences in nitrite reductase activities (Table 4.2).

TABLE 4.2 Nitrite reductase in IR22 rice seedling grown in NH_4^+ and NO_3^-

N source	Days germinated	Nitrate reductase Shoot	(n mol NO_2^- min^{-1} gfw^{-1}) Root
NH_4^+	5	41	34
	7	21	Trace
	10	24	Trace
	14	Trace	Trace
NO_3^-	5	109	202
	7	377	325
	10	764	311
	14	639	158
LSD (5%)		48	44

Nitrite reductase activity was higher in the seedling grown in NO_3^- than in that grown in NH_4^+. It was higher in the shoot than in the root. Peak activity occurred 10 days after germination in the shoot and 7 to 10 days after germination in the root of NO_3^- grown seedling, thus following closely the trend of nitrate reductase. Nitrite reductase activity was lower in the root than in the shoot of the NO_3^- grown seedling, except in the five-days old sample. The presence of higher levels of NO_3^-, nitrate reductase and nitrite reductase in the shoots than roots reflect the relatively higher rate of oxidation of NH_4^+ in the leaves (Marwaha and Juliano, 1976).

4.7. Nitrite in Cell Regeneration

The nitrite ion content and the activity of nitrate reductase and nitrite reductase in scutellum derived calluses of rice varieties using a modified R2 medium (medium A) and a medium derived from modified R2 medium (medium B). In medium A, marked differences were observed in callus growth between the varieties. The calluses of the poor–growth varieties were injured by toxic nitrite ions, which lead to browning and inhibited growth. The calluses of poor growth varieties had significantly lower levels of nitrite reductase activity than good growth varieties. On the other hand, no differences between groups were observed in the nitrate reductase activity. These results indicate that the higher nitrite ion levels observed in the poor growth varieties resulted from a lower ability to reduce nitrite and that nitrite reductase activity is one of the physiological factors that correlates with differences between varieties in rice cell cultures. In medium B, the calluses of the poor-growth varieties grew as well as the good–growth varieties, but also had significantly lower levels of nitrite reductase. Nitrate reductase activity was repressed in the calluses of both varieties in medium B compared to culture in medium A. The results suggest that repressed nitrate reductase activity causes the calluses of poor, growth varieties to accumulate only trace amounts of nitrite ions despite lower nitrite reductase activity and as a result, callus growth improved in medium B (Ogawa et al. 1999).

4.8. *Nitrite Reductase and Stress*

Salinity increased nitrite reductase activity in seedlings of sensitive cultivars whereas in tolerant suppression in root nitrite reductase activity was observed due to salinity. The activity of nitrite reductase was higher in roots than shoots (Katiyar and Dubey, 1992). Inhibitive effect of cadmium on activity of enzymes such as nitrate reductase and nitrite reductase is reported by Boussama (1999a, b).

The influence of thiobencarb at 1500 ppm on nitrite reductase did not show any appreciable change and nither nitrite is accumulated in the rice leaf sections.

Observation that synthesis of nitrite reductase is dependent on the synthesis of chloroplast protein appeared surprising since it is known that nitrite reductase is a monomeric protein of about 63 KDa and is nuclear encoded like nitrate reductase. However, in subsequent studies it has been shown that the synthesis of nitrite reductase also requires a 'plastidic factor' produced by functional chloroplasts. This explains the sensitivity of nitrite reductase to the addition of chloramphenicol, besides cyclohexamide.

In contrast to the large numbers of plant mutants lacking various activities of nitrate reductase, until recently no mutants lacking the enzyme have been isolated. This may be due in part to the toxic nature of nitrite, which would have a deleterious effect on metabolism. A mutant of barley has been isolated, that lacks nitrite reductase activity in the leaves and roots and can only be maintained by growth on ammonia or glutamine. The mutation is in a single nuclear gene Nir 1 and the plants have been shown to contain elevated levels of nitrate reductase activity (Duncanson et al. 1993). Vaucher et al. (1992) have expressed antisense nitrite reductase in RNA in tobacco plants. One transformed plant was shown to lack nitrite reductase activity and accumulated nitrite. The plant had a greatly reduced level of ammonia, glutamine and protein in the leaf tissue.

4.9. *Genetic Variations*

By conventional crosses of low regeneration rice strain Koshikari with high regeneration rice strain Kabalath, it was identified some quantative trait loci, which control the regeneration ability in rice. Using a map-based cloning, nitrite reductase determines regeneration ability in rice. Molecular analyses revealed that the poor regeneration ability of Koshikari is caused by lower expression than in Kasalath and the specific activity of nitrite reductase.

Using the nitrite reductase gene as a selection marker, it was succeeded in selectively transforming a foreign gene into rice without exogenous marker genes. It is demonstrated that nitrate assimilation is an important process in rice regeneration and also provide an additional selectable marker for rice transformation.

A ferredoxin–nitrite reductase (ECI.7.7.1) cDNA was isolated and sequenced from a λ gt11 cDNA library constructed from nitrate induced greening shoot of rice (*Oryza sativa* L) seedlings. The nucleotide sequence of the DNA clone contains an open reading frame of 1788 nucleotides. There exists a strong bias for the third codon usage of GIC (95.5%) as in the case of maize enzyme. The deduced amino acid sequence shows an overall homology to the maize (81%) and the dicot enzymes (70–74%) suggesting that primary structure of ferredoxim nitrite reductase is highly conserved in higher plants (Terada et al. 1995).

Transgenic plants of *Arabidopsis* bearing the spinach (*Spinacia olecacea*) nitrite reductase gene that catalyzes the six-electron reduction of nitrite to ammonia in the second step of the nitrate assimilation pathway were produced by using the cauliflower mosaic virus 355 promoter and nopalive synthase terminator. Integration of gene was confirmed by a genomic polymerase chain reaction (PCR) and southern–blot analysis. Its expression by a reverse transcriptase–PCR and two dimensional polyacryl amide gel electrophoresis western blot analysis; total (Spinach + Arabidopsis) nitrite reductase mRNA content by competitive reverse transcriptase–PCR; localization of nitrite reductase activity in the chloroplast by fractionation analysis, and NO_2 assimilation by analysis of reduced nitrogen derived from NO_2 (NO_2–RN). Twelve independent transgenic plant lines were characterized in depth. Three positive correlations were found for nitrite reductase gene expression: between total nitrite reductase mRNA and total nitrite reductase protein contents ($r = 0.74$), between total nitrite reductase protein and nitrite reductase activity ($r = 0.71$); and between nitrite reductase activity and NO_2^- RN ($r = 0.65$). Of these twelve lines, four had significantly higher nitrite reductase activity than the wild–type control ($P < 0.01$) and three had significantly higher NO_2^- RN ($P < 0.01$). Each of the latter three had one to two copies of spinach nitrite reductase cDNA per haploid genome. The nitrite reductase flux control coefficient for NO_2 assimilation was estimated to be about 0.4. A similar value was obtained for an nitrite reductase antisense tobacco (Cv Xanthix HFD8). The flux control coefficients of nitrate reductase and glutamine synthetase were much smaller than this value. These indicate the nitrite reductase as a controlling enzyme in NO_2^- assimilation by plants (Takahashi et al. 2001).

References

Ali, A., S. Sivakami and N. Raghuram. 2007. Effect of nitrate, nitrite, ammonium, glutamate, glutamine and 2-oxoglutarate on RNA levels and enzyme activities of nitrate reductase and nitrite reductase in rice. *Physiol. Mol. Boil. Plant,* **13:** 17-25.

Barlaan, E.A. and M. Ichii .1996. Genotypic variability in nitrate assimilation in rice. *In:* Rice Genetics (ed. G.S. Khush). IRRI, Philippines pp. 434-440.

Boussama, N., O. Ouartiti and M.H. Ghorbel. 1999a. Changes in growth and nitrogen assimilation in barley seedlings under cadmium stress. *J. Plant Nutr.,* **22:** 731-752.

Boussama, N., O. Ouartiti, A. Suzuki and M.H. Ghorbel. 1999b. Cd stress on nitrogen assimilation. J. *Plant Physiol* ., **155:** 310-317.

Duncanson, E., A.F. Gilkes, D.W. Kirk, A. Sherman and Jt Wray. 1993. nir 1, a conditional-lethal mutation in barley causing a defect in nitrite reduction. *Mol. Gen. Genet.,* **236:** 275-282.

Forde, B.G. 2000. Nitrate transporters in plants: Structure, function and regulation. *Biochem. Biophys. Acta,* **1465:** 219-235.

Hase, T., P. Schurmann and D.B. Knaff. 2006. The interaction of ferredoxin with ferredoxin-dependent enzymes. pp.477-498. *In:* Golbeck J. (editor) Photosystem 1. springer, Dordrecht, The Netherlands.

Katiyar, A. and R.S. Dubey. 1992. Influence of NaCl salinity on behaviors of nitrate reductase and nitrite reductase in rice seedlings differing in salt tolerance. *J. Agron. Crop Sci.,* **169:** 289-297.

Krueger, R.J. and F.M. Siegel. 1982a. Evidence for siroheme-FeS interaction in spinach ferredoxin-sulfite reductase. *Biochemistry,* **21:** 2905-2909.

Krueger, R.J. and L.M. Siegel. 1982b. Spinach siroheme exzymes: Isolation and characterization of ferredoxin-sulfite reductase and comparison of properties with ferredoxin-nitrite reductase. *Biochemistry,* **21:** 2892-2904.

Kuznetsova, S., D.B. Knaff, M. Hirasawa, R. Lagoutte and P. Setif. 2004a. The mechanism of spinach chloroplast ferredoxin dependent nitrite reductase-spectroscopic evidence for intermediate states. *Biochemistry,* **43:** 510-517.

Kuznetsova, S., D.B. Knaff, M. Hirasawa, P. Setif and P.A. Mattioli. 2004b. Reactions of spinach nitrite reductase with its substrate nitrite and a putative intermediate, hydroxylamine. *Biochemistry,* **43:** 10765-10774.

Marwaha, R.S. and B.O. Juliano. 1976. Aspects of nitrogen metabolism in the rice seedlings. *Plant Physoil.,* **57:** 923-927.

Naik, M.S., K.V. Sardhambal and Shiv Prakash. 1976. Non-exzymatic photochemical reduction of nitrate in rice seedlings. *Nature,* **262:** 396-397.

Ogawa, M. and S. Ida. 1987. Biosynthesis of ferredoxin-nitrite reductase in rice seedlings. *Plant Cell Physiol,* **28:** 1501-1508.

Ogawa, T., H. Fuknoka, H. Yano and Y. Ohkawa. 1999. Relationships between nitrite reductase activitiy and genotype dependent callus growth in rice cell cultures. *Plant Cell Reports,* **18:** 576-581.

Pecsvaradi, A. and F. Zsoldos. 1996. Nitrate reductase and nitrite reductase activity in nitrite and chlorate stressed rice seedings. *Plant Physiol. Biochem.,* **34:** 659-663.

Ramirez, J.M., F.F. Delcampo, A. Paneque and M. Losada. 1966. Ferredoxin-nitrite reductase from spinach. *Biochem. Biophys. Acta,* **118:** 58-71.

Swamy, U., M. Wang, J.N. Tripathy, S.K. Kim, M. Hirasawa, D.B. Knaff and J.P. Allen. 2005. Structure of spinach nitrite reductase: Implications for multi-electron reactions by the iron-sulpur : siroheme cofactor. *Biochemistry,* **44:** 16054-16063.

Takahashi, M., Y. Sasaki, S. Ida and H. Morikawa. 2001. Nitrite reductase gene enrichment improves assimilation of NO_2 in Arabidopsis. *Plant Physiol.,* **126:** 731-741.

Terada ,Y., H. Aoki, T. Tanaka, H. Morikawa and S. Ida. 1995. Cloning and nucleotide sequence of a leaf ferredoxin nitrite reductase cDNA of rice. *Biocio. Biotech. And Biochem.,* **59:** 2183-2185.

Vaucher, H., J. Kronenberger, A. Lepingle, E. Vilaine, J.P. Boutin and M. Caboche. 1992. Inhibition of tabocco nitrite reductase activity by expression of antisense RNA. *Plant J.,* **2:** 559-569.

chapter five

Ammonia Assimilation

The pathway for ammonium assimilation in higher plants has been well established. Ammonium, whether resulting from nitrate assimilation or from other secondary sources, is first incorporated into glutamine in a reaction catalyzed by glutamine synthetase.

$$\text{Glutamate} + NH_3 + \text{ATP} \rightarrow \text{Glutamine} + H_2O + \text{ADP} + \text{Pi}.$$

Glutamate synthase catalyzes the combination of glutamine with 2-oxoglutarate to form two molecules of glutamate, one of which serves as substrate for glutamine synthetase, while the other glutamate is available for transport, storage or further metabolism. The reductant may be either reduced ferredoxin or NADH, depending on the GOGAT species.

$$\text{Glutamine} + \text{2-oxoglutarate} + \text{Fdred or NADH} \rightarrow \text{2 Glutamate} + \text{Fdox or NAD}$$

These two reactions form a cycle with the conversion of one molecule of 2-oxoglutarate and ammonium to one molecule of glutamate (Miflin and Lea, 1980).

Ammonia is liberated by photorespiration, protein and amino acid metabolism, and the breakdown of nitrogen transport compounds. The rate of ammonia generated during photorespiration is about 10 times the rate of nitrate assimilation.

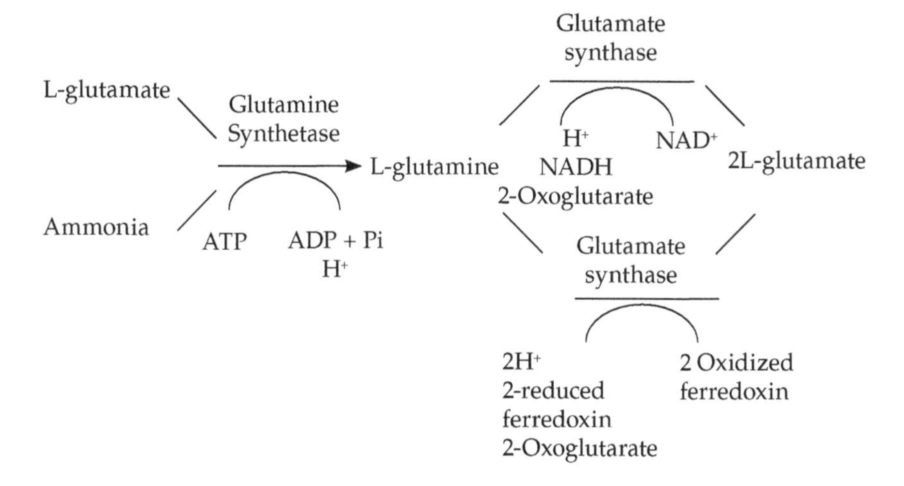

The ammonia released from photorespiration is immediately re-assimilated via the ammonia assimilation cycle. The net result of the cycle is the conversion of one molecule of ammonia to one molecule of glutamate at the expense of one ATP molecule.

The first step in the assimilation of ammonia is the adenosine triphosphate (ATP) dependent reaction with glutamate to form glutamine, catalyzed by glutamine synthetase. The ammonia may have been generated by direct primary nitrate assimilation, or from secondary metabolism such as photorespiration (Leegood et al. 1995) and the catabolism of aminoacids, in particular asparagine (Lea et al. 2007) and through uptake of ammonium ion. Glutamine synthetase activity is located in both the cytoplasm and chloroplasts/plastids in most but not all higher plants. The GS enzyme proteins can be readily separated by standard chromatographic, localization and Western blotting techniques into cytoplasmic (GS1) and plastidic (GS2) forms.

The subunit of cytosolic GS1 has a molecular mass of 38-40 KDa whilst the plastid GS2 form is larger at 44-45 KDa, and the proteins can be usefully separated by simple SDS-PAGE. The quaternary molecular structure of plant GS has proved difficult to establish due to major differences between the eukaryotic and prokaryotic proteins. Initially, it was thought that the mammal and plant enzymes were octamers; however, there is now strongcrystallographic evidence that both native enzyme proteins exist as decamers comprising two pentameric rings (Krajewski et al. 2008). The first crystal structure of a plant GS protein was obtained by Unno et al. (2006), using the stable maize GS1a protein. The protein is composed of two face to face pentameric rings of identical subunits with a total of ten active sites, each formed between neighbouring subunits within each ring. The first step in the GS reaction is the transfer of the terminal phosphoryl group of ATP to the g-carboxyl group of the glutamate to produce the activated intermediate g-glutamyl phosphate. In the second step, a bound ammonium ion is de-protonated, forming ammonia, which attacks the carbonyl carbon to form glutamine with the release of phosphate (Lea and Miflin, 2011). Experiments clearly show that individual GS1 proteins have a non-reductant role in plant metabolism.

5.1. *Synthesis of Glutamine*

Glutamine synthetase (EC 6.3.1.2), an enzyme catalyzing formation of glutamine from glutamate and ammonium ion, is one of the most important enzymes in nitrogen metabolism. Due to glutamine synthetase activity, inorganic nitrogen is incorporated in the cell metabolism and is further used in biosynthesis of several highly important metabolites. The first part started with the discovery of methionine sulfoxime in 1949. Since that time several inhibitors of this enzyme, classified in groups, derivatives of methionine sulfoxime, phosphorus containing analogues of glutamic acid, bisphosphonates and miscellaneous inhibitors, have been developed and described (Berlicki, 2008). Various L-amino acids and nucleotides were used to inhibit the Mg^{2+} supported biosynthetic activity of rice leaf glutamine synthetases GS1 and GS2. The most potent inhibitors are glycine, alanine, aspartic acid, serine, AMP and ADP. When increasing concentration of the six inhibitors to saturating levels, the remaining percent activity of GS1 and GS2 decreased to a finite value indicating partial inhibition. The results ware further confirmed by fractional inhibition analysis and inhibition kinetics. The combined effects obtained with all 15 possible pairs of these six inhibitors on the activities of GS1 and GS2 showed that six pairs are cumulative inhibitors and nine pairs are antagonistic inhibitors on GS1, whereas 14 pairs are cumulative inhibition and one pair is antagonistic inhibition on GS2. The results indicated that both GS1 and GS2 possess separate binding sites for each of these six inhibitors, but interactions of these inhibitor sites on the surface of both enzymes are not identical. When glutamate was the varied substrate, glycine and alanine were partial uncompetitive inhibitors, while serine and aspartic acid were partial noncompetitive inhibitors of GS1. However, glycine, alanine and aspartic acid were partial mixed type inhibitors and serine was partial noncompetitive inhibitor of GS2. When ATP was the varied substrate, AMP and ADP were partial competitive inhibitors of both GS1 and GS2 (Yuan and Hou, 1989).

Ammonia is assimilated into glutamine and glutamate which serve to translocate organic nitrogen from sources to sinks in higher plants. The major enzymes involved are glutamate synthase and glutamate dehydrogenase. However, glutamine synthetase has a vital role in ammonium assimilation and the activity of the enzyme is considered to be a critical and possibly the rate limiting step in ammonium assimilation. In subsequent studies using radioisotope labeled (^{13}N) and stable (^{15}N) nitrogen isotopes, enzyme inhibitors and mutants of plant nitrogen metabolism, indicated that the primary assimilation of ammonium into amino acids occurs through the joint action of GS and GOGAT. In rice cytosolic GS (GS1) protein was distributed homogenously through all cells of the root. NADH GOGAT protein was strongly induced and its cellular location altered by ammonium treatment, becoming concentrated within the epidermal and exodermal cells. Fd-GOGAT protein location changed with root development from a widespread distribution in young cells to becoming concentrated within the central cylinder as cells matured. Plastid GS protein was barely detectable in rice roots, but was the major isoform in leaves, being present in the mesophyll and parenchyma sheath cells. GS1 was specific to the vascular bundle, as was NADH GOGAT, whereas FdGOGAT was primarily found in mesophyll cells (Tobin and Yamaya, 2001).

Nitrogen assimilation is all important physiological process for plant growth and development. Inorganic nitrogen could be assimilated by plants into forms of glutamine and glutamic acid. Glutamine synthetase is a key enzyme for nitrogen assimilation, which regulates nitrogen metabolism. Synthesis and transformation of amino acid could be improved by enhancing the activity of glutamine synthetase, thus enhancing nitrogen movement in the metabolism. The activity of glutamine synthetase is positively correlated with the amount of protein nitrogen in single rice grain at significant level (Yang et al. 2005). In previous studies, some researchers had studied the effect of chemicals on the activity of glutamine synthetase. For example, carboxymethyl chitosan could enhance the activities of glutamine synthetase in rice leaves at heading stage so as to increase the amounts of total nitrogen and protein nitrogen in rice grain (Li et al. 2001). Uniconazole could increase the glutamine synthetase activity and protein content in rice grain at the grainfilling stage (Yang et al. 2005). Applying more phosphate or potash fertilizer could improve the activities of glutamine synthetase in leaves and grains, thereby enhanced protein synthesis and metabolism in rice leaves (Tang and Yu, 2002). More nitrogen application in the whole growth period or at the booting stage could increase the amounts of total nitrogen, protein nitrogen, non-protein nitrogen and the activities of glutamine synthetase in leaves and grains of rice (Tang, 2000). Moreover, increasing nitrogen application could also improve the activity of glutamine synthesis in flag leaf of wheat.

Chromatographic, kinetic, and regulatory properties of glutamine synthetase in rice by DEAE Sephacel column chromatography, two forms GS1 and GS2 were identified in leaves and one form (Glutamine synthetase R) was identified in roots. Purification on hydroxyapatite and gel electrophoresis showed that glutamine synthetase R was distinct from the leaf enzymes. The three isoforms were purified to similar specific activity. Their properties suggest that heat lability, pH optimum about 8 km for L-glutamate of 20 mM and inhibition by glucosamine-6-phosphate were the main characteristics of GS2. Heat stability, pH optimum about 7.5, km for L-glutamate of 2 mM and no effect of glucosamine-6-phosphate was differentiated GS1 from GS2. Glutamine synthetase R was also a labile protein but its kinetic and regulatory properties were quite similar to those of GS1 (Hirel and Gadal, 1980).

Previous studies using ion-exchange chromatography reported that one isoform of glutamine synthetase (GS) existed in cytosol of rice (*Oryza sativa* L) roots. Further, report suggested that two isoforms of GS were observed in rice roots using native PAGE, GS activity staining and immunoblotting. One isoform (GSra) had a mobility similar to that of cytosolic GS in rice leaves (GS1). This isoform existed in seminal roots of germinating seeds and in the roots of seedlings grown in N-free and N-containing culture solutions. Another isoform (GSrb) moved more slowly than GSra and occurred only in the presence of external N. Both ammonium-N and nitrate-N induced GSrb, but the activity of GSrb was higher under ammonium-N than nitrate-N. After the removal of N-source from the solution, GSrb activity and GSrb protein disappeared as judged by native PAGE and immunoblotting but GSra remained relatively constant. It was estimated that GSrb contributed about 80% of the total GS activity of the rice roots grown in the solution containing NH_4^+. SDS-PAGE and immunoblotting test showed that subunits of GSra, GSrb and GS1 had the same molecular weight. The results suggest that GSra was a constitutive enzyme and GSrb was an enzyme induced by external nitrogen (Zhang et al. 1997). Glutamine synthetase is a key enzyme in nitrogen metabolism; it catalyses the crucial incorporation of inorganic ammonium into glutamine. Two full length cDNAs that encode the rice cytosolic glutamine synthetase genes (OsGS1:1 and OsGS1:2) were isolated from a Minghui 63 normalized cDNA library, and gluA encoding GS in E. Coli was isolated by PCR amplification. Transformants for GS gene (GS1:2 and gluA) in rice were produced by an *Agrobacterium tumefaciens*-mediated transformation method, and transcripts of GS gene accumulated at higher levels in the primary transgenic plants. The results indicated an increased metabolic level in GS-over expressed plants, which showed higher total GS activities and soluble protein concentrations in leaves and higher total amino acids and total nitrogen content in the whole plant. Decreases in both grain yield production and total amino acids were observed in seeds of GS-over expressed plants compared with wild type plants. In addition, GS1:2-over expressed plants exhibited resistance to Basta selection and higher sensitivity to salt, drought and cold stress conditions, whereas the other two types of GS-over expressed plants failed to show any significant changes for these stress conditions compared with the wild type plants (Cai et al. 2009).

Of the three genes encoded cytosolic GS in rice, OsGS1:1 is critical for normal growth and grainfilling. However, the basis of its physiological function that may alter the rate of nitrogen assimilation and carbon metabolism within the context of metabolic networks remain unclear. A rice mutant lacking OsGS1:1 and its background wild type (WT) was studied. The mutant plants exhibited severe retardation of shoot growth in the presence of ammonium compared with the WT. Over accumulation of free ammonium in the leaf sheath and roots of the mutant indicated the importance of OsGS1:1 for ammonium assimilation in both organs. The metabolic profiles of the mutant revealed (i) an imbalance in levels of sugars, amino acids and metabolites in the tricarboxylic acid cycle, and (ii) over accumulation of secondary metabolites, particularly in the roots under continuous supply of ammonium. Metabolite-to-metabolite correlation analysis revealed the presence of mutant-specific networks between tryptamine and other primary metabolites in the roots (Kusaro et al. 2011).

During the greening of etiolated rice leaves, total glutamine synthetase activity increases about two fold, and after 48h the level of activity usually observed in green leaves is obtained. A density-labeling experiment with deuterium demonstrated that the increase in enzyme activity is due to a synthesis of the enzyme. The enhanced activity obtained upon greening is the result of two different phenomena : there is a fivefold increase of chloroplastic glutamine synthetase content accompanied by a concomitant decrease (two fold) of the cytosolic glutamine synthetase. The increase of chloroplastic glutamine synthetase (GS2) is only inhibited by Cycloheximide and not by Lincomycin. The result indicates a cytosolic synthesis of GS2. The synthesis of GS2 was confirmed by a quantification of protein by an immunochemical method. It was demonstrated the GS2 protein context in green leaves is fivefold higher than in etiolated leaves (Hirel et al. 1982).

Heat shock on the subsequent cd-induced decrease in the activity of glutamine synthetase (GS) and increase in specific activity of protease in the rice leaves. Heat shock exposure of rice seedlings for 3h in the dark was effective in reducing subsequent cd-induced decrease in the activity of glutamine synthetase and increase in the specific activity of protease (Lin et al. 2010). Glutamine synthetase (GS) is an important enzyme in nitrogen metabolism, as it catalyzes the assimilation of all inorganic nitrogen into organic compounds.

5.2. *Glutamine Synthetase in Rice Genotypes*

Fifty varieties of rice genotypes were evaluated for the leaf GS enzyme activities at Hyderabed. The results suggested that maximum enzyme activity was observed in N-0 as compared to N-100. The genotypes RPHR-1005, 111-3, China 9998 and RPHR-101096 showed low, and IR 40750R, NQ1-38, IR40750R, SC5-2-2-1 and GQ-58 showed moderate, whereas 524-2, 628-2, C-28, B-95 and PNR-2-49 were showing high GS enzyme activity in N-0 whereas in N-100 very amount of enzyme activity was detected. There were significant differences between genotypes and treatments (Table 5.1). The GS activity was one of the selection criteria to identify nitrogen use efficient cultivars and their use in developing mapping population for high NUE.

TABLE 5.1 Glutamine synthetase activity (absorbance at 540 nm gfw^{-1}h^{-1}) in leaves of rice genotypes

Genotype	N_0	N_{100}	Genotype	N_0	N_{100}
EPLT – 109	ND	ND	PNR - 3158	0.039	0.008
IBL – 57	ND	ND	517	0.040	0.008
KMR – 3	ND	ND	611-1	0.040	0.008

SG27 – 77	ND	ND	GQ - 70	0.043	0.008
AJAYA – R	ND	ND	MTU - 9992	0.047	0.009
SC 5-9-3	ND	ND	GQ – 86	0.048	0.009
695-1	ND	ND	NDR – 3026	0.051	0.010
TG – 70	ND	ND	SGRT – 131	0.058	0.011
RPHR – 1005	0.009	0.003	B-95-91	0.060	0.011
111-3	0.014	0.003	Salivahava	0.069	0.011
China 9998	0.014	0.004	IR40750 – R	0.070	0.012
RPHR – 101096	0.017	0.004	1163	0.085	0.013
IR40750 (H39)	0.026	0.004	RCW – 56	0.087	0.014
GQ – 58	0.028	0.004	SG22-289-3	0.103	0.014
IR40750R(H5)	0.028	0.004	ICRD 16-1-4-2-1	0.090	0.014
SC5-2-2-1	0.028	0.004	SG22-2-3-1	0.121	0.016
NRI – 38	0.029	0.005	B-95-91 (H31)	0.135	0.018
TG – 164	0.033	0.006	Shrabani	0.143	0.019
EPLT - 104	0.032	0.005	IBL – 52 – 1	0.155	0.019
DR714 – 1 – 2R	0.034	0.006	B95 – 16	0.219	0.022
B-95-12	0.035	0.007	PNR - 2-49	0.238	0.024
TG – 23	0.036	0.007	B - 95-95	0.335	0.024
ICRD-19-9-2-1	0.037	0.007	C - 28	0.358	0.025
GQ – 54	0.038	0.007	628-2	0.495	0.025
			524-2	0.650	0.031

ND – Not detected

Four japonica rice varieties different in cooking qualities were considered in a pot experiment to know the relationship between the activities of glutamine synthetase during grain filling and rice quality. The activities of glutamine synthetase gradually increased and then declined as a single peak curve in the course of grain filling. The 15th day after heading was a turning point, before which the enzymatic activities in the inferior rice varieties with high protein content were higher than those in the superior rice varieties with low protein content, and after which it was converse. The activity of glutamine synthetase in grain was correlated with the taste meter value, peak viscosity and breakdown negatively at the early stage of grain filling whereas positively at the middle and late stages. Moreover, it was correlated with the protein content of rice grain and setback positively at the early stage and negatively at the middle and late stages. The correlation degree varied with the course of grainfilling.

From 15 days to 20 days after heading was a critical stage, in which the direction of correlation between the activity of glutamine synthetase and taste meter value and RVA properties of rice changed (Zheng-xun et al. 2007) (Tables 5.2 and 5.3).

TABLE 5.2 Multiple comparison of glutamine synthetase activities (OD/grain) in rice grains at the filling stage

Variety	10 DAH	15 DAH	20 DAH	25 DAH	30 DAH	35 DAH
Shuiludoo 1	0.0271	0.0301	0.0243	0.0184	0.0143	0.0115
Touker 180	0.0156	0.0211	0.0255	0.0255	0.0187	0.0151
Fujihikan	0.0184	0.0224	0.0252	0.0202	0.0160	0.0126
Dongnong 415	0.0202	0.0241	0.0199	0.0160	0.0128	0.0102
Mean	0.0178	0.0244	0.0237	0.0193	0.0155	0.0148

DAH – Days after heading

TABLE5.3 Coefficients of Correlation between glutamine synthetase activity and protein content, taste meter value, RVA properties at different grain filling stages.

Days after heading	Protein Content	Taste meter Value	Peak Viscosity	Breakdown	Setback
10	0.4061	-0.9201	-0.6365	-0.7511	0.7614
15	0.3885	-0.9219	-0.5656	-0.6817	0.6894
20	-0.9610*	0.5047	0.7375	0.5899	-0.5237
25	-0.9204	0.7536	0.9598*	0.8943	-0.8546
30	-0.8616	0.7864	0.9870*	0.9493	-0.9201
35	-0.8316	0.7512	0.9954**	0.9600	-0.9347

* and ** are significant at 5% and 1% respectively.

5.3. *Abiotic Stress*

Under stress conditions such as drought and high salinity, stomatal closure triggered by ABA limits CO_2 supply to the leaf leading to over reduction of the photosynthetic electron transport chain. Therefore, enhancement of the enzyme activity involved in active oxygen scavenging systems may be a potent strategy to increase salt tolerance. An alternative strategy to cope with oxidative damage under salt stress might be the suppression of active oxygen production. Photorespiration may function as a possible route for the dissipation of excess light energy or reducing power. Although photorespiration includes many metabolic steps which are performed across chloroplasts, mitochondria and peroxisomes, several studies suggest that the rate limiting step is the reassimilation of ammonia catalyzed by chloroplastic glutamine synthetase (GS2). Previously, it was demonstrated that a transgenic rice plant over expressing GS2 constitutively had increased photorespiration capacity and increased tolerance to high salinity and high temperature.

Ammonium and nitrate assimilation is curtailed seriously during salinity stress. Therefore, N status in the plant is also significantly influenced by salinity. In aerobic soils, nitrate is the main dominant species and it is converted to ammonium by the sequential action of two enzymes, nitrate reductase and nitrite reductase. In higher plants NH_4^+ is mainly assimilated through the concerted action of glutamine synthetase (GS) and glutamate synthase (GOGAT). Salinity curtailed NO_3^- uptake by decreasing the activities of nitrate reductase and nitrite reductase in plants also decreased NH_4^+ assimilation seriously by influencing GS activity. It had also been reported that both NR and GS were repressed by the salinity.

Rice plants would more frequently suffer from high temperature (HT) stress at the grain filling stage in future. A japonica rice variety Koshihikari and an indica rice variety IR72 were used to study the effect of high temperature on dynamic changes of glutamine synthetase (GS), glutamate synthase (GOGAT), glutamic oxalo-acetic transaminase (GOT), and glutamate pyruvate transaminase (GPT) activities in grains. Under HT, the activities of GOGAT, GOT, GPT and soluble protein contents in grains significantly increased, whereas GS activity significantly decreased at the grain filling stage. In addition to the increase of protein and amino acids contents, it was suggested that GOGAT, GOT and GPT in grains played important roles in nitrogen metabolism at the grain filling stage. Since, the decrease of GS activity in grains did not influence the accumulations of amino acid and protein, it is implied that GS might not be the key enzyme in regulating glutamine content in grains.

The regulation of GS isoforms by water deficit (WD) was organ specific. Two GS isoforms i.e. OSGS1 and OSGS2 were differentially regulated in IR64 (drought sensitive) and Khitish (drought tolerant) cultivars of rice. Water deficit (WD) has adverse effect on rice and acclimation requires essential reactions of primary metabolism to continue. Rice plants utilize ammonium as major nitrogen source, which is assimilated into glutamine by the reaction of glutamine synthetase. Rice plants possess one gene (OsGS2) for chloroplastic GS2 and three genes (OsGS1:1, OsGS1:2 and OsGS1:3) for cytosolic GS1. Here, it is reported the effect of WD on the regulation of GS isoforms in drought sensitive (IR64) and drought tolerant (Khitish) rice cultivars. Under WD total GS activity in root and leaf decreased significantly in IR64 seedlings in comparison to Khitish seedlings. The reduced GS activity in IR64 leaf was mainly due to decrease in

GS2 activity, which correlated with decrease in corresponding transcript and polypeptide contents. GS1 transcript and polypeptide accumulated in leaf during WD; however, GS1 activity was maintained at a constant level. Total GS activity in stem of both the varieties was insensitive to WD. Among GS1 genes, OSGS1:1 expression was differently regulated by WD in the two varieties. Its transcript accumulated more abundantly in IR64 leaf than in Khitish leaf following WD, OSGC1:1 mRNA level in stem and root tissues declined in IR64 and enhanced in Khitish. A steady OsGS1:2 expression patterns were noted in leaf, stem and root of both the cultivars. Results suggest that OsGS2 and OsGS1:1 expression may contribute to drought tolerance of Khitish cultivar under WD conditions (Singh and Ghosh, 2013).

5.4. Rice Root System

Rice plants in paddy fields prefer to utilize ammonium as a major nitrogen source. Glutamine synthetase serves for assimilation of ammonium in rice root, and ameliorates the toxic effect of ammonium excess. Among the three isoenzymes of the cytosolic GS1 gene family in rice, OsGS1:1 and OsGS 1:2 were abundantly expressed in roots. Analysis of the purified enzymes showed that OsGLN1:1 and OsGLN1:2 can be classified into high affinity subtypes with relatively high Vmax values, as compared with the major high affinity isoenzyme GLN1:1 in Arabidopsis (GLN1:2 and GLN1:3) were absent in rice roots. The OsGLN1:1 and OsGLN1:2 transcripts showed reciprocal responses to ammonium supply in the surface cell layers of roots. OsGLN1:1 accumulated in dermatogens, epidermis and exodermis under nitrogen limited condition. By contrast, OsGLN1:2 was abundantly expressed in the same cell layers under nitrogen sufficient conditions, replenishing the loss of OsGLN1:1 following ammonium treatment. Within the central cylinder of elongating zone, OsGLN1:1 and OsGLN1:2 were both induced by ammonium, which was distinguishable from the response observed in the surface cell layers. The high capacity Gln synthetic activities of OsGLN1:1 and OsGLN1:2 facilitate active ammonium assimilation in specific cell types in rice roots (Ishiyama et al. 2004).

A direct correlation was reported between an enhanced GS activity in transgenic plants in some cases, which is noted by an increase in biomass or yield by transforming novel GS1 construct. Similarly, Kozaki and Takeba (1996) constructed transgenic tobacco plants enriched or reduced in plastidic glutamine synthetase. Ectopic expression of GS1 has been shown to alter plant and over expression of GS1 in transgenic could cause the enhancement of photosynthetic rates, higher rates of photorespiration, and enhanced resistance to water stress. The over expression of soyabean cytosolic GS1 in the shoots of lotus corniculatus was reported to accelerate plant development leading to early senescence and premature flowering particularly when plants were grown in conditions of high ammonium. Additional emperical evidence for enhanced nitrogen–assimilation efficiency in GS1 transgenic lines had also been noted. However, differences in the degree of ectopic GS1 expression have been reported and attributed the positional effects, effectiveness of chimeric constructs, or differences in growth conditions. This may be due to lack of correlation between the enhanced expression of GS1 and concomitant growth (Cheng-gang et al. 2011).

The potential role of photorespiration in the protection against salt stress was examined with-transgenic rice plants. *Oryza sativa* L. Cv. Kinuhikari was transformed with a chloroplastic glutamine synthetase (GS2) gene from rice. Each transgenic rice plant line showed a different accumulation level of GS2. A transgenic plant line, G39-2, which accumulated about 1.5 fold more GS2 than the control plant, had an increased photorespiration capacity. In another line G241-12, GS2 was about lost and photorespiration activity could not be detected. Fluorescence quenching analysis revealed that photorespiration could prevent the over-reduction of electron transport systems. When exposed to 150mM NaCl for two weeks, the control rice plants completely lost photosystem II activity, but G39-2 plants retained more than 90% activity after the two week treatment, where G 241-12 plants lost these activities within one week. In the presence of isomicotinic acid hydrazide, an inhibitor of photorespiration G 39-2 showed the same salt tolerance as the control plants. The intracellular contents of NH_4^+ and Na^+ in the stressed plants correlated well with the levels of GS2. Thus, the enhancement of photorespiration conferred resistance to salt in rice plants. Preliminary results also suggest chilling tolerance in the transformant (Hoshida et al. 2000).

5.5. *Glutamate Synthesis*

Two forms of GOGAT are present in plastids from higher plants, one that uses Fd as a source reductant (Fd-GOGAT) and the other that uses NADH-GOGAT. Fd-GOGAT is the major enzyme for glutamate synthesis in photosynthetic tissues, whereas the NADH-GOGAT is predominant in non-photosynthetic tissues. Fd-GOGAT was first isolated from pea leaves and can represent upto 1% of total leaf protein. The enzyme has been shown to be dimeric in rice. Its molecular mass is about 115 KDa for the enzyme from rice. Using immunogold localization, Fd-GOGAT was found in chloroplast stroma of mesophyll, xylem paranchyama and epidermal cells. Fd is present in roots, and mechanism for supply of reductant via the oxidative pentose phosphate have been proposed. In rice, Fd-GOGAT activity was highest in the youngest roots cells in the tips and then decreased as the cell matured towards the root base. In the younger tissue, the enzyme was present in all cell types, but in older tissues it was only in the central cylinder. The application of ammonium ions did not affect the distribution of the Fd-GOGAT protein. An involvement of thioredoxin in the activation of Fd-GOGAT has been proposed.

There are two rice varieties grown under humid tropic conditions: (i) Piaui, a land race adopted to low N availability and reduced light supply, and (ii) IAC-47, an improved variety were conducted in controlled conditions of 24°C and 12/12 hour light/dark periods (200 μE $m^{-2}s^{-1}$) using a growth chamber, simulating humid tropic environment. Rice plants were grown under 0.1 and 1.0mM NH_4^+-N in a Hougland and Arnon nutrient solution, pH5.5 upto 26 days after germination. The activity of PMH^+-ATPases in the plasma membrane, vaccuole VH^+-ATPases, H^+-ATPases as well as GS, GOGAT and GDH were determined. PMH^+-ATPases presented higher activity in IAC-47 roots. On the other hand Piaui roots showed an enhanced microsomal protein context and VH^+-ATPases activity, which apparently allows the variety to absorb N as well as the improved one. Increase in cytosolic NH_4^+ would result in high GDH deamination in Piaui variety at 1.0mM NH_4^+. Meanwhile, GS increases were observed in IAC-47 shoots, together with GOGAT increases at same treatment (Garrido et al. 2012).

The presence of glutamate synthase (GOGAT) in rice root extracts and the relationship among electron donors, nitrogen donors and the activity were studied using 15N amido labeled glutamine, asparazine, ^{14}C-2-oxoglutarate and inhibitors. The high molecular fraction of rice root extracts prepared by Sephadex-50 column showed ferredoxin-dependent GOGAT activity, but pyridine nucleotide dependent activity could not be detected in it. Asparagine did not act as a nitrogen donor for rice root GOGAT. Methy viologen could be a substitute for ferredoxin, but GOGAT activity with it was about ¼ of that with ferredoxin. Accordingly, rice root GOGAT was considered to be the same type as that observed in leaves of many higher plants, but different from that discovered in pea roots and cultural carrot tissues (Arima, 1978). In developing young tissues of rice plants, NADH-dependent glutamate synthase (NADH-GOGAT) has been suggested to play an important role in the utilization of glutamine, which is transported from senesing tissues and roots (Hayakawa et al. 1994). Because NADH-GOGAT protein is located in specific cells in both the leaf blade and grain of rice and because its protein content and activity change dramatically during the early stage of the seed ripening (Hayakawa et al. 1993), the expression of NADH-GOGAT gene might be regulated as a cell specific and age specific manner. It was recently shown that the mRNA, protein and activity of NADH-GOGAT in roots of rice seedlings increased more than 10-fold within a day of the start of a supply of nitrogen (Yamaya et al. 1995). This increase was specific to NADH-GOGAT in the roots; little change was observed in the activity of protein content of Fd-GOGAT or GS. A supply of NO_4^+ was most effective on the apparent inducible increase in NADH-GOGAT but adding NO_3^+, glutamine or asparazine also caused rapid increases to a lesser extent in both rice roots (Yamaya et al. 1995) and rice cell culture (Watanbe et al. 1996). After growth for 26 days in water, rice seedlings were transferred to a medium containing 1mM NH_4Cl. The mRNA for NADH-glutamate synthase was markedly increased in the roots within three hours. Methionine sulfoxime completely inhibited this accumulation, suggesting that NH_4^+ is not a direct inducer of this process (Hirose et al. 1997).

Ferredoxin dependent glutamate synthase from rice leaves (*Oryza sativa* L.) cv Delta was purified 206-fold with a final specific activity of 35.9 mmoles glutamate formed per min per milligram protein by a procedure including ammonium sulfate fractionation, DEAE-cellulose chromatography, Sephacyrl S-300 gel filtration, and ferredocin-sepharose affinity chromatography. The purified enzyme yielded a single protein band on polyacrylamide gel electrophoresis. Molecular weight of the native enzyme was estimated to be 224 kDa by Sepharose 6B gel filtration. Electrophoresis of the dissociated enzyme in sodium dodecyl sulfate-polyacrylamide gel gave a single protein band which corresponds to the subunit molecular weight of 115 kDa. Thus, it is concluded that the glutamate synthase is composed of two polypeptide chains exhibiting the same molecular mass. Spectrophotometric analysis indicated that the enzyme is free of iron sulfide and flavin. The pH optimum was 7.3. The enzyme had a negative cooperativity (Hill number of 0.70) for glutamine, and its km value increased 270 to 570 μM at a glutamine concentration higher than 800 μM, respectively. Asparagine and oxaloacetate could not be substituted for glutamine and a-ketoglutarate, respectively. Enzyme activity was not detected with pyridine nucleotides as electron donors. Azaserine and several divalent cations were potent inhibitors. The purified enzyme was stabilized by dithiorthreitol (Suzuki and Gadal, 1982).

High levels of NADH-GOGAT protein (33.1 mg protein/g fresh weight) and activity were detected in the 10th leaf blade before emergence. The unexpanded nongreen protein of the 9th leaf blade contained more than 50% of the NADH-GOGAT protein and activity per gram fresh weight when compared with the 10th leaf. The expanding, green portion of the 9th leaf blade outside of the sheath contained a slightly lower abundance of NADH-GOGAT protein than the non-green portion of the 9th leaf blade on a fresh weight basis. The fully expanded leaf blades at positions lower than the 9th leaf had decreased NADH-GOGAT levels as a function of increasing age, and the oldest, 5th blade contained only 4% of the NADH-GOGAT protein compared with the youngest 10th leaf blade. Fd-GOGAT protein, on the other hand, was the major form of GOGAT in the green tissues, and the highest amount of Fd-GOGAT protein. The content of plastidic glutamine synthetase polypeptide was also the highest in the 7th leaf blade (429 mg/g fresh weight) and lowest in non-green blades and sheath. On the other hand, the relative abundance of cytosolic glutamine synthetase polypeptide was the highest in the oldest leaf blade, decreasing 10 to 20% of that value in young, non-green leaves (Yamaya et al. 1992) (Table 5.4).

TABLE 5.4 Activities of NADH and Fd-dependent GOGAT at various leaf blade and sheath positions of rice plants

Leaf position from base	NADH-GOGAT (m unit/g fresh weight)	Fd-GOGAT (m unit/g fresh weight)	Ratio of Fd-GOGAT : NADH-GOGAT
Leaf blade			
5th	25.2	7250	288
6th	37.0	7350	199
7th	31.6	8350	264
8th	45.6	5660	124
9th inside (non-green)	89.2	349	3.9
9th outside (green)	31.7	5190	164
10th	155	297	1.9
Leaf Sheath			
5th	28.2	658	23.3
6th	19.5	705	36.2
7th	29.0	570	19.7
8th	21.7	278	12.8
9th	21.8	127	5.8

5.6. *Glutamate Synthase in Rice*

Nitrogen accumulation in the apical spiekelets on the primary branches of the main stem of rice plants have been studied during the ripening process (0-35 days after flowering). The level of NADH dependent glutamate synthase (GOGAT) protein and activity increased 4 and 6-fold, respectively, in the first 15 days after flowering. Maximum levels of NADH-GOGAT were found at the time when the spikelets had just began to increase in dry weight and to accumulate storage proteins. Subsequently, both the level of NADH-GOGAT protein and its activity of spikelets declined rapidly. Although changes in ferredoxin-dependent GOGAT paralleled changes in NADH-GOGAT, the relative abundance of NADH-GOGAT protein in the spikelets was about three times higher than that of Fd-GOGAT from 5 to 15 days after flowering. When the chaff (lemna and palea) was separated from the spikelets 10 days after flowering, 16% of the NADH-GOGAT protein was found in the chaff and 84% in the young grain tissues (endosperm, tastae, aleurone tissues and embryo). On the other hand Fd-GOGAT protein was distributed 52% in the chaff and 48% in the young grain tissues in spikelets of the same age. Activity of NADP-isocitrate dehydrogenase, which may generate the 2-oxoglutarate required for the GOGAT reactions, was much higher than that of total GOGAT activities on a spikelet basis during the ripening process. The results suggest that in rice plants NADH-GOGAT is responsible for the synthesis of glutamate from the glutamine that is transported from senescing tissues to the spikelets (Hayakawa et al. 1993; Watanbe et al. 1996).

OSGlt1 and OSGlt2 are expressed primarily in roots when N is limiting, but in leaves when non-limiting. Transcription of OSGlt2 is decreased, but OSGlt2 is increased by increasing N level in roots and leaves. When rice roots are exposed to NO_2^- and NH_4^+ for two hours after N starvation, transcriptions of OSGln1; OSGlt1, OSGlt2 and OSGlu1 are all repressed by NO_2^- and NH_4^+ (Zhao and Shi, 2006)

In the first crop season both the glutamate synthase and glutamate dehydrogenase in leaves and sheaths showed the highest activity at tillering and a higher activity at harvesting. In the second crop season, the glutamate synthase in leaves and sheaths showed higher activities at the beginning of tillering, booting and ripening to harvesting. The effects of water stress on nitrogen forms, activities of some enzymes related to nitrogen metabolism and photosynthetic properties in rice were studied using a hydroponic experiment. Water stress was caused by adding PEG-6000 into the solution. Slight water stress (PEG ≤ 5%, water potential ≥ – 0.05 MPa) has little effects on the contexts of amino acid nitrogen, soluble protein and activities of nitrate reductase, glutamine synthetase, glutamate synthase and glutamate dehydrogenase, and moreover the absorption and accumulation of nitrate nitrogen is stimulated. However, photosynthetic rate and dry matter accumulation are significantly inhibited. The dry matter accumulation change is significantly correlated with the regulation of photosynthetic rate more than positive effect of resisting badness effect from environment. Therefore, photosynthetic rate has much more effect on dry matter accumulation than metabolism *in vivo* of rice. When PEG concentration ≥ 10% (water potential ≤ –0.15 MPa), the concentrations of different nitrogen forms, some key enzymes of nitrogen metabolism and photosynthetic rate are significantly decreased. These effects on roots under the water stress are significantly larger than those on leaves (Yuan etal., 2009).

Okadaic acid (OKA), a potent and specific inhibitor of protein serine/threonine phosphatases 1 and 2A, induced the accumulation of NADH-glutamate synthase (GOGAT) mRNA within four hours in rice (*Oryza sativa* L.) cell cultures. In contrast to the transient accumulation of NADH-GOGAT mRNA by NH_4^+, OKA caused a continuous accumulation for at least 24 hour. The induction of NADH-GOGAT mRNA by OKA was not inhibited in the presence of methionine sulfoxime, which inhibited the NH_4^+ induced accumulation of mRNA. These results suggest that the OKA-sensitive protein phosphatase is involved in the regulation of NADH-GOGAT gene expression and probably plays a role in the regulation in the signal transduction pathway down stream from NH_4^+, although a signal transduction pathway, other than that of nitrogen sensing, could be possible. Nuclear run-on assays demonstrated that the accumulation of NADH-GOGAT mRNA induced by the supply of either NH_4^+ or OKA was mainly regulated at the transcription level. OKA effects were synergistic to the NH_4^+ induced expression of the NADH-GOGAT gene. In the presence of K-252a, a protein kinase inhibitor, the accumulation of NADH-GOGAT mRNA induced by either NH_4^+ or OKA was reduced. The possible roles of protein phosphatases in the regulation of NADH-GOGAT gene expression is noted (Hirose and Yamaya, 1999).

NADH-glutamate synthase was inhibited by abscisic acid (ABA) at 10^{-6}M. Addition of either gibberellic acid (GA3) or dichloropenoxy acetic acid (2, 4-D) at 10^{-6}M to ABA counteracted its inhibitory effect. Kinetin at the same concentration increased inhibition of the enzyme by ABA, cycloheximide, rifampicin, cordycepin and chloramphenicol at 0.3M reduced mediated-increase in the enzyme activity by 2, 4-D particularly *in vivo.* The enzyme was purified to homeogeneity as determined by SDS-PAGE. The subunit matter of GOGAT as determined by SDS-PAGE was 200KDa. The specific activity was 131.2 μmg^{-1} protein. Group specific modifiers like N-bromoguccinimide (NBs), Densylchloride (DC), Tetranitromethane (TNM) and 2-ethoxy-l-ethoxy-1, 2-dihydroquinoline (EEDO) inactivated the purified exzyme. The results suggested the presence of tryptophanyl, lysyl, tyrosyl and carboxyl groups essential for the enzyme catalysis. Double logarithmic plots of the observed pseudo first order rate constants against modifier concentration revealed modification of only one residue of each group (El-Shora, 2001).

It is noted in maize that glutamate synthase (GOGAT) activity increase with increase in concentration of putrescine upto 100μM in root and upto 50μM in shoot and further increase in concentration resulted in decline of enzymatic activity. Protein and total nitrogen content increased upto 10mM concentration both in root and shoot of maize seedlings (Awasthi et al. 2013).

GOGAT is found in all types of organisms and its amino acids sequence is remarkably well conserved. To examine the evolutionary relationships among the eubacterial and eukaryotic GOGAT proteins, a phylogenetic tree can be constructed based on the aminoacid sequences of regions common to all eubacterial and eukaryotic GOGAT proteins. With the exception of the *Synechocysus* sp. gltB gene product, all of the Fd-GOGAT protein cluster together. This group contains Fd-GOGATs from higher plants (Maize, Rice and Arabidopsis) two red alga species and two cyanobacteria. This analysis shows that eukaryotic Fd-GOGAT is closely related to bacterial Fd-GOGATs and suggests that the gene encoding these enzymes are derived from the eubacterial precursors of chloroplasts. These data are consistent with an endosymbiotic origin of plastids. Additional support for this conclusion comes from the finding that the Fd-GOGAT gene is found in the plastid genomes of red algal. Higher plant GOGAT genes are located in the nuclear genome, into which they were presumably transferred from the symbiont genome. The phylogenetic analysis also suggests that the gltB gene which has been found in all cyanobacteria examined is generally more similar to eukaryotic and eubacterial NAD(P)H-GOGATs than Fd-GOGATs and did not functionally replace the NADH-GOGAT genes found in higher plants.

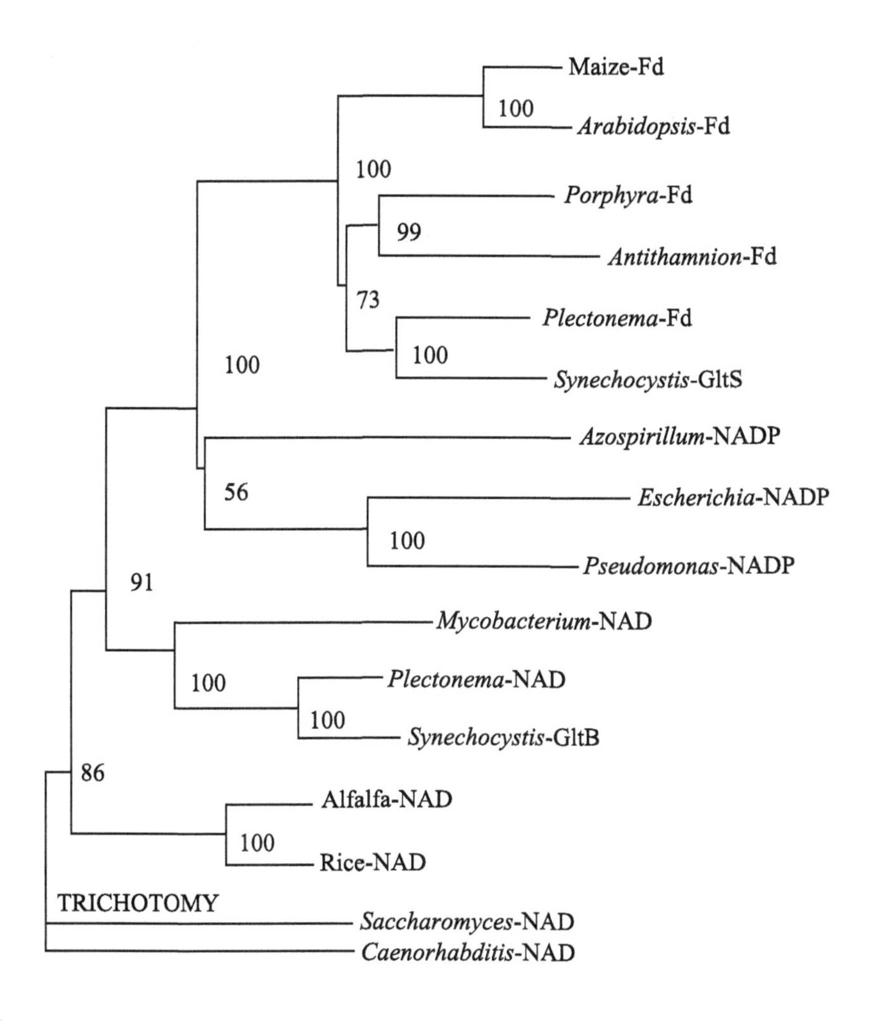

Analysis of transcriptional profiles of rice GOGAT genes using a genome-wide microarray database, and investigation on the effects of suppressions of glutamate synthase genes on carbon and nitrogen metabolism using GOGAT co-suppressed rice plants had been conducted. Transcriptional profiles showed that rice GOGAT genes were expressed differently in various tissues and organs, which suggested that they have different roles *in vivo*. Compared with the wild type, tiller number, total shoot dry weight and yield of GOGAT co-suppressed plants were significantly decreased. Physiological and biochemical studies showed that the contents of nitrate, several kinds of free amino acids, chlorophyll, sugars, sugar phosphates, and pyridine nucleotides were significantly decreased in leaves of GOGAT co-suppressed plants, but the contexts of free ammonium, 2-oxoglutarate and isocitrate in leaves were increased. It is concluded that GOGATs play essential roles in carbon and nitrogen metabolism and that they are indispensable for efficient nitrogen assimilation in rice (Yong En et al. 2011) (Tables 5.5 and 5.6).

TABLE 5.5 Morphological characteristics of field grown GOGAT co-suppressed plants.

Trait	WT	NAI – 4	NAI – 7	NAI – 8
Tillers per plant	10.04	2.44	2.52	5.00
Plant height (cm)	99.10	79.97	81.86	79.38
Total shoot weight (g)	48.48	7.01	7.40	14.32
100 grain weight (g)	2.46	2.32	2.35	2.28
Yield per plant (g)	27.77	4.17	4.56	7.63

TABLE 5.6 Enzyme activities in co-suppressed lines and wild type (n mol g FW^{-1} min^{-1})

Enzyme	Leaves				Roots			
	WT	NAI–4	NAI–7	NAI–8	WT	NAI–4	NAI–7	NAI-8
NADH–GOGAT	157.7	32.9	23.9	53.9	121.7	58.2	68.6	79.4
Fd–GOGAT	111.8	64.9	81.3	50.7	-	-	-	-
NADP–MDH	291.1	295.8	260.9	288.7	-	-	-	-
GS	3298.0	3187.9	3467.5	2917.0	548.9	607.7	714.1	722.5
NR	191.0	276.1	313.6	211.1	150.3	84.7	41.8	71.2
NiR	261.6	295.5	302.0	358.0	366.5	338.0	333.3	364.2
NAD-GDH	2039.7	2019.5	449.7	1921.5	2969.6	3338.5	2621.4	2143.2
NADP-GDH	193.0	191.4	155.5	158.7	459.4	555.2	434.6	547.0
PEP Case	1503.4	884.5	1104.5	1236.6	343.8	616.4	315.4	367.0
NADP–ICDH	1076.6	1106.3	1066.3	1020.6	1165.1	1280.1	1171.7	1301.2
FB Pase	2707.5	2629.4	2691.8	2584.7	9096	937.8	929.8	842.7

Fd-GOGAT mRNA accumulated 4flods and the enzyme polypeptide (3-fold) and activity (3-fold) also increased in leaf cells, while NADH-GOGAT activity remained constantly low. Leaf specific induction of Fd-GOGAT mRNA (3-fold) occurred in etiolated leaves by low fluence red light, and far red light reversibly repressed the mRNA accumulation. Red/far red reversible induction also occurred for Fd-GOGAT polypeptide (2-fold) and activity (2-fold), implicating the phytochrome dependent induction of Fd-GOGAT. In contrast, NADH-GOGAT activity remained constant, irrespective of red/far red light treatments. Fd-GOGAT showed diurnal changes under light/dark cycles with the maximum early in the morning and the minimum in the afternoon at the levels of mRNA, enzyme polypeptide and activity. Gln diurnaly changed in parallel with Fd-GOGAT mRNA. The indication of Fd-GOGAT provides evidence that light and metabolites are the major signal for the Gln and Glu formation in maize leaf cells (Suzuki et al. 2001).

Glutamine synthetase (GS) and glutamate synthase (GOGAT) serve for primary assimilation of nitrogen in higher plants. When NH_4^+ quickly taken up in roots, through GS/GOGAT cycle to ameliorate the toxic effect of excess NH_4^+. The sequence of the rice genome is almost complete, facilitating the identification of the GS and GOGAT gene families in this species. Thus it was revealed that a distinct expression pattern for these genes, OsGln 1:1 and OsGln 1:2 mainly functions in roots and OsGln 2 and OsGln 1 are preferentially expressed in leaves. However, transcriptions of OsGln 1:1, OsGln 1:2, OsGln2 and OsGln1 in leaves are all increased by increased nitrogen level while those in the roots not influenced or even decreased. OSGlt2 are expressed primarily in roots when nitrogen is limiting but in leaves when nitrogen is non-limiting. Transcription of OsGlt1 is decreased but OsGlt2 is increased by increased nitrogen level in roots and leaves. When rice roots are exposed to NO_3^- and NH_4^+ for two hours after N starvation transcriptions of OsGln 1:1, OsGlt2 and OsGlu1 are all repressed by NO_3^- and NH_4^+. OsGln 1:2 expression shows significant up regulation by NH_4^+ and down regulation by NO_3^-, while OsGln2 down regulation by NH_4^+ and up regulation by NO_3^- (Zhao and Shi, 2006).

A5822 bp long cDNA clone encoding the full length ferredoxin-dependent glutamate synthase (Fd-GOGAT) protein was isolated from roots of rice (*Oryza sativa* L. Cv. Sasanishiki). Its sequence was identical to those of partial cDNAs for Fd-GOGAT from leaves and shoots of rice. The predicted open reading frame (4848 bp) encodes a 1616 amino acids protein with a molecular mass 175034 Da that includes a 96 amino acid presequence. The combined nucleotide sequence of genomic clones for Fd-GOGAT isolated from rice was 20899 bp long and contained an entire structural gene, a 5672 bp 5′-upstream region from the first methionine and a 779 bp 3′ downstream region from the stop codon. The predicted transcribed region (15.4 kb) consisted of 33 exons separated by 32 introns (Hayakawa et al. 2003).

Transcriptional profiles showed that rice GOGAT genes were expressed differently in various tissues and organs, which suggested that they have different role *in vivo*. Compared with the wild type, tiller number, total shoot dry weight, and yield of GOGAT co-suppressed plants were significantly decreased.

Genomic clones for NADH-dependent glutamate synthase (NADH-GOGAT) were obtained from a genomic library of rice *(Oryza sativa* L Cv. Sasaniskhi). A genomic clone (λ Nos 42, 14kb) covered entire structural gene and a 3.7 kb 5′-upstream region from the first methionine. Another clone (λ Nos 23, 14kb) contained a 1.8 kb 3′-downstream region from the stop codon. A7047 bp long clone (λ Nos R51) consisting of full length cDNA for NADH GOGAT was isolated from a cDNA library prepared using mRNA from roots of rice seedlings treated with 1mM NH_4Cl for 12 hours. The presumed transcribed region (11.7 kb) consisted of 23 exons separated by 22 introns. Rice NADH-GOGAT is synthesized as a 2166 amino acid protein with a molecular mass of 236.7 kDa that includes a 99 amino acid presequence. DNA gel blot analysis suggested that NADH-GOGAT occurred as a single gene in rice. Primer extension experiments map the transcription start of NADH-GOGAT to identical positions. The 3.7 kb 5′-upstream region was able to transiently express a reporter gene in cultured rice cells. Putative motifs related to the regulation of NADH-GOGAT gene expression were looked for within the 5′-upstream region by database (Goto et al. 1998).

References

Arima, Y. 1978. Glutamate synthase in rice root extracts and the relationship among electron donors, nitrogen donors and its activity. *Plant Cell Physiol,* **19:** 955-961.

Awasthi, V., I.K. Gautam, R.S. Sengar and S.K. Garg. 2013. Influence of putrescine on enzymes of ammonium assimilation in maize seedling. *Am. J. Plant Sci.,* **4:** 297-301.

Berlicki, L. 2008. Inhibitors of glutamine synthetase and their potential application in medicine. *Mini Rev. Med. Chem.* **8:** 869-878.

Cai, H., Y. Zhou, J. Xiao, X. Li, Q. Zhang and X. Lian. 2009. Overexpressed glutamine synthetase gene modifies nitrogen metabolism and abiotic stress responses in rice. *Plant Cell Rep.* **28:** 527-537.

Cheng-gang, L., C. Liping, W. Yan, L. Jia, X. Guang-li and L. Tian. 2011. High temperature at grain filling stage affects nitrogen metabolism enzyme activities in grains and grain nutritional quality in rice. *Rice Sci.* **15:** 210-216.

El-Shora, H.M. 2001. Effect of growth regulators and group modifiers on NADH glutamate synthase of marrow cotyledons. *Online J.Biol Sci.,* **1:** 597-602.

Goto, S.., T. Akagawa, S. Kojima, T. Hayakawa and T. Yamaya. 1998. Organization and structure of NADH-dependent glutamate synthase gene from rice plants. *Biochem. Biophysic. Acta,* **1387:** 298-308.

Garrido, R.G., F.S.R.G. Garrido, S.R. desouza and M.S. Fernandes. 2012. *Trop. Subtrop. Agroeco.*, **15:** 699-706.

Hayakawa, T., T. Yamaya, T. Mae and K. Ojima. 1993. Changes in the content of two glutamate synthase proteins in spikelets of rice (*Oryza sativa*) plants during ripening. *Plant Physiol.*, **101:** 1257-1262.

Hayakawa, T., T. Nakamura, F. Hattori, T. Mae, K. Ojima and T. Yamaya. 1994. Cellular localization of NADH dependent glutamate synthase protein in vascular bundles of unexpanded leaf blades and young grains of rice plants. *Planta,* **193:** 455-460.

Hayakawa, T., T. Sakai, K. Ishiyama, N. Hirose, H. Nakajima, M. Takezawa, K. Naito, M. Nakayama, T. Akagawa, S. Goto and T. Yamaya. 2003. Organisation and structure of ferredoxin dependent glutamate synthase and intracellular localization of enzyme protein in rice plants. *Plant Biotec.,* **20:** 43-55.

Hirose, N., T. Hayakawa and T. Yamaya. 1997. Inducible accumulation of mRNA for NADH-dependent glutamate synthase in rice roots in response to ammonium ions. *Plant Cell Physiol.,* **38:** 1295-1297.

Hirel, B. and P. Gadal. 1980. Glutamine synthetics in rice — A comparative study of the enzymes from roots and leaves. *Plant Physiol.,* **60:** 619-623.

Hirel, B., J. Vidal and P. Gadal. 1982. Evidence for a cytosolic-dependent light inductionof chloroplastic glutamine synthetase during greening of etiolated rice leaves. *Planta,* **155:** 17-23.

Hirose, N. and T. Yamaya. 1999. Okadaic acid mimics nitrogen stimulated transcriptionof the NADH-glutamate synthase gene in rice cell cultures. *Plant Physiol.,* **121:** 805-812.

Hoshida, H., Y. Tanaka, T. Hibino, Y. Hayashi, A. Tanaka and T. Takabe. 2000. Enhanced tolerance to salt stress in transgenic rice that over expressed chloroplast glutamine synthetase. *Plant Mol. Biol.,* **43:** 103-111.

Ishiyama, K., E. Inoue, M. Tabuchi, T. Yamaya and H. Takahashi. 2004. Biochemical background and compartmentalized functions of cytosolic glutamine synthetase for active ammonium assimilation. *Plant cell Physiol.* **45:** 1640-1647.

Kozaki, A. and G. Takeba. 1996. Photorespiration protects C_3 plants from photooxidation. *Nature,* **384:** 557-560.

Krajewski, W.W., R. Collins and L. Holmberg-Schiavone. 2008. Crystal structure of mainniation glutamine synthetases illustrate substrate-induced conformational changes and provides opportunities for drug and herbicide design. *J. Mol. Biol.* **375:** 217-228.

Kusaro, M., M. Tabuchi, A. Fukushima, K. Funayama, C. Diaz, M. Kobayashi, N. Hayashi, Y.N. Tsuchiya, H. Takahashi, A. Kamata, T. Yamaya and K. Saito. 2011. Metabolomics data reveal a crucial role of cytosolic glutamine synthetase 1: 1 in coordinating metabolic balance in rice. *The Plant J.* **66:** 456-466.

Lea, P.J., L. Sodek and M.A.J. Parry. 2007. Aspaazine in plants. Ann. *Applied Biol.,* **150:** 126-150.

Lea, P.J. and B.J. Miflin. 2011. Nitrogen assimilations and its relevance to crop improvement. *Ann Plant Reviews,* **42:** 1-30.

Leegood, R.C., P.J. Lea, M.D. Adcock and R.E. Hauster. 1995. The regulation and control of protorespiration. *J. Exp. Bot.,* **46:** 1397-1414.

Lin, Y.L., Y.Y. Chao and C.H. Kao. 2010. Exposure of rice seedlings to heat shock protects subsequent Cd induced decrease in glutamine synthetase activity and increase in specific protease activity in leaves. *J. Plant Physiol.* **167:** 1061-1065.

Miflin, B.J. and P.J. Lea. 1980. Ammonia Assimilation. pp.169-202 In: Miflin, B.J.(ed) The Biochemistry of plants. Vol. 5. Academic Press, New York.

Singh, K.K. and S. Ghosh. 2013. Regulation of glutamine synthetase isoforms in two differentially drought tolerant rice (L) cultivars under water deficit conditions. *Plant cell Re,.* **32:** 183-193.

Suzuki, A. and P. Gadal. 1982. Glutamate synthetase from rice leaves. *Plant Physiol,* **69:** 848-852.

Suzuki, A., S. Rional, S. Lemarchand, N. Godfroy, Y. Roux, J.P. Boutin and S. Rothstein. 2001. Regulation by light and metabolites of ferredoxin – dependent glutamate synthase in maize. *Physiol. Plant.* **112:** 524-530.

Tang, X.R. and T.Q. Yu. 2002. Effects of P and K fertilizers on yield and protein content in fodder rice and their mechanisms. *Sci. Agric. Sinica,* **35:** 372-377.

Tang, X.R. 2000. Effect of N supply on yield and protein content and its mechanism in fodder hybrid rice. *Hybrid Rice,* **15:** 34-37.

Tang, W.Y., Z.F. Xiang, W.J. Reu and X.C. Wang. 2005. Effect of S-3307 on nitrogen metabolism and grain protein content in rice. *Chinese J. Rice Sci,* **19:** 63-67.

Tobin, A.K. and T. Yamaya. 2001. Cellular compartmentation of ammonium assimilation in rice and barley. *J. Exp. Bot.,* **52:** 591-604.

Unno, H., T. Uchida, H. Sugawara, G. Kurisu, T. Sugiyama, T. Yamaya, H. Sakakibara, T.Hase and M. Kusunoki. 2006. Atomic structure of plant glutamine synthetase. *J. Biol. Chem.* **281:** 29287-29296.

Watanbe, S., T. Sakai, S. Goto, T. Yagimura, T. Hayakawa and T. Yamaya. 1996. Expression of NADH-dependent glutamate synthase in response to the supply of nitrogen in rice cells in suspension culture. *Plant Cell Physiol.,* **37:** 1034-1037.

Yamaya, T., Y. Hayakawa, K. Tanasawa, K. Kamachi, T. Mae and K. Ojimka. 1992. Tissue distribution of glutamate synthase and glutamine synthetase in rice leaves. *Plant Physiol.,* **100:** 1427-1432.

Yamaya, T., H. Tanno, N. Hirose, S. Watanbe and T. Hayakawa. 1995. A supply of nitrogen causes increase in the levels of NADH dependent glutamate synthase protein and the activity of the enzyme in roots of rice seedings. *Plant Cell Physiol.* **36:** 1197-1204.

Yong En, L.U., Luo Feng, Y. Meng, L. l. Xiang Hua and L. Xing Ming. 2011. Suppression of glutamate synthase genes significantly affects carbon and nitrogen metabolism in rice (*Oryza sativa* L). *Sci. China,* **54:** 651-663.

Yuan, S., S. Young-jian, W.'He-Zhou and M. Jun. 2009. Effects of water stress on activities of nitrogen assimilation enzymes and photosynthetic characteristics of rice seedlings. *Plant Nutr. Fert. Sci,* **15:** 1016-1022.

Yuan, H.F. and C.R. Hou. 1989. Regulatory properties of rice leaf glutamine synthetase by amino acids and nucleotides. *Bot. Bull. Academia Sinica,* **30:** 71-89.

Zhao, X.Q. and W.M. Shi. 2006. Expression analysis of the glutamine synthetase and glutamate synthase gene families in young rice *(Oryza sativa)* seedlings. *Plant Sci.,* **170:** 748-754.

Zhang, C., S. Peng, X. Peng, A.Q. Charez and J. BenneH. 1997. Response of glutamine synthetase isoforms to nitrogen sources in rice (*Oryza sativa* L.) roots. *Plant Sci.,* **125:** 163-170.

Zheng-xun, J., Q. Chun-rong, Y. Jing, L. Hai-Ying and P. Zhong-ze. 2007. Changes in activities of glutamine synthetase during grain filling and their relation to rice quality. *Rice Science,* **14:** 211-216.

chapter six

Polyamines

Polyamines are small polycationic molecules found ubiquitously in all organisms and function in a wide variety of biological processes. In the past decade, molecular and genetic studies using mutants and transgenic plants with an altered activity of enzymes involved in polyamine biosynthesis have contributed much to a better understanding of the biological functions of polyamines in plants.

In contrast to animals and fungi, in which ornithine decarboxylase (ODC) is the first and rate-limiting enzyme in the synthesis of polyamines, plants typically use arginine decarboxylase (ADC). The Arabidopsis thaliana genome lacks a gene encoding ODC (Hanfrey et al. 2005). Putrescine is converted to spermidine and spermine by successive activities of spermidine synthase and spermine synthase with the use of decarboxylated S-adenosyl methionine (dcSAM) as an aminopropyl donor. The dcSAM is produced by S-adenosyl methionine decarboxylase (SAMDC) from SAM. SAMDC is thought to be a major regulatory enzyme in the synthesis of spermidine and spermine and is also known to influence the rate of ethylene production in plants given that the precursor of ethylene, 1-aminocyclopropane-1-carboxylic acid, is derived from SAM. Polyamines are further metabolized by oxidation and conjugation with

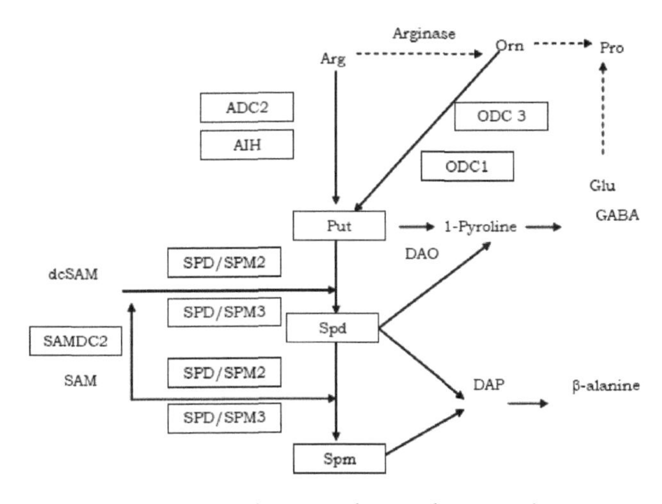

FIGURE 6.1 Polyamines biosynthesis in plants.

other molecules (Bagni and Tassoni, 2001; Cona et al. 2006; Moschou et al. 2008). Plant polyamines are preferentially detected in actively growing tissues and under stress conditions and have been implicated in the control of cell division, embryogenesis, root formation, fruit development and ripening, and responses to biotic and abiotic stresses (Kumar et al. 1997). The molecular mechanism of how polyamines act in these processes had remained unclear. In the past decade, however, molecular and genetic studies with mutants and transgenic plants having no or altered activity of enzymes involved in the biosynthesis of polyamines have contributed much to a better understanding of the biological functions of polyamines in plants.

6.1. *Estimation of Polyamines*

A sensitive (0.01-1 nmol) method has been developed for the analysis of polyamines in higher plant extracts based on high performance liquid chromatography (HPLC) of their benzoyl derivatives Putrescine, cadaverine, agmatine, spermidine, spermine, and the less common polyamines nor-spermidine and homospermidine can be completely resolved by reverse phase HPLC, isocratic elution with methanol:water (64%, v/v) through a 5-µm C_{18} column, and detection at 254 nm. The method can be directly applied to crude plant extracts, and it is not subject to interference by carbohydrates and phenolics. A good quantitative correlation was found between HPLC analysis of benzoylpolyamines and thin layer chromatography of their dansyl derivatives. With the HPLC method, polyamine titers have been reproducibly estimated for various organs of amaranth, *Lemna*, oat, pea, *Pharbitis*, and potato. The analyses correlate well with results of thin layer chromatography determinations.

Rice (*Oryza sativa* L.) is one of the most important food crops in the world. Almost half of the world's population depend on rice as their staple food. Rice is particularly susceptible to soil water deficit (Inthapan and Fukai, 1988). For upland rice drought is a major constraint on productivity and for rain fed lowland rice drought is the major environmental factor with a reduction of productivity up to 35%. Most high-yielding rice cultivars developed for irrigated conditions are highly susceptible to drought stress as well. The estimated average annual loss of rice production due to drought conditions world-wide is about 18 million tons, or 3.6 billion US$. Drought delays the development of the rice plant, and strongly affects morphology as well as physiological processes like transpiration, photosynthesis, respiration and translocation of assimilates to the grain. Leaf and root phenology of rice cultivars are known to influence their vegetative response to water deficit. The development of drought tolerant rice varieties is one of the challenges of the next decades.

6.2. *Biological Importance of Polyamines*

Polyamines, especially putrescine (Put), spermidine (Spd) and spermine (Spm) have been implicated in a wide range of biological processes, including growth, development and apoptosis (Evans and Malmberg, 1989; Galston, 1983; Galston and Sawheny, 1990; Kuehn and Phillips, 2005). Polyamines are also associated with responses of plants to environmental stresses, including mineral nutrient deficiencies, osmotic and drought stress, salinity, heat, chilling, hypoxia and environmental pollutants. Treatment with inhibitors of polyamine biosynthesis reduces stress tolerance whereas addition of exogenous polyamines restores successful stress acclimation (Liu et al. 2004). Therefore, polyamines are thought to play an essential role in the environmental stress tolerance of plants.

However, the physiological function of polyamines under abiotic stress conditions is not clear (Capell et al. 2004; Ma et al. 2005). Polyamines are positively charged at physiological pH and are therefore able to interact with negatively charged molecules, such as nucleic acids, acidic phospholipids, proteins and cell wall components such as pectin (Bouchereau et al. 1999; Kakkar and Sawhney, 2002; Martin-Tanguy, 2001). The multiple suggested roles of polyamines encompass involvement in protein phosphorylation, conformational transition of DNA (Martin-Tanguy, 2001), maintenance of ion balance, prevention of senescence, radical scavenging, membrane stabilization (Kakkar and Sawheny, 2002) and regulation of gene expression by enhancing the DNA-binding activity of transcription factors (Panagiotidis et al. 1995).

Also the specific functions of the different polyamines are unclear. Since Flores et al. reported massive accumulation of Put in leaf cells and protoplasts of oat in response to osmotic stress (Flores and Galston, 1982), a similar increase has been shown under osmotic or drought stress in rice (Yang et al. 2007) and other plant species (Flores and Galston, 1984; Galiba et al. 1993; Turner and Stewart, 1986; Legocka and Kluk, 2005; Aziz et al. 1997). Also, Spd and Spm accumulate under osmotic stress in some plants (Tiburcio et al. 1986), while a reduction of Put or Spd levels was observed in others (Turner and Stewart, 1986; Maiale et al. 2004; Mo and Pua, 2002). In more detailed analyses of rice Put, Spd and Spm were found to accumulate under osmotic stress (Capell et al. 2004) depending on stress intensity and duration (Le. fevre et al. 2001).

6.3. *Putrescine*

Put is synthesized either directly from ornithine by ornithine decarboxylase (ODC; EC 4.1.1.17) or indirectly from arginine via agmatine. The pathway is initiated by the arginine decarboxylase reaction (ADC; EC 4.1.1.19). Agmatine is sequentially converted to N-carbamoylputrescine

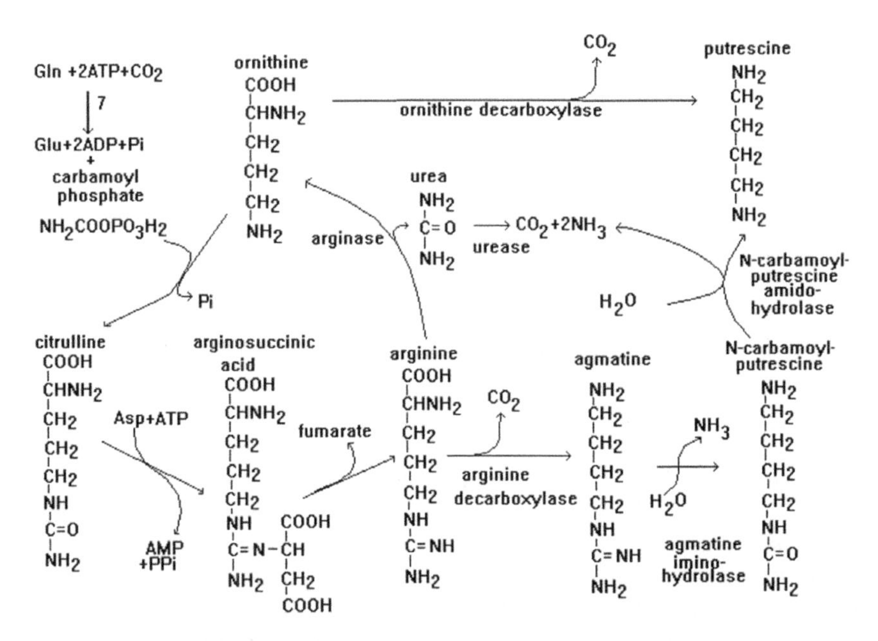

FIGURE 6.2 Details of putrescine biosynthesis.

by agmatine iminohydrolase (AIH; EC 3.5.3.12) and finally to Put by N-carbamoylputrescine amidohydrolase (CPA; EC 3.5.1.53). Spd and Spm are synthesized from Put by the transfer of aminopropyl groups from decarboxylated S-adenosylmethionine (SAM). These reactions are catalysed by Spd synthase (SPD; EC 2.5.1.16) and Spm synthase (SPM; EC 2.5.1.22). The decarboxylated SAM precursor is produced from SAM by S-adenosylmethionine decarboxylase (SAMDC; EC 4.1.1.50). The details of polyamine biosynthesis and integration of polyamine metabolic pathways into the surrounding metabolic network has been recently published for *Arabidopsis*.

The accumulation of Put during drought stress is thought to be primarily the result of increased ADC activity that may be controlled by transcript levels and/or enzyme activity in *Arabidopsis* (Alcazar et al. 2006; Urano et al. 2005), rice (Yang et al. 2007) and other species. Abiotic stress tolerance of plants was improved by constitutive over-expression of various genes encoding polyamine biosynthesis enzymes.

Polyamines (PAs), spermidine (Spd), spermine (Spm), and their diamine obligate precursor putrescine (Put) have been frequently described as endogenous plant growth regulators or intracellular messengers mediating physiological responses (Galston, 1983; Davies, 2004). In higher plants, Put can be directly synthesized from ornithine via ornithine decarboxylase (ODC; EC 4.1.1.17) or indirectly from arginine via arginine decarboxylase (ADC; EC4.1.1.19) (Gemperlová et al. 2006). Spd and Spm are synthesized via Spd synthase (EC 2.5.1.16) and Spm synthase (EC 2.5.1.22), respectively, by sequential addition of aminopropyl groups to Put. This aminopropyl group is provided by decarboxylated S-adenosyl-L-methionine, which is a product of S-adenosyl-L-methionine decarboxylase (SAMDC; EC 4.1.1.50) (Maiale et al. 2004). Because of their polycationic nature at physiologically relevant ionic and pH conditions, PAs occur in cells not only in the free form but also in the soluble-conjugated and insoluble-conjugated forms (Martin-Tanguy, 2001; Gemperlová et al. 2006). Soluble-conjugated PAs are mostly linked to hydroxycinnamic acid monomers, and insoluble-conjugated PAs to hydroxycinnamic acid dimers and trimers, and to macromolecules such as proteins (Kasukabe et al. 2004). PAs are implicated in many physiological processes such as cell division, morphogenesis, and development.

A number of investigators have used PA inhibitors to modulate the cellular PA titer in order to determine their role in various plant processes. Four commonly used inhibitors of PA synthesis are: 1. Difluoromethylornithine (DFMO), an irreversible inhibitor of ODC; 2. Difluoromethylarginine (DFMA), an irreversible inhibitor of ADC (Bitonti et al. 1987); 3. Methylglyoxyl-bis guanylhydrazone (MGBG), a competitive inhibitor of S-adenosyl-methionine decarboxylase (SAMDC) (Williams-Ashman and Schenone, 1992); and 4. Cyclohexylamine (CHA), a competitive inhibitor of spermidine synthase (Hibasami et al. 1980). Common oxidases are diamine oxidase and polyamine oxidase (PAO), as reviewed by Smith and Marshall (1988). Each PA has been found to be catabolized by a specific oxidase.

Several investigations have dealt with localization of PAs and their biosynthetic enzymes in plants (Slocum, 1991). However, paucity of information regarding the exact cellular and subcellular localization of these entities remains one of interest concern.

6.4. *Abiotic Stress*

Saline (NaCl) stress in germinated seedlings of rice Cv. Rupsail causes an increase in polyamine content and concomitantly an enhancement of arginine decarboxylase (ADC) activity; in the case of putrescine, the increase is 100%. The responsiveness of coleoptiles and roots is different; ADC activity is maximum in roots and coleoptiles at 50 mM and 100 mM respectively. Difluoromethyl arginine, a potent inhibitor of ADC, reduces both polyamine accumulation and ADC activity. NaCl can enhance the activity of ADC by 1.6-fold *in vitro*. [^{14}C]leucine incorporation into protein at 400 mM NaCl is decreased 4- and 10-fold in coleoptiles and roots, respectively. Efflux of sugar, amino acid, polyamines and total electrolytes increased gradually with the increase in concentration of salt. Influx of Na^+ and Cl^- and efflux of K^+ in coleoptiles and roots are directly proportional to the concentration of NaCl applied (Basu et al. 1988).

Enhanced production and accumulation of free and conjugated polyamines as well as increased activities of their biosynthetic enzymes in plants have been associated with heat stress. Perchloric acid-soluble free, as well as conjugated polyamines, and their metabolic enzymes were studied under 45°C heat stress in callus raised from heat-tolerant and -sensitive rice cultivars. The levels of free and conjugated polyamines, as well as arginine decarboxylase (EC 4.1.1.19) and polyamine oxidase (EC 1.4.34) activities were higher in tolerant than in sensitive callus under non-stressed conditions. Heat stress caused greater accumulation of free and conjugated polyamines in callus of the heat-tolerant cultivar N22 than in that of the heat-sensitive cultivar IR8. In particular, the uncommon polyamines norspermidine and norspermine were detected in Cv. N22, which increased appreciably during stress, but they were not detected in callus of Cv. IR8. Arginine decarboxylase and polyamine oxidase activities increased to a larger extent in N22 than in IR8 callus during stress, activities that were well correlated with the increased levels of common and uncommon polyamines. Increased levels of transglutaminase activity indicated the high titre of conjugated polyamines (Roy and Ghosh, 1996).

Environmental changes, irrespective of source, cause a variety of stresses in plants. These stresses affect the growth and development and trigger a series of morphological, physiological, biochemical and molecular changes in plants. Abiotic stress is the primary cause of crop loss worldwide. The most challenging job before the plant biologists is the development of stress tolerant plants and maintenance of sufficient yield of crops in this changing environment. Polyamines can be of great use to enhance stress tolerance in such crop plants. Polyamines are small organic polycations present in all organisms and have a leading role in cell cycle, expression of genes, signaling, plant growth and development and tolerance to a variety of abiotic stresses. High accumulation of polyamines (putrescine, spermidine and spermine) in plants during abiotic stress has been well documented and is correlated with increased tolerance to abiotic stress. Genetic engineering of PA biosynthetic genes in crop plants is the way to create tolerance against different stresses. The injury due to cold causes alteration in the membrane structure, and the chilling injury involves phase transition in the molecular ordering of membrane lipids (Raison and Lyons, 1971). This can cause several deleterious effects like increased membrane permeability and alteration of the activity of membrane proteins.

Cold treatment has been reported to increase the levels of Put, and this correlates with the increase in the induction of arginine decarboxylase (ADC) genes (ADC1, ADC2 and SAMDC2) (Urano et al. 2003; Cuevas et al. 2008, 2009). On the other hand, levels of free Spd and Spm remain constant or even decrease in response to cold treatment (Alcazar et al.

2010). The absence of correlation between enhanced SAMDC2 expression and the decrease Spm levels may be a result of increased Spm catabolism (Cuevas et al. 2008; Alcazar et al. 2010). Boucereau et al. (1999) reported that in the chilling-tolerant-cultivar, chilling induced an increase in free abscisic acid (ABA) levels first, then ADC activity and finally free Put levels. Fluridone, an inhibitor of ABA synthesis, inhibited the increase of free ABA levels, ADC activity and free Put levels in chilled seedlings of a chilling-tolerant cultivar. These effects resulted in a reduced tolerance to chilling and could be reversed by the pre-chilling treatment with ABA. All these results suggest that Put and ABA are integrated in a positive feedback loop, in which ABA and Put reciprocally promote each other's biosynthesis in response to abiotic stress. This highlights a novel mode of action of polyamines as regulators of ABA biosynthesis (Alcazar et al. 2010).

Incubation of 3-d-old seedlings of *Oryza sativa* L. Cv Arborio, under anaerobic conditions, leads to a large increase in the titer of free putrescine while aerobic incubation causes a slight decrease. After two days, the putrescine level is about 2.5 times greater without oxygen than in air. The rice coleoptile also accumulates a large amount of bound putrescine and, to a lesser extent, spermidine and spermine (mainly as acid-soluble conjugates). Accumulation of conjugates in the roots is severely inhibited by the anaerobic treatment. Feeding experiments with labeled amino acids showed that anoxia stimulates the release of (14)CO_2 from tissues fed with [(14)C]arginine and that arginine is the precursor in putrescine biosynthesis. After two days of anoxia, the activity of arginine decarboxylase was 42% and 89% greater in coleoptile and root, respectively, than in the aerobic condition (Reggiani et al., 1989).

A selection of 21 rice cultivars (*Oryza sativa* L. ssp. *indica* and *japonica*) was characterized under moderate long-term drought stress by comprehensive physiological analyses and determination of the contents of polyamines and selected metabolites directly related to polyamine metabolism. To investigate the potential regulation of polyamine biosynthesis at the transcriptional level, the expression of 21 genes encoding enzymes involved in these pathways were analyzed by qRT-PCR. Analysis of the genomic loci revealed that 11 of these genes were located in drought-related QTL regions, in agreement with a proposed role of polyamine metabolism in rice drought tolerance. The cultivars differed widely in their drought tolerance and parameters such as biomass and photosynthetic quantum yield were significantly affected by drought treatment. Under optimal irrigation free putrescine was the predominant polyamine followed by free spermidine and spermine. When exposed to drought, putrescine levels decreased markedly and spermine became predominant in all cultivars. There were no correlations between polyamine contents and drought tolerance (Do et al. 2013).

TABLE 6.1 Correlations of the relative contents of polyamines in the flag leaf with the yield maintenance ratio of rice

Polyamines	8 DAWW	16 DAWW	24 DAWW	32 DAWW
Free polyamines				
Putrescine	0.97**	-0.29	-0.85*	-0.96**
Spermidine	0.94**	0.89*	0.97**	0.99**
Spermine	0.96**	0.98**	0.95**	0.98**
Soluble-conjugated polyamines				
Putrescine	0.39	-0.11	0.28	-0.49
Spermidine	0.48	0.59	0.78	0.41
Spermine	0.13	0.41	0.07	0.04
Insoluble-conjugated polyamines				
Putrescine	0.95**	0.84*	0.96**	0.98**
Spermidine	0.03	-0.32	0.19	0.01
Spermine	0.35	0.21	0.52	0.43

DAWW = Days after withholding water
** significant at 1% level, * significant at 5% level

Plants were pot-grown, and well-watered (WW) and water-stressed (WS) treatments were conducted from complete elongation of the flag leaf to grain maturity.

GC-MS analysis revealed drought-induced changes of the levels of ornithine/arginine (substrate), substrates of polyamine synthesis, proline, product of a competing pathway and GABA, a potential degradation product. Gene expression analysis indicated that ADC-dependent polyamine biosynthesis responded much more strongly to drought than the ODC-dependent pathway. Nevertheless the fold change in transcript abundance of ODC1 under drought stress was linearly correlated with the drought tolerance of the cultivars.

New rice genotype was developed with West African Rice Development Association (WARDA) in 1990s. NERICA rice shows both vigorous growth and tolerance of stressors such as drought and disease. The purpose of this study was to clarify the physiological and biochemical responses to salt stress of NERICA rice seedlings. The degree of growth inhibition caused by salt stress was small in NERICA rice varieties as compared with *japonica* Nipponbare. Na accumulation in leaf blades was high in salt-sensitive varieties. Accumulation of proline, a known compatible solute, was also induced by salt stress, especially in salt-sensitive varieties; it was thought that this accumulation was brought on salt-stress injury. The contents of polyamines, especially spermidine, were high in the pre-stressed leaf blades of NERICA rice seedlings. After the salt-stress treatment, the polyamine content of leaf blades differed with the degree of salt tolerance of the NERICA rice seedlings. These results suggested that the salt tolerance of NERICA rice seedlings might be associated not only with the regulation of Na absorption and translocation but also with their ability to maintain leaf polyamine levels under salt-stress conditions.

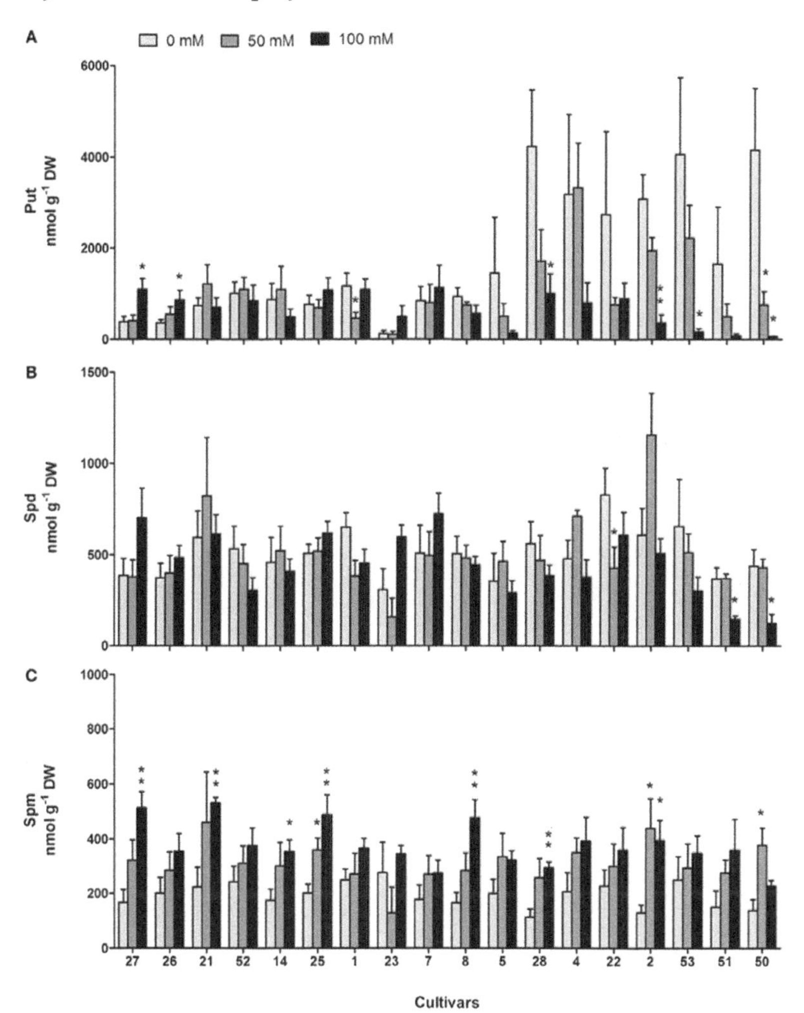

FIGURE 6.3 Variation of polyamines in rice genotypes with salinity.

This study tested the hypothesis that the interaction between polyamines and ethylene may mediate the effects of soil drying on grain filling of rice (*Oryza sativa* L.). Two rice cultivars were pot grown. Three treatments, well-watered, moderate soil drying (MD), and severe soil drying (SD), were imposed from 8 d post-anthesis until maturity. The endosperm cell division rate, grain-filling rate, and grain weight of earlier flowering superior spikelets showed no significant differences among the three treatments. However, those of the later flowering inferior spikelets were significantly increased under MD and significantly reduced under SD when compared with those which were well watered. The two cultivars showed the same tendencies. MD increased the contents of free spermidine (Spd) and free spermine (Spm), the activities of *S*-adenosyl-L-methionine decarboxylase and Spd synthase, and expression levels of polyamine synthesis genes, and decreased the ethylene evolution rate, the contents of 1-aminocylopropane-1-carboxylic acid (ACC) and hydrogen peroxide, the activities of ACC synthase, ACC oxidase, and polyamine

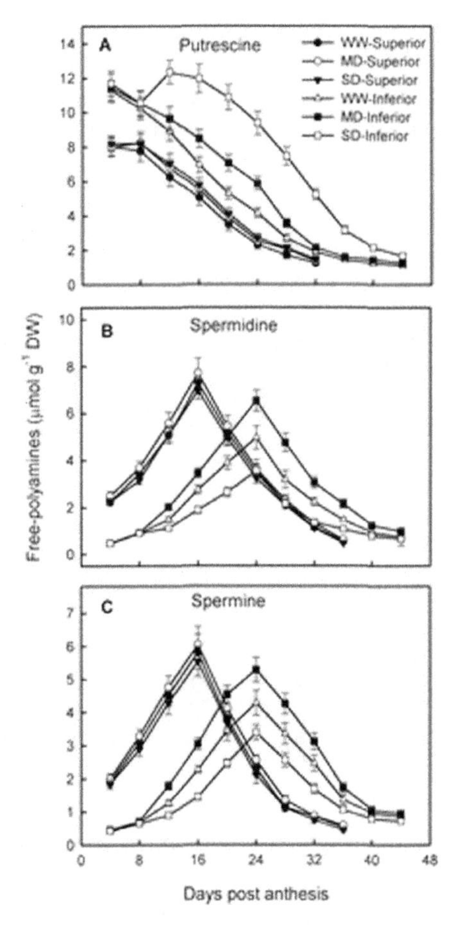

FIGURE 6.4 Changes of polyamines in superior and inferior grains under different water levels during grain filling.

oxidase, and the expression levels of ethylene synthesis genes in inferior spikelets. SD exhibited the opposite effects. Application of Spd, Spm, or an inhibitor of ethylene synthesis to rice panicles significantly reduced ethylene and ACClevels, but significantly increased Spd and Spm contents, grain-filling rate, and grain weight of inferior spikelets. The results were reversed when ACC or an inhibitor of Spd and Spm synthesis was applied. The results suggest that a potential metabolic interaction between polyamines and ethylene biosynthesis responds to soil drying and mediates the grain filling of inferior spikelets in rice.

Highly cold tolerant cultivars such as HSC-55 and Plovdiv-22 seedlings grown at low temperature (18/13 °C) produced about 33 and 28%, respectively of dry matter of their controls grown at normal temperature (28/23 °C). In contrast, Amaroo and Doongara at low temperature produced only about 17% of dry matter of their controls. The cultivars with higher seedling growth at low temperature also accumulated higher amount of spermine in shoots. There was a significant correlation between spermine content and seedling growth at low temperature (R^2 = 0.85).

An induction of the transcript level and/or activity of ADC could be shown for rice (Chattopadhyay et al. 1997) as well as for other species (Mo and Pua., 2002; Urano et al. 2004; Hao et al. 2005a; Legocka and Kluk, 2005; Liu et al. 2006) under salinity. Transcript levels of other polyamine biosynthesis-related genes are also increased under salt stress, e.g., SAMDC in rice (Li and Chen, 2000), soybean (Tian et al. 2004), wheat (Li and Chen, 2000), *Arabidopsis* (Urano et al. 2003) and apple (Hao et al. 2005), SPD and SPM in *Arabidopsis* (Urano et al. 2003), and maize (Rodríguez-Kessler et al. 2006). Tolerance to drought was improved by constitutive over-expression of oat ADC in rice (Capell et al. 1998), with a simultaneous effect on plant development. When polyamine accumulation was induced by over-expression of oat ADC or *Tritodermum* SAMDC under the control of an ABA-inducible promoter, rice plants were more resistant to high salinity (Roy and Wu, 2001, 2002). Furthermore, over-expression of the *Datura stramonium* ADC gene under the control of the stress activated maize ubiqitin-1 promoter conferred tolerance to osmotic stress in rice (Capell et al. 2004).

Soil salinity affects a large proportion of rural area and limits agricultural productivity. To investigate differential adaptation to soil salinity, salt tolerance of 18 varieties of *Oryza sativa* using a hydroponic culture system were studied. Based on visual inspection and photosynthetic parameters, cultivars were classified according to their tolerance level. Additionally, biomass parameters were correlated with salt tolerance. Polyamines have frequently been demonstrated to be involved in plant stress responses and, therefore, soluble leaf polyamines were measured. Under salinity, putrescine (Put) content was unchanged or increased in tolerant, while dropped in sensitive cultivars. Spermidine (Spd) content was unchanged at lower NaCl concentrations in all, while reduced at 100 mM NaCl in

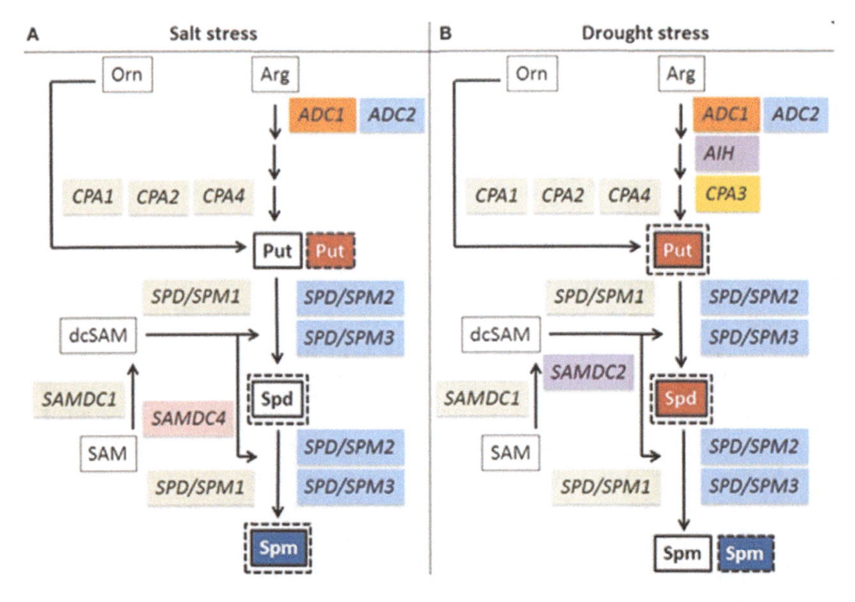

FIGURE 6.5 Changes in polyamines and enzymes under salt and drought stress.

sensitive cultivars. Spermine (Spm) content was increased in all cultivars. A comparison with data from 21 cultivars under long-term, moderate drought stress revealed an increase of Spm under both stress conditions. While Spm became the most prominent polyamine under drought, levels of all three polyamines were relatively similar under salt stress. Put levels were reduced under both, drought and salt stress, while changes in Spd were different under drought (decrease) or salt (unchanged) conditions. Regulation of polyamine metabolism at the transcript level during exposure to salinity was studied for genes encoding enzymes involved in the biosynthesis of polyamines and compared to expression under drought stress. Based on expression profiles, investigated genes were divided into generally stress-induced genes (ADC2, SPD/SPM2, SPD/SPM3), one generally stress-repressed gene (ADC1), constitutively expressed genes (CPA1, CPA2, CPA4, SAMDC1, SPD/SPM1), specifically drought-induced genes (SAMDC2, AIH), one specifically drought-repressed gene (CPA3) and one specifically salt-stress repressed gene (SAMDC4), revealing both overlapping and specific stress responses under these conditions.

6.5. Transport and Localization of Polyamines

The results of the study suggested that

- Computational analysis resulted in a distinct group of ten candidate polyamine transporter genes from rice and *Arabidopsis.*

- Heterologous expression of (Polyamines transports PUTs) in the *agp2_* yeast mutant showed that all seven PUTs analyzed were polyamine transporters. Except for OsPUT3.2, all the PUTs are high affinity spermidine importers and low affinity putrescine transporters.
- Analysis of PUTs by RT-PCR revealed that they have diverse tissue expression.
- GUS–reporter analysis showed that AtPUT2 and AtPUT3 are expressed mainly in the vascular bundles indicating that they have a role in long distance transport of polyamines.
- Transient expression of PUTs in onion epidermal cells showed no clear organullar localization among the PUTs except OsPUT3.1. OsPUT3.1 appears to be localized to the mitochondria.
- Transient expression in rice protoplasts revealed that OsPUT3.1 might be localized to chloroplast.
- OsPUT3.1 and OsPUT3.2 are alternatively spliced forms of the same rice gene and appear to have different affinities for spermidine uptake. Transient expression showed OsPUT3.1 localized to the mitochondria (in onion) and chloroplast (in rice protoplast).
- This work has shown that use of GFP-gene fusions might not be a good strategy to resolve the subcellular localizations of PUTs (Mulangi, 2011).

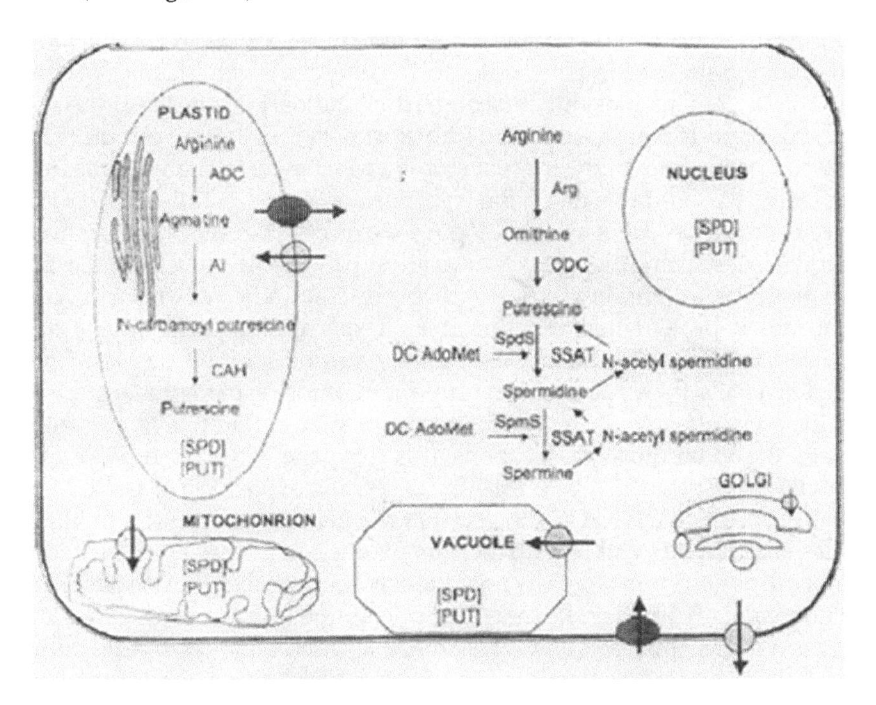

FIGURE 6.6 Localization and transport of polyamines in a cell.

6.6. Engineering Polyamine Biosynthesis

Engineering of the plant polyamine biosynthetic pathway has concentrated mostly on two species, tobacco and rice. It has generated a diverse rice germplasm with altered polyamine content. Transgenic rice plants expressing the *Samdc* cDNA accumulated spermidine and spermine in seeds at two to three-fold higher levels compared to wild type. In a different set of experiments, we were will be it was able to measure a ten-fold putrescine accumulation in transgenic rice plants harboring oat *adc* cDNA compared to wild type. Reduction in endogenous *adc* transcript levels in rice resulted in depletion of putrescine and spermidine pools, with no concomitant changes in expression of downstream genes in the pathway.

In general, studies focusing on spatial expression of these transgenes demonstrate that more dramatic changes in polyamine content occur in storage compared to vegetative tissues, such as leaves and roots. Therefore, we believe that the polyamine biosynthetic pathway in plants is regulated strongly in a spatial manner. In tomato, enhanced fruit juice quality and prolonged vine life of fresh fruits with increased lycopene was achieved by expression of yeast *Samdc* driven by the ripening-inducible E8 promoter.

Polyamines are nitrogenous compounds found in all eukaryotic and prokaryotic cells and absolutely essential for cell viability. In plants, they regulate several growth and developmental processes and the levels of polyamines are also correlated with the plant responses to various biotic and abiotic stresses. In plant cells, polyamines are synthesized in plastids and cytosol. This biosynthetic compartmentation indicates that the specific transporters are essential to transport polyamines between the cellular compartments. In the present study, a phylogenetic analysis was used to identify candidate polyamine transporters in rice. A full-length cDNA rice clone AK068055 was heterologously expressed in the Saccharomyces cerevisiae spermidine uptake mutant, agp2Δ. Radiological uptake and competitive inhibition studies with putrescine indicated that rice gene encodes a protein that functioned as a spermidine-preferential transporter. In competition experiments with several amino acids at 25-fold higher levels than spermidine, only methionine, asparagine and glutamine were effective in reducing uptake of spermidine to 60% of control rates. Based on those observations, this rice gene was named polyamine uptake transporter 1 (OsPUT1). Tissue-specific expression of OsPUT1 by semi quantitative RT-PCR showed that the gene was expressed in all tissues except seeds and roots. Transient expression assays in onion epidermal cells and rice protoplasts failed to localize to a cellular compartment. The characterization of the first plant polyamine transporter sets the stage for a systems approach that can be used to build a model to fully define how the biosynthesis, degradation, and transport of polyamines in plants mediate developmental and biotic responses.

6.7. *Stress Tolerance*

ADC is a key enzyme in the synthesis of putrescine in plants (ADC: ADC1 and ADC2). In contrast to ADC1, which is constitutively expressed in all tissues, ADC2 is responsive to abiotic stresses such as drought and wounding (Soyka and Heyer, 1999; Pérez-Amador *et al.* 2002). A loss-of-function mutant of ADC2 fails to show the osmotic-stress-induced increase in ADC activity observed in the wild-type, but exhibits no obvious phenotype under normal growth conditions (Soyka and Heyer, 1999). This was the first report on a genetically mapped mutant allele of a polyamine biosynthetic gene in plants. Another mutant allele of ADC2 was later shown to be more sensitive to salt stress than wild-type plants (Urano et al. 2004). This mutant has reduced levels of putrescine, but not of spermidine or spermine, and its stress tolerance is restored by exogenously supplied putrescine. These findings suggest a direct protective role of putrescine in abiotic stress tolerance. It is also clear that putrescine is important as a precursor for the biosynthesis of higher polyamines.

According to a threshold model based on studies using transgenic plants with altered putrescine levels (Capell et al. 2004), the putrescine level must exceed a certain threshold to enhance the synthesis of spermidine and spermine under stress, such synthesis being necessary for recovery from the stress. The double mutant of ADC1 and ADC2, which could not produce polyamines, dies at the embryo stage (Urano et al. 2005). Putrescine also serves as a precursor for the biosynthesis of pyridine and tropane alkaloids such as nicotine in tobacco and hyoscyamine in *Datura stramonium*. The first committed step is catalyzed by putrescine N-methyltransferase, which converts putrescine into N-methylputrescine and is structurally related to spermidine synthase (Hashimoto et al. 1989).

6.8. *Spermidine and Plant Growth*

Spermidine-deficient mutants in *Escherichia coli* are viable with no growth defects, while the yeast spe3 mutant, which has no spermidine synthase activity, requires spermidine for growth (Tabor and Tabor, 1999). There are two genes encoding spermidine synthase, SPDS1 and SPDS2, in the *Arabidopsis*genome. Each single mutant of these genes shows no growth defects but embryo development of the double mutant is arrested at the heart stage, indicating a requirement for spermidine during the course of embryogenesis (Imai et al. 2004a). On the other hand, as described below, spermine is not essential for viability of *Arabidopsis* (Imai et al. 2004b). Although it remains to be examined whether spermine can functionally substitute for spermidine or not, spermidine may be specifically required for some aspects of development. For instance, spermidine is a substrate for deoxyhypusine synthesis. The unusual amino acid deoxyhypusine is produced by deoxyhypusine synthase (DHS),which transfers the butylamine moiety of spermidine to the1-amino group of highly conserved lysine-50 of inactive eukaryotic translation initiation factor 5A (eIF5A) precursor.

The deoxyhypusine-eIF5A is further converted to the active hypusine-eIF5A, which functions in the transport of newly transcribed mRNAs from the nucleus to cytoplasm. The fact that activated eIF5A is essential for eukaryotic cell growth and proliferation is consistent with the absolute requirement for spermidine in plant embryo development. A mutation in the DHS gene results in the arrest of embryo-sac development in *Arabidopsis* (Pagnussat et al. 2005) probably because deoxyhypusine-eIF5A cannot be supplied by maternal tissues. Plant polyamines are present not only as free molecules but also as conjugates to cinnamic acids such as *p*-coumaric, the ferulic and caffeic acids (Bagni and Tassoni, 2001). The resulting conjugates are known as hydroxycinnamic acid amides. Recently, two genes encoding spermidine disinapoyl transferase (SDT) and spermidine di coumaroyl transferase (SCT), which mediate the production of spermidine conjugates, were identified in *Arabidopsis* (Luo et al. 2009). SDT and SCT are highly expressed in developing embryos and root tips, respectively. In addition, an anther tapetum-specific gene encoding spermidine hydroxycinnamoyl transferase was cloned from *Arabidopsis* (Grienenberger et al. 2009). The spermidine conjugates produced are implicated in protecting against pathogens, detoxifying phenolic compounds, and/or serving as a reserve of polyamines that are available to actively proliferating tissues, although not always essential for survival. In most organisms, polyamines can also be covalently bound to glutamine residues of certain proteins by the action of transglutaminase (TGase). The widespread occurrence of TGase activity in all plant tissues suggests the significance of inter- or intra-molecular cross-link formation of the proteins by polyamines (Serafini-Fracassini and Del Duca, 2008).

In some plant species, an aminobutyl group of spermidine is transferred to putrescine by the action of homospermidine synthase (HSS), which catalyses the formation of an uncommon polyamine, homospermidine. Homospermidine is the first intermediate of the biosynthesis of the pyrrolizidine alkaloids that serve as defence compounds in such families as *Asteraceae*, *Boraginaceae* and *Orchidaceae*. The HSS gene from *Senecio vernalis*was cloned and shown to be derived from the DHS gene (Ober and Hartmann, 1999).

6.9. Spermine in Stress Response

The *E. coli* genome does not contain a gene for spermin synthase. The yeast spe4 mutant, which lacks spermine synthase activity, shows normal growth in the absence of spermine (Tabor and Tabor, 1999). Furthermore, there are many reports that polyamines, in particular spermine, interact with functionally diverse ion channels and receptors. In general, the efficacy of polyamines in modulating or blocking these types of proteins decreases in the order spermine spermidine putrescine. In plants, no requirement for spermine under normal growth conditions has been demonstrated in a loss-of-function mutant of SPMS in *Arabidopsis* (Imai et al. 2004b) but the existence of a polyamine metabolon, a large protein complex containing both spermidine synthase and spermine synthase in *Arabidopsis*, is probably responsible for the efficient production of spermine in plant cells (Panicot et al. 2002). The SPMS mutant appears to be more sensitive to drought and salt stresses than the wild-type (Yamaguchi et al. 2007). This phenotype might be related to the fact that inward potassium currents across the plasma membrane of guard-cells are blocked by intracellular polyamines (Liu et al. 2000).

Blocking of ion channels by polyamines in plants has also been reported for vacuolar cation channels in barley and red beet (Dobrovinskaya et al. 1999), and for non-selective cation channels in pea mesophyll cells (Shabala et al. 2006). Although increasing evidence supports a modulating role for spermine in the control of ion channel and receptor activities, one of the most important roles of spermine, which occurs in millimolar concentrations in the nucleus, has been thought to be in protecting DNA from free radical attack and subsequent mutation (Ha et al. 1998). On the other hand, spermine plays a role as a mediator in defence signalling against plant pathogens (Yamakawa et al. 1998; Takahashi et al. 2003). This 'spermine

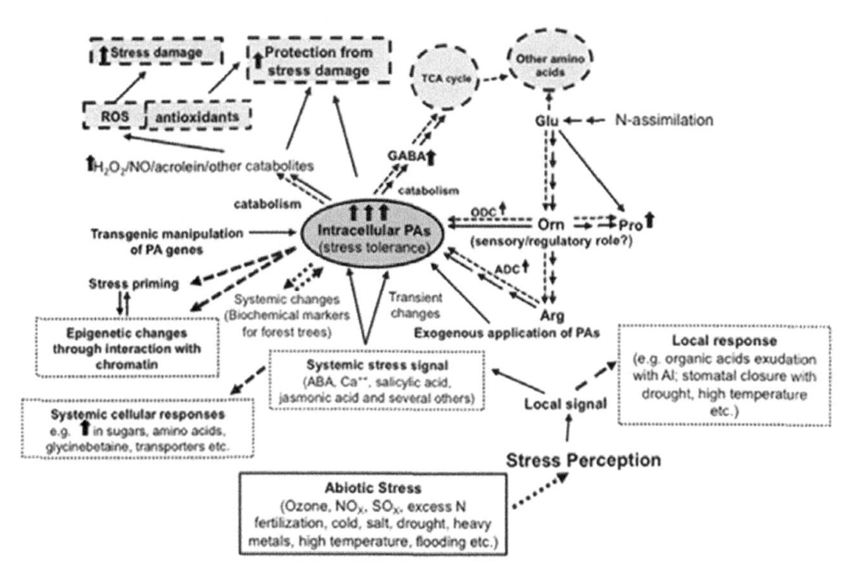

FIGURE 6.7 Polyamines in stress.

signalling pathway' involves accumulation of spermine in the apoplast, upregulation of a subset of defence-related genes such as those encoding pathogens in related proteins and mitogen-activated protein kinases, and a type of programmed cell death known as the hypersensitive response. This response is triggered by spermine-derivedH_2O_2, produced through the action of polyamine oxidase (PAO) localized in the apoplast (Cona et al. 2006; Kusano et al. 2008; Moschou et al. 2008b). Taken together, these data indicate double-edged roles of spermine in cell survival: ylogenetic tree of the amino acid sequences of spermidine/spermine synthase-related proteins. The full-length amino acid sequences were aligned by use of Clustal X and the tree was drawn by the neighbour-joining method with the software TreeView. ACL5 represents thermospermine synthase or its putative orthologues, which may have been acquired by an ancestor of the plant lineage through horizontal gene transfer from archaea or bacteria, as suggested by the absence of ACL5-like sequences in animals and fungi (Minguet et al. 2008) as a free radical scavenger in the nucleus and as a source of free radicals in the apoplast, although the possibility cannot be ruled out that the interaction of spermine with other molecules is involved in the cell death.

6.10. *Thermospermine*

Thermospermine is a structural isomer of spermine first discovered in thermophilic bacteria (Oshima, 2007). Thermospermine is synthesized from spermidine in a similar reaction to spermine synthesis. The *Arabidopsis* gene encoding thermospermine synthase, ACL5, was previously identified as encoding spermine synthase from the acaulis5 (acl5) mutant, which shows a severely dwarfed phenotype with over-proliferation of xylem tissues (Hanzawa et al. 2000), but a detailed biochemical study of ACL5 and its orthologue cloned from the diatom *Thalassiosirap seudonana*reveals that they encode thermospermine synthase (Knott et al. 2007). A study of the evolutionary pathways of genes involved in polyamine biosynthesis suggests that plants acquired the ability to synthesize thermospermine at an early stage of evolution by horizontal gene transfer from a prokaryote (Minguet et al. 2008). Exogenous application of thermospermine but not spermine to the acl5 mutant partially rescues plant height and reduces the acl5 transcript level that is increased in the mutant (Kakehi et al. 2008). Reduction in ACL5 expression by thermospermine is also observed in wild-type plants and indicates a negative feedback control of thermospermine synthesis. ACL5 shows a preferential expression in developing xylem vessel elements, suggesting a role in preventing premature cell death in xylem differentiation (Clay and Nelson, 2005; Muñiz et al. 2008). It was confirmed that the dwarf phenotype of acl5 was partially overcome by transgenic expression of a moss ACL5 orthologue in acl5 plants.

An interesting question is how the role of thermospermine was integrated into the regulation of stem elongation during the evolution of higher plants. In experiments designed to elucidate the mode of action of thermospermine in stem elongation, suppressor mutants of acl5, named sac, that show a recovery from the dwarf phenotype in acl5 have been isolated. One of the responsible genes, SAC51, encodes a bHLH-type (basic-helix-loop-helix) transcription factor and the sac51-d allele has a point mutation in the 4th upstream open reading frame (uORF) of the five uORFs present in the SAC51 50 leader sequence (Imai et al. 2006). The dominant nature of the sac51-d phenotype in acl5 can be attributed to an increased translational efficiency of the SAC51 main ORF because the 4th uORF of SAC51 has an inhibitory effect on the translation of the main ORF. The gene for another mutant, sac52-d, encodes ribosomal protein L10 (RPL10A) and the sac52-d allele also increases the translational efficiency of the SAC51 main ORF (Imai et al. 2008). In addition, we have found that a newly identified mutant, sac56-d, has a defect in a gene encoding RPL4 and shows a similar effect to that of sac52-d on SAC51 translation . These results suggest that SAC51 is one of the key transcription factors controlling stem elongation and that thermospermine plays a crucial role in its uORF-mediated translational regulation.

6.11. *Polyamines in Translation*

It is well known that spermidine and spermine regulate the translation of SAMDC with a small uORF of its mRNA that encodes a conserved hexapeptide (MAGDIS) in mammals (Ruan et al. 1996). The MAGDIS peptide causes ribosome stalling at its own termination codon in the presence of elevated levels of polyamines through polyamine-dependent interaction of the peptide with a component of the translational machinery; consequently, MAGDIS blocks translation of the SAMDC main ORF and the synthesis of polyamines. A similar autoregulatory circuit is found in the SAMDC translation in *Arabidopsis*but its regulation involves two uORFs, the first and shorter of which is responsible, at reduced levels of polyamines, for repressing translation of the second and longer uORF, which in turn blocks SAMDC translation constitutively irrespective of polyamine levels (Hanfrey et al. 2005).

The predicted role of spermidine and spermine in bypassing the inhibitory effect of the first uORF on the translation of the second uORF seems to be opposite to that in mediating the effect of the MAGDIS uORF in mammals. Phenotype of the *Arabidopsis* acl5-1 mutants, which are defective in the synthesis of thermospermine. Forty-day-old plants of acl5-1 and transgenic acl5-1 expressing a full-length cDNA of the moss ACL5B (GenBank accession number EDQ54752) are noted and, rather, to be related to that of thermospermine in enhancing the SAC51 main ORF translation. For polyamines to bypass the inhibitory effect of uORFs, several possible mechanisms are conceivable, including translational frameshifting and ribosome shunting.

6.12. Polyamines Oxidation

While putrescine is catabolized by the copper-containing diamine oxidase to 4-aminobutanal with concomitant production of NH_3 and H_2O_2, spermidine, spermine and probably thermospermine are catabolized by PAOs (Bagni and Tassoni, 2001). PAOs are classified as those de-aminating spermidine and spermine to 4-aminobutanal and N-(3-aminopropyl)-4-aminobutanal, respectively, along with 1,3-diaminopropane and H_2O_2, and those back-converting spermine to spermidine and spermidine to putrescine with concomitant production of 3-aminopropanal and H_2O_2 (Moschou et al. 2008). These de-amination reactions occur mainly in the apoplast and the resulting H_2O_2 plays a role in mediating a complex array of defence responses to microbial pathogens as described above. Furthermore, polyamine-derived H_2O_2 has been implicated in cell-wall maturation and lignification during development as well as in wound-healing and cell-wall reinforcement during pathogen invasion (Cona et al. 2006). During abiotic stress responses, spermidine is secreted into the apoplast and oxidized to produce H_2O_2 (Moschou et al. 2008b). PAOs possessing back-conversion activity are sorted into peroxisomes in *Arabidopsis* as in mammals but their physiological roles remain unclear (Moschou et al. 2008b).

The physiological roles of polyamines in plants are gradually being elucidated at the molecular level. Among them, the role of extracellular polyamines as a source of H_2O_2 and that of thermospermine in stem elongation seem to be unique to higher plants. Further knowledge of polyamine catabolism, including degradation and conjugation, will shed light on the maintenance of polyamine homeostasis in plant cells. Although the export system of polyamines in plants needs to be investigated, their import mechanism will also need to be addressed, in view of the presumed occurrence in the soil of sufficient polyamines for the root to utilize. On the other hand, thermospermine synthesis is maintained through a feedback regulation of ACL5 gene expression in *Arabidopsis*, the mechanism of which remains unknown. Detailed characterization of a series of sac mutants and their responsible genes should aid in deciphering the precise mode of action of thermospermine in stem elongation. Further identification of polyamine-responsive genes might also help us to understand the role of each polyamine. The molecular mechanisms of the action of polyamines are obviously different from those of plant hormones, in which ubiquitin–proteasome pathways play a central part (Santner and Estelle, 2009), and undoubtedly provide a new understanding of plant growth regulation and stress responses.

Polyamines (PAs) in rice (*Oryza sativa* L.) plants are involved in drought resistance. Six rice cultivars differing in drought resistance were used and subjected to well watered and water-stressed treatments during their reproductive period. The activities of arginine decarboxylase, S-adenosyl-L-methionine decarboxylase, and spermidine (Spd) synthase in the leaves were significantly enhanced by water stress, in good agreement with the increase in putrescine (Put), Spd, and spermine (Spm) contents there. The increased contents of free Spd, free Spm, and insoluble-conjugated Put under water stress were significantly correlated with the yield maintenance ratio (the ratio of grain yield under water stressed conditions to grain yield under well-watered conditions) of the cultivars. Free Put at an early stage of water stress positively, whereas at a later stage negatively, correlated with the yield maintenance ratio. No significant differences were observed in soluble conjugated PAs and insoluble-conjugated Spd and Spm among the cultivars. Free PAs showed significant accumulation when leaf water potentials reached 20.51 MPa to 20.62 MPa for the drought-resistant cultivars and 20.70 MPa to 20.84 MPa for the drought susceptible ones. The results suggest that rice has a large capacity to enhance PA biosynthesis in leaves in response to water stress. The role of PAs in plant defence to water stress varies with PA forms and stress stages. In adapting to drought it would be good for rice to have the physiological traits of higher levels of free Spd/free Spm and insoluble-conjugated Put, as well as early accumulation of free PAs, under water stress.

PA and ETH reportedly share the same S-adenosylmethionine biosynthetic precursor, and increasing PA biosynthesis has a notable effect on ETH synthesis rates. Exogenous PA represses ETH synthesis in oat (*Avena sativa* L.) leaves and rice panicles. In addition, exogenous ABA increased the Put content in chickpeas (*Cicer arietinum* L.). This reduced endogenous ABA content led to a decrease in the PA levels in maize. These studies provided clear evidence that there is a close relationship between PA and hormones in the regulation of plant growth. However, little is known about the relationship between PA and hormones in the regulation of wheat grain filling.

S-adenosylmethionine decarboxylase (SAMDC) is a key enzyme involved in the biosynthesis of the polyamines, viz. spermidine and spermine. SAMDC cDNA from Tritordeum was introduced into the rice (*Oryza sativa* L.) genome using an *Agrobacterium*-mediated transformation method. Transgenic rice plants showed normal growth and development. DNA blot analysis confirmed stable integration of the transgene. ABA-induced SAMDC gene expression was observed in RI transgenic plants under NaCl stress and controls. There was a three- to four-fold increase in spermidine and spermine levels in transformed plants under the NaCl stress.

References

Alcazar, R., F. Marco, J. Cuevas, M. Patron, A. Ferrando and P. Carrasco. 2006. Involvement of polymines in plant response to abiotic stress. *Biotechnos Letters*, **28:** 1867-1876.

Alcazar, R., T. Altabella, F. Marco, C. Bortolotti, M. Reymond, C. Koncz, P. Carrasso and A.F. Tiburcio 2010. Polymines: molecules with regulatory functions in plant abiotic stress tolerance. *Planta*, **231:** 1237-1249.

Aziz, A., J. Martin-Tanguy and F. Larher. 1997. Plasticity of polyamine metabolism associated with high osmotic stress in rape leaf discs and with ethylene treatment. *Plant Growth Regul.*, **21:** 153-163.

Bagni, N. and A. Tassoni. 2001. Biosynthesis, oxidation and conjugation of aliphatic polymines in higher plants. *Amino Acids*, **20:** 301-317.

Basu, R., N. Maitra and B. Ghosh. 1988. Salinity results in polyamine accumulation in early rice (*oryzas satival*) seedlings. *Aust. J. Plant Physiol.*, **15:** 777-786.

Bey, P., C. Danzn and M. Jung, 1987. Inhibition of basic amino acid decarboxylases involved in polyamine biosynthesis, pp.1-32. *In:* Inhibition of polyamine metabolism. McCann, P.P. Pegg, A.E. and Spperdsma (Ed.), Academic press, Orlando.

Bitonti, A.J., P. Carara, P.P. McCann and P. Bey. 1987. Catalytic irreversible inhibition of bacterial and plant arginine decarboxylase activities by novel substrate and product analogues. *Biochem J.*, **242:** 69-74.

Bouchereau, A., A. Aziz, F. Larher and J. Martin-Tanguy. 1999. Polyamines and environmental challenges : recent development. *Plant Sci.*, **140:** 103-125.

Capell, T.C., Escobar, H. Liu, D. Burtin, O. Lepri and P. Christon. 1998. Over expression of the oat arginine decarboxylase cDNA intransgenic rice (*Oryza sativa*) affects normal development patterns *in vitro* and results in putrescine accumulation in transgenic plants. *Threo. Appl. Genet.*, **91:** 246-254.

Capell, T., L. Bassie and P. Christon. 2004. Modulation of the polyamine biosynthetic pathway in transgenic rice confers tolerance to drought stress. *Proc. Natl. Acad. Sci. USA.*, **101:** 9909-9914.

Chattopadhyay, M.K., S. Gupta, D.N. Sengupta and B. Ghosh. 1997. Expression of organine decarboxylase in seedlings of indica rice (*Oryza sativa* L) cultivars as affected by salinity stress. *Plant Mol. Boil*, **34:** 477-483.

Chen, T., Y. Xu, Thang, Z. Nang, J. Tana and Zhang. 2013. Polymines and ethylene interact in rice grains in response to soil drying during grain filling. *J.Exp. Bot.* **64:** 2532-2538.

Clay, N.R. and T. Nelson. 2005. *Arabidopsis* thick vein mutation affects vein thickness and organ vascularization, and resides in a provascular cell specific spermine synthase involved in vein definition and in polar auxin transport. *Plant Physiol.*, **138:** 767-777.

Cona, A., G. Rea, R. Angelini, R. Federico and P. Tauladoraki. 2006. Functions of amine oxidases in plant development and defense. *Trends in Plant Sci.*, **11:** 80-88.

Cuevas, J.C., R. Lopez-Cobollo, R. Alcazar, X. Zarza, C. Koncz, T. Altabella, J. Salinas, A.F. Tiburcio and A. Ferrando. 2008. Putrescine is involved in *Arabidopsis* freezing tolerance and cold acclimation by regulating abscisic acid levels in response to low temperature. *Plant Physiol.*, **148:** 1094-1105.

Cuevas,. J.C., R. Lepoz-Cobollo, R. Alcazar, R. Zarza, C. Koncz, T. Altabella, J. Salinas, A.F. Tiburcio and A. Ferrando. 2009. Putrescine as a signal to modulate the indispensable ABA increase under cold stress. *Plant Signal. Behav.*, **4:** 219-220.

Davies, P.J. 2004. The plant hormones: Their nature, occurence and function. pp.1-15. In: Davies, P.J. (ed) Plant hormones, biosynthesis, signal transduction, action. Academic Press, Dordrecht.

Do, P.T., J. Degenkolbe, A. Erban, A.G. Heyer, J. Kopka, K.I. Kohl, D.K. Hincha and E. Zuther. 2013. Dissecting rice polyamine metabolism under controlled long term drought stress. *Plos ONE*, **8:** 325. doi : 10.1371/journal. pone. 0060325

Do, P.T., O. Drechsel, A.G. Heyer, D.K. Hincha and E. Zuther. 2014. Changes in free polyamine levels, expression of polyamine biosynthesis genes, and performance of rice cultivars under salt stress a comparison with responses to drought. *Front. Plant Sci.*, **5:** 49-63.

Dobrovinskaya, O.R., J. Muniz and I.L. Pottosin. 1999. Inhibition of vacuolar ion channels by polyamines. *J. Memb. Biol.* **167:** 127-140.

Evans, P.T. and R.L. Malmberg. 1989. Do polyamines have roles in plant development ? *Annu. Rev. Plant Physiol.*, **40:** 235-269.

Flores, H. and A.W. Galston. 1982. Polyamines and plant stress: Activation of putrescine biosynthesis by osmotic shock. *Science*, **217:** 1257-1261.

Flores, H. and A.W. Galston. 1984. Osmotic stress-induced polyamine accumulation in cereal leaves. I. Physiological parameters of the response. *Plant Physiol.* **75:** 102-109.

Fukai, S. and M. Cooper. 1995. Development of drought-resistant cultivars using physio-morphological traits in rice. *Field Crops Res.* **40:** 67-86

Galiba, G., G. Kocsy, R. Kaur-Sawhney, J. Sutka and A.W. Galston. 1993. Crromosomal localization of osmotic and salt stress induced differential alterations in polyamine content in wheat. *Plant Sci.*, **92:** 203-211.

Galston, A.W. 1983. Green Wisdom., Penguin Group, USA.

Galston, A.W. and R.K. Sawhney. 1990. Polyamines in plant physiology. *Plant Physiol.*, **94:** 406-410.

Gemperlova, L., M. Novakova, R. Vankova, J. Eder and M. Cvikrova. 2006. Diurnal changes in polyamine content, arginine and ornithine decarbosylase, and diamine oxidase in tobacco leaves. *J. Exp. Bot.*, **57:** 1413-1421.

Grienenberger, E., S. Besseau, P. Geoffroy, D. Debayle, D. Heintz, C. Lapierre, B. Pollet, T. Heitz and M. Legrand. 2009. A BAHD acyltransferase is expressed in the tapetum of Arabidopsis anthers and is involved in the synthesis of hydroxycinnamoyl spermidines. *Plant J.* **58:** 246-259

Ha, H.C., N.S. Sirisoma, P. Kuppusamy, J.L. Zewier, P.M. Woster and R.A. Casero Jr. 1998. The natural polyamine spermine functions directly as a free radical scavenger. *Proc. Nalt. Acad. Sci.*, **95:** 11140-11445.

Hanfrey, C., K.A. Elliott, M. Franceschetti, M.J. Mayer, C. Illiugworth and A.J. Michael. 2005. A dual upstream open reading frame-based autoregulatory circuit controlling polyamine responsive translation. *J. Biol. Chem.*, **280:** 39229-39237.

Hanzawa, Y., T. Takahashi, A.J. Michael, D.Burtix, D.Long, M. Pineiro, G. Coupland and Y. Komeda. 2000. ACAULIS5 an Arabidopsis gene required for stem elongation, encodes a spermine synthase. *EMBO J.* **19:** 4248-4256.

Hao, Y.J., H. Kitashiba, C. Honda, K. Nada and T. Moriguchi. 2005a. Expression of arginine decarboxylase and ornithine decarboxylase genes in apple cells and stressed shoots. *J. Exp. Bot.*, **56:** 1105-1115.

Hao, Y.J., Z. Zhang, H. Kitashiba, C. Honda, B. Ubi, M. Kita and T. Moriguchi. 2005b. Molecular Cloning and functional characterization of two apple S-adenosyl methionine decarboxylase genes and their different expression in fruit development, cell growth and stress response. *Gene*, **350:** 41-50.

Hashimoto, T., Y. Yukimune and Y. Yamada. 1989. Putrescine and putrescine N. Methyl transferase in the biosynthesis of tropane alkaloids in cultured roots of Hyoscyamus albus: 11. Incorporation of labeled precursors. *Planta,* **178:** 131-137

Hibasami, H., M. Tanaka, J. Nagai and T. Ikeda. 1980. Dicyclohexylamine, a potent inhibitor of mommalion and yeast S-adenosylmethionine. *Biochem. Biophys. Res. Commu.,* **46:** 288-295.

Imai, A., T. Matsuyama, Y. Hanzawa, T. Akiyama, M., Tamaoki, H. Saji, Y. Shirano, T. Kato, H. Hayashi, D. Shibata, S. Tabata, Y. Komeda and T. Takahashi. 2004a. Spermidine synthase genes are essential for survival of Arabidopsis. *Plant Physiol.,* **135:** 1565-1573.

Imai, A. T. Akiyama, T. Kato, S. Sato, S. Tabata, K.T. Yamamoto and T. Takahashi. 2004b. Spermine is not essential for survival of Arabidopsis. *FEBS Letters,* **556:** 148-152.

Imai, A., Y. Hanzawa, M. Komura, K.T. Yamamoto, Y. Komeda and T. Takahash. 2006. The dwarf phenotype of the Arabidopsis ac 15 mutant is suppressed by a mutation in an upstream ORF of a bHL Hgene. *Development,* **133:** 3575-3585.

Imai, A., M. Komura, E. Kawano, Y. Kuwashiro and T. Takahashi. 2008. A semi-dominant mutation in the ribosomal protein LIOgene suppresses the dwarf phenotype of the ac 15 mutant in Arabidopsis thaliara. *Plant J.* **56:** 881-890.

Ivanov, I.P., S. Matsufuji, Y. Murakami, R.F. Gesteland and J.F. Atkins. 2000. Conservation of polymine regulation by translational frame shifting from yeast to mammals. *EMBO J.,* **19:** 1907-1917.

Kakehi, J., Y. Kuwashiro, M. Niitsu and T. Takahashi. 2008. Thermospermine is required for stem elougation in Arabidopsis thaliana. *Plant Cell Physiol.,* **49:** 1342-1349.

Kakkar, R.K. and V.K. Sawhney. 2002. Polyamine research in plants – A changing perspective. *Physiol. Plant,* **116:** 281-292.

Kasukabe, Y., L. He, K. Nada, S. Misawa, I. Ihara and S. Tachibana. 2004. Over expression of spermidine synthase tolerance to multiple environmental stresses and upregulates the expression of various stress regulated genes in transgenic Arabidopsis. *Plant Cell Physiol.,* **45:** 712-722.

Kaur-Sawhney, R., A.F. Tiburcio, T. Altbella and A.W. Galston. 2003. Polyamines in plants: An overview. *J. Cell Mol. Biol.,* **2:** 1-12.

Knott, J.M., P. Romer and M. Sumper. 2007. Putative spermine synthases from Thalassiosira pseudonana and Arabidopsis thaliana synthesize thermospermine rather than spermine. *FEBS Letters,* **581:** 3081-3086.

Kusano, T., T. Berberich, C. Tateda and Y. Takahashi. 2008. Polyamines: Essential factor for growth and survival. *Planta,* **228:** 367-381.

Kumar, A., T. Altabella, M. Taylor and A.F. Tiburcio. 1997. Recent advances in polyamine research. *Trends in Plant Sci.,* **2:** 124-130.

Kuehn, G.D. and G.C. Phillips. 2005. Role of polyamines in apoptosis and other recent advances in plant polyamines. *Crit. Rev. Plant Sci.,* **24:** 123-130.

Lefevre, I., Gratia and S. Lutts. 2001. Discrimination between the inonic and osmotic components of salt stress in relation to free polyamine level in rice (*Oryza sativa*). *Plant Sci.,* **161:** 943-952.

Legocka, J. and A. Kluk. 2005. Effect of salt and osmotic stress on changes in polyamine content and arginine decarboxylase activity in Lupinus luteus seedlings. *J. Plant Physiol.,* **162:** 662-668.

Li., Z.Y. and S.Y. Chen. 2000. Differential accumulation of the S-adenosyl methionine decarboxylase transcript in rice seedlings in response to salt and drought stresses. *Theor. Appl. Genet.*, **100:** 782-788.

Liu, H.P., B.H. Dong, Y.Y. Zhang, Z.P. Liu and Y.L. Liu. 2004. Relationship between osmotic stress and the levels of free, conjugated and bound polyamines in leaves of wheat seedlings. *Plant Sci.*, **166:** 1261-1267.

Liu, G.F., J. Yang and J. Zhu. 2006. Mapping QTL for biomass yield and its components in rice (*Oryza sativa* L). *Acta Gen. Sinica.*, **33:** 607-616.

Luo, J., C. Fuell, A. Parr, L. Hill, P. Baikey, K. Elliott, S.A. Fairhurst, C. Martin and A.J. Michael. 2009. A movel polyamine acyltransferase responsible for the accumulation of spermidine conjugates in Arabidopsis seed. *Plant cell.*, **21:** 318-333.

Ma, N., H. Tan, X. Liu, J. Xue, Y. Li and J. Gao. 2005. Transcriptional regulation of ethylene receptor and CTR genes involved in ethylene induced flower opening in cut rose (Rosa hybrid) Cv. Samantha. *J. Exp. Biol.*, **57:** 2763-2773.

Maiale, S., D.H. Sanchez, A. Guirado, A. Vidal and O.A. Ruiz. 2004. Spermine accumulation under salt stress. *J. Plant Physiol.*, **161:** 35-42.

Martin-Tanguy, J. 2001. Metabolism and function of polyamines in plants – Recent development (new approaches). *Plant Growth Regul.*, **34:** 135-148.

Minguet, E.G., F. Vera-Sirera, A. Marina, J. Carbonell and M.A. Blazquez. 2008. Evolutionary diversification in polyamine biosynthesis. *Mol. Biol. Evol.*, **25:** 2119-2128.

Mo, H. and E.C. Pua. 2002. Upregulation of arginine decarboxylase gene expression and accumulation of polyamines in mustard (Brassica juncea) in response to stress. *Physiol Plant.*, **114:** 439-449.

Moschou, P.N., K.A. Paschaldis and K.A. Roubelakis-Angelakis. 2008b. Plant-polyamine catabolism : The start of the art. *Plant Signal. Behav.*, **3:** 1061-1066.

Mulangi, G.R.V. 2011. Characterization of polyamine transporters from rice and Arabidopsis. Ph.D. Thesis. Bowling Green State University.

Muniz, L., E.G. Minguet, S.K. Singh, E. Pesquet, F. Vera-Sirera, C.L. Moreau-Courtois, J. Carbonell, M.A. Blazquez and H. Tauminen. 2008. ACAULIS5 controls Arabidopsis Xylem specification through the prevention of premature cell death. *Development*, **135:** 2573-2582.

Nishimura, K., H. Okudaira, E. Ochiai, K. Higashi, M. Kaneko, I. Ishii, T. Nishimura, N. Dohmae, K. Kashiwagi and K. Igarashi. 2009. Identifications of proteins whose synthesis is preferentially enhanced by polyamines at the level of translation in mammalian cells. *Int. J. Biochem. Cell Biol.*, **41:** 2251-2261.

Ober, F. and T.H. Hartman. 1999. Homospermidine synthase, the first pathway specific enzyme of pyrrolizidine alkaloid biosynthesis, evolved from deoxyhypusine synthase. *Prox. Natt. Acad. Sci. USA.*, **96:** 14777-14782.

Oshima, T. 2007. Unique polyamines produced by an extreme thermophile, Thermus thermophilus. *Amino Acids*, **33:** 367-372.

Pagnussat, G.C., H.J. Yu, Q.A. Ngo, S. Rajani, S. Mayalagu, C.S. Johnson, A. Capron, L.F. Xie, D. Ye and V. Sundaresan. 2005. Genetic and molecular identification of genes required for female gametophyte development and function in Arabidopsis. *Development*, **132:** 603-614

Panagiotidis, C.A., S. Artandi, K. Calame and S.J. Silverstein. 1995. Polyamines alter sequence specific DNA-protein interactions. *Nucleic Acids Res.* **23:** 1800-1809.

Perez-Amador, M.A., J. Leon, P.J. Green and J. Carbonell. 2002. Induction of arginine decarboxylase ADC2 gene provides evidence for the involvement of polyamines in the wound response in Arabidopsis. *Plant Physiol.,* **130:** 1454-1463.

Raison, J.K. and J.M. Lyons. 1971. Hibernation: Alternation of mitochondrial membranes as a requisite for metabolism at low temperature. *Proc. Natl. Acad. Sci. USA,* **68:** 2092-2094.

Reggiani, R., A. Hochkoeppler and A. Bertani. 1989. Polyamines in rice seedlings under oxygen-deficit stress. *Plant Physiol.,* **91:** 1197-1201.

Rodriguez-Kessler, M., A. Alpuche-Solis, O. Ruiz and J. Jimenez-Bremont. 2006. Effect of salt stress on the regulation of maize (Zea mays L.) genes involved in polyamine biosynthesis. *Plant Growth Regul.,* **48:** 175-185.

Roy, M. and B. Ghosh. 1996. Polyamines, both common and uncommon under heat stress in rice (*Oryza sativa*) Callus. *Physiol. Plant.,* **98:** 196-200.

Roy, M. and R. Wu. 2001. Arginine decarboxylase transgene expression and analysis of environmental stress tolerance in transgenic rice. *Plant Sci.,* **160:** 869-875.

Roy, M. and R. Wu. 2002. Over expression of S-adenosylmethionine decarboxylase gene in rice increases polyamine level and enhanced sodumum chloride stress tolerance. *Plant Sci.,* **163:** 987-992.

Ruan, H., L.M. Shantz, A.E. Pegg and D.R. Morris. 1996. The upstream open reading frame of the mRNA encoding S-adenosyl methionine decarboxylase is a polyamine-responsive translational control element. *J. Biol. Chem.,* **271:** 29576-29582.

Santner, A. and M. Estelle. 2009. Recent advances and emerging trends in plant hormone signaling. *Nature,* **459:** 1071-1078.

Serafini-Fracassini, D. and S. Del Duca. 2008. Transglutaminases: widespread cross-linking enzymes in plants. *Ann. Bot.,* **102:** 145-152.

Shabala, S., V. Demidchik, L. Shabala, T.A. Cuin, S. J. Smith, A.J. Miller, J.M. Davies and I.A. Newman. 2006. Extracelluian Ca^{2+} ameliorates NaCl induced K^{+} loss from Arabidopsis root and leaf cells by controlling plasma membrane channels. *Plant Physiol.* **141:** 1653-1665.

Slocum, R.D. 1991b. Polyamine biosynthesis in plants. In: Biochemistry and physiology of polyamines in plants. Slocum R.D. and Flores, H.E. (Ed). CRC Press, FL, USA. pp. 22-40.

Smith, T.A. and J.H.A. Marshall. 1988. The di and polyamine oxidases of plant. pp.573-587. In: Progress in polyamine Research (Advances in Experimental Biology and Medicine 250). Plenum Press, New York.

Soyka, S. and A.G. Heyer. 1999. *Arabidopsis* Knockout mutation of ADC2 gene reveals inducibility by osmotic stress. *FEBS Letters.,* **458:** 219-223.

Tabor, C.W. and H. Tabor. 1999. It all started on a streetcar in Boston. *Annu. Rev. Biochem.,* **68:** 1-32.

Takahashi, Y., T. Berberich, A. Miyazaki, S. Seo, Y. Ohashi and T. Kusano. 2003. Spermine signalling in tobacco: Activation of mitogen-activated protein kinases by spermine is mediated through mitochondrial dysfunction. *Plant J.,* **36:** 820-829.

Tian. A.G., J.Y. Zhao, J.S. Zhang, J.Y. Gai and S.Y. Chen. 2004. Genomic characterization of the S-adenosylmethionine decarboxylase genes from soybean. *Theor. Appl. Genet.,* **108:** 842-850.

Tiburcio, A.F., R. Kaur-Sawhney and A.W. Galston. 1986. Polyamine metabolism and osmotic stress. II. Improvement of Oat protoplasts by an inhibitor of putrescine biosynthesis. *Plant Physiol.,* **82:** 375-378.

Turner, L.B. and G.R. Stewart. 1986. The effect of water stress upon polyamine levels in barley (Hordeum vulgave L.) leaves. *J. Exp. Bot.,* **37:** 170-177.

Urano, K., Y. Yoshida, T. Nanjo, Y. Igarashi, M. Seki, F. Sekiguchi, K. Yamaguchi-Shinozaki and K. Shinozaki. 2003. Characterization of Arabidopsis genes involved in biosynthesis of polyamines abiotic stress responses and developmental stages. *Plant Cell Environ.,* **26:** 1917-1926.

Urano, K., Y. Yoshida, T. Nanjo, T. Ito, K. Yamaguchi-Shinozaki and K. Shinozaki. 2004. Arabidopsis stress-inductible gene for arginine decarboxylase AtADC2 is required for accumulation of putrescine in salt tolerance. *Biochem. Biophys. Res. Commu.,* **313:** 369-375.

Urano, K., T. Hobo and K. Shinozaki, 2005. Arabidopsis ADC genes involved in polyamine biosynthesis are essential for seed development. *FEBS Letters,* **579:** 1557-1564.

Williams–Ashman, H.G. and A. Schenone. 1992. A methyl-glyoxyl-bis (guanylhydrazone) as a potent inhibitor of mammalian and yeast S-adexosylmethionine. *Biochem. Biophys. Res. Commu.,* **46:** 288-295.

Yamakawa, H., H. Kamada, M. Satoh and Y. Ohashi. 1998. Spermine is a salicylate-independent endogenous inducer for both tobacco acidic pathogenesis related proteins and resistance against tobacco mosaic virus infection. *Plant Physiol.,* **118:** 1213-1222.

Yamaguchi, K., Y. Takahashi, T. Berberich, A. Imai, A., Miyazaki, T. Takahashi, A. Michael and T. Kusano. 2007. A Protective role for the polyamine spermine against drought stress in Arabidopsis Biochem. *Biophys. Res. Commun.* **352:** 486-490

Yang, J., J. Zhang, K. Liu, Z. Wang and L. Liu. 2007. Involvement of polyamines in the drought resistance of rice. *J. Exp. Bot.* **58:** 1545-1555

chapter seven

Amino Acids

Amino acids are biologically important organic compounds composed of amine (-NH_2) and carboxylic acid (-COOH) functional groups, along with a side-chain specific to each amino acid. The key elements of an amino acid are carbon, hydrogen, oxygen and nitrogen, though other elements are found in the side-chains of certain amino acids.

Amino acids are the structural units (monomers) that make up proteins. They join together to form short polymer chains called peptides or longer chains called either polypeptides or proteins. These polymers are linear and unbranched, with each amino acid within the chain attached to two neighboring amino acids. The process of making proteins is called translation and involves the step-by-step addition of amino acids to a growing protein chain by a ribozyme that is called a ribosome(Rodnina et al. 2007). The order in which the amino acids are added is read through the genetic code from an mRNA template, which is a RNA copy of one of the organism's genes.

Twenty-two amino acids are naturally incorporated into polypeptides and are called proteinogenic or natural amino acids. Of these, 20 are encoded by the universal genetic code.

The 20 amino acids encoded directly by the genetic code can be divided into several groups based on their properties. Important factors are charge, hydrophilicity or hydrophobicity, size and functional groups (Creighton, 1993). These properties are important for protein structure and protein–protein interactions. The water-soluble proteins tend to have their hydrophobic residues (Leu, Ile, Val, Phe and Trp) buried in the middle of the protein, whereas hydrophilic side-chains are exposed to the aqueous solvent. The integral membrane proteins tend to have outer rings of exposed hydrophobic amino acids that anchor them into the lipid bilayer. In the case part-way between these two extremes, some peripheral membrane proteins have a patch of hydrophobic amino acids on their surface that locks onto the membrane. In similar fashion, proteins that have to bind to positively charged molecules have surfaces rich with negatively charged amino acids like glutamate and aspartate, while proteins binding to negatively charged molecules have surfaces rich with positively charged chains like lysine and arginine. There are different hydrophobicity scales of amino acid residues (Urry, 2004).

Some amino acids have special properties such as cysteine, that can form covalent disulfide bonds to other cysteine residues, proline that forms a cycle to the polypeptide backbone, and glycine that is more flexible than other amino acids.

Many proteins undergo a range of post translational modifications, when additional chemical groups are attached to the amino acids in proteins. Some modifications can produce hydrophobic lipoproteins (Magee and Seabra, 2005), or hydrophilic glycoproteins (Pilobello and Mahal, 2007). These types of modification allow the reversible targeting of a protein to a membrane. For example, the addition and removal of the fatty acid palmitic acid to cysteine residues in some signaling proteins causes the proteins to attach and then detach from cell membranes (Smotrys and Linder, 2004).

Glutamate occupies a central position in amino acid metabolism in plants. The acidic amino acid is formed by the action of glutamate synthase, utilizing glutamine and 2-oxoglutarate. However, glutamate is also the substrate for the synthesis of glutamine from ammonia, catalysed by glutamine synthetase. The α-amino group of glutamate may be transferred to other amino acids by the action of a wide range of multispecific aminotransferases. In addition, both the carbon skeleton and a-amino group of glutamate form the basis for the synthesis of 4-aminobutyric acid, arginine and proline. Finally, glutamate may be deaminated by glutamate dehydrogenase to form ammonia and 2-oxoglutarate. The possibility that the cellular concentrations of glutamate within the plant are homeostatically regulated by the combined action of these pathways. Evidence that the well known signalling properties of glutamate in animals may also extend to the plant kingdom. The existence in plants of glutamate-activated ion channels and their possible relationship to the GLR gene family that is homologous to ionotropic glutamate receptors (iGluRs) in animals are known. Glutamate signaling might have an evolutionary

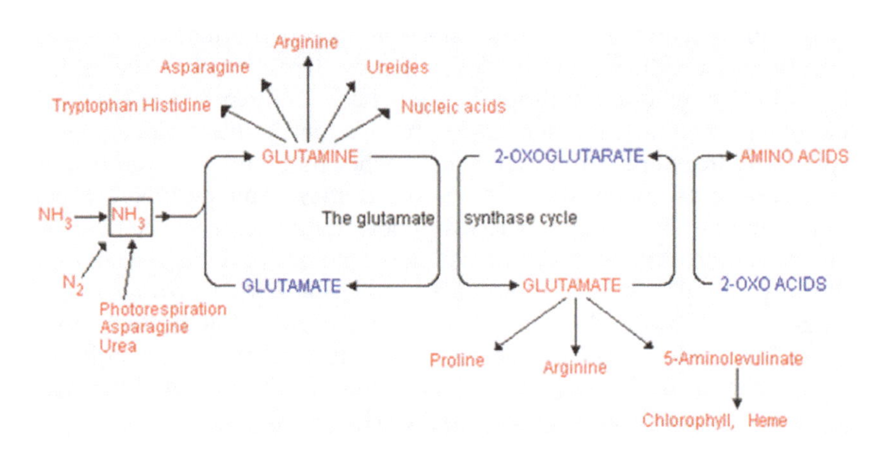

FIGURE 7.1 Formation of amino acids.

perspective, and the roles it might play in plants, both in endogenous signaling pathways.

7.1. Amino Acid Families

The molecules we got to know when discussing glycolysis and the citric acid cycle were carbon-containing but nitrogen-free. Some of these intermediates are starting points of amino acid biosynthetic pathways. Based on chemical similarities and only few starting compounds, all amino acids can be regarded as members of five families:

1. The glutamate family starting with *alpha*-ketoglutarate
2. The aspartate family with the starting compound oxaloacetate
3. The alanine-valine-leucine group (pyruvate)
4. The serine-glycine group (3-phosphoglycerate)
5. The family of aromatic amino acids (phosphoenolpyruvate and erythrose-4-phosphate, an intermediate in the pentose phosphate pathway).

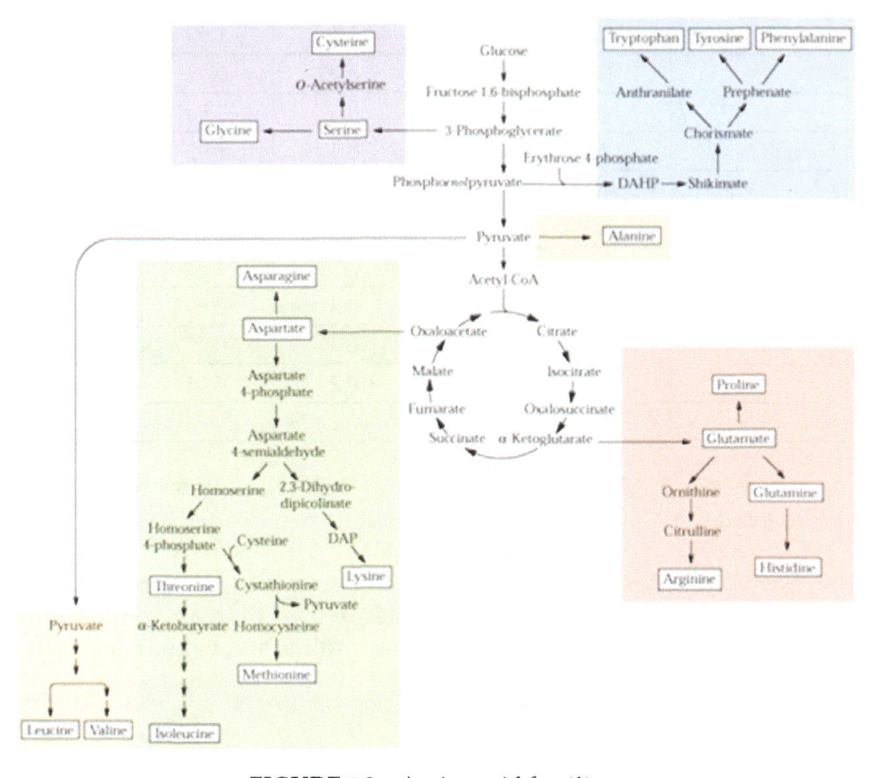

FIGURE 7.2 Amino acid families.

7.2. Amino Acids in Rice

A study was conducted with four genotypes under ammonium and nitrate nutrition with or without phosphorus and free amino acids were analyzed in xylem and phloem sap, leaf and stem. Results indicate that free amino acid content was maximum in phloem sap. It had also been revealed that nitrate nutrition accumulated less amino acid than ammonium. In general phosphorus nutrition enhanced free amino acid appreciably (Table 7.1).

TABLE 7.1 Variation of free amino acids in different parts of rice genotypes.

Genotype	Treatment	Leaf (mg g^{-1})	Stem (mg g^{-1})	Xylemsap (mg ml^{-1})	Phoemsap (mg ml^{-1})
Shanyou63	A	0.97	0.48	1.09	1.49
	AP	1.08	0.62	1.96	2.27
	N	0.70	0.36	1.16	1.27
	NP	0.72	0.39	1.84	1.28
Yangdao6	A	0.79	0.41	1.54	1.27
	AP	1.15	0.46	1.99	1.76
	N	0.53	0.23	0.93	1.18
	NP	0.67	0.37	1.80	1.34
86you8	A	0.56	0.30	1.92	1.16
	AP	0.68	0.42	1.57	1.20
	N	0.85	0.28	0.75	1.22
	NP	o.76	0.28	1.55	1.21
Wuyunjing7	A	0.75	0.31	0.85	1.14
	AP	0.87	0.34	0.70	1.15
	N	0.44	0.23	0.64	1.13
	NP	0.61	0.22	0.63	1.12

A = Ammonium, AP = Ammonium + Phosphate, N = Nitrate, NP = Nitrate + Phosphate

Bacilio-Jimenez et al. (2003) studied the amino acid composition of root exudates from rice at different stages of growth in an axenic hydroponic system and observed that histidine, proline and valine come out in significantly high amount (Table 7.2).

TABLE 7.2 Changes of amino acids in rice root exudates

Amino acid	Culture (days)			
	7	14	21	28
Asx	0.12	0.03	0.01	0.04
Glx	0.22	0.09	0.03	0.07
Ser	0.27	0.11	0.03	0.09
Gly	0.74	0.30	0.08	0.25
His	1.58	1.38	1.21	ND
Arg	0.11	0.08	ND	ND
Thr	0.24	0.10	0.03	0.09
Ala	0.48	0.17	0.04	0.16
Pro	1.33	0.20	0.06	0.24
Tyr	0.04	0.01	0.001	ND
Val	1.73	0.57	0.26	0.83
Met	0.02	0.02	0.01	ND
Cys	ND	ND	ND	ND
ILe	0.21	0.08	0.02	0.06
Leu	0.28	0.11	0.03	0.09
Phe	0.18	0.06	0.02	0.04
Lys	0.29	0.07	0.01	0.05
Total	7.84	3.38	1.84	2.01

Values denoted by different letters differ significantly $P < 0.05$ using one-way ANOVA. Values are the mean of six replicates with 12 seedlings in each culture system. Standard error of the mean was lower than 10%. ND – not detected.

Das Gupta and Basuchaudhuri (1974) observed a significant increase in protein content of grains associated with a corresponding increase in most of the protein-bound amino acids, viz. leucine, phenylalanine, methionine and valine, alanine, threonine, glutamic acid, serine and glycine, aspartic acid, lysine, arginine and histidine, asparazine and proline when nitrogen fertilization was applied.

The improvement of grain quality, such as protein content (PC) and amino acid composition, has been a major concern of rice breeders. It was constructed a population of 190 recombinant inbred lines (RILs) from a cross between Zhenshan 97 and Nanyangzhan to map the quantitative trait locus or loci (QTL) for amino acid content (AAC) as characterized by each of the AACs, total essential AAC, and all AAC. Using the data collected from milled rice in 2002 and 2004, it was identified 18 chromosomal regions for 19 components of AAC. For 13 of all the loci, the Zhenshan 97 allele increased the trait values. Most QTL were co-localized, forming ten QTL clusters in 2002 and six in 2004. The QTL clusters varied in both effects and locations, and the mean values of variation explained by individual QTL in the clusters ranged from 4.3% to 28.82%. A relatively strong QTL cluster, consisting of up to 19 individual QTL, was found at the bottom of chromosome.

The major QTL clusters identified for two different years were coincident. A wide coincidence was found between the QTL detected and the loci involved in amino acid metabolism pathways, including N assimilation and transfer, and amino acid or protein biosynthesis. The results will be useful for candidate gene identification and marker-assisted favorable allele transfer in rice breeding programs (Table 7.3).

TABLE 7.3 Descriptive statistics of the 19 components of AAC in the parents and RIL population observed in rice grain

Amino acid	Zhenshan 97	Nanyangzhan	Population range 2002	Zhenshan 97	Nanyangzhan	Population range2004
Asp	12.95	9.44	8.08-15.08	12.81	7.93	7.52-15.05
Thr	4.86	3.70	3.13-5.38	4.77	3.20	3.01-5.73
Ser	6.54	5.14	4.28-7.80	6.48	4.41	4.06-8.48
Glu	26.90	20.25	16.55-30.93	27.21	16.91	15.84-34.05
Gly	6.01	4.50	3.88-6.63	6.12	3.85	3.63-7.16
Ala	7.71	5.64	4.83-9.23	7.61	4.77	4.56-8.85
Cys	3.03	2.60	2.40-3.93	2.87	2.39	1.83-3.61
Val	7.59	5.74	4.95-8.53	8.19	5.04	4.86-9.66
Met	2.81	2.70	2.03-3.55	2.26	2.06	1.42-3.49
Ile	5.37	4.05	3.40-6.15	6.17	3.52	3.42-7.28
Leu	11.39	8.51	7.20-12.53	11.76	7.43	6.82-13.94
Tyr	5.35	3.91	3.55-6.78	5.15	3.95	2.48-6.90
Phe	7.77	5.76	4.85-8.70	8.11	4.81	4.60-9.04
Lys	4.73	3.42	3.10-5.60	5.09	3.16	3.08-5.71
His	3.20	2.42	2.05-3.78	3.86	2.16	2.10-4.33
Arg	11.52	8.33	7.03-12.68	11.00	7.13	5.95-13.66
Pro	5.87	4.38	2.93-6.38	5.73	3.50	3.36-6.99
Eaa	59.23	44.62	37.93-65.03	61.21	38.43	35.26-71.50
Total	133.59	100.48	85.80-154.13	135.20	86.14	78.85-160.20

Salinity is one of the major problems to increasing rice production in paddy field. The role of free amino acids as compatible solutes is controversial and the different salt compositions effect on rice response to salinity stress is not completely clear. Therefore, a glasshouse experiment was carried out to determine free amino acid/acids which are involved in defense mechanism under salt stress conditions and different salt compositions effects on free amino acids status at seedling stage. Two rice genotypes differing in salinity resistance were grown hydroponically. The rice seedlings were exposed to salinity stress at EC 7 $dS.m^{-1}$ by NaCl and mixture of $NaCl:Na_2SO_4$ at 1:2 and 2:1 molar ratios. Free amino acids in shoot and root tissue of rice seedlings were measured using high performance liquid chromatography (HPLC). The rice seedlings showed significantly amino acids accumulation in their shoots greater than in their roots. Khazar, a salt sensitive genotype, demonstrated higher total free amino acids and aspargine than Fajr, a salt tolerant genotype. A positive correlation between aspargine accumulation and water content percentage in shoot tissue was recorded. Likewise the results revealed that there was a significant difference across salt compositions. It was suggested that SO_4 decreased Cl toxicity effects on rice seedlings growth (Table 7.4).

TABLE 7.4 Effects of salt stress on the content of proline and total amino acids in the 2nd leaf blade of NERICA rice seedlings

Variety	Proline (µmol/gDW) Control	Proline (µmol/gDW) Salt stressed	TAA (µmol/gDW) Control	TAA (µmol/gDW) Salt stressed
NERICA 1	0.42	3.29	52.67	80.94
NERICA 2	0.51	11.47	58.72	163.20
NERICA 3	0.43	5.73	43.49	110.46
NERICA 4	0.43	5.68	55.01	97.14
NERICA 5	0.45	3.75	58.40	121.73
NERICA 6	0.46	6.30	52.38	109.81
NERICA 7	0.44	2.45	49.09	97.64

Sekhar and Reddy (1982) studied twelve scented (basmati) and one non-scented variety and analyzed for their amino acid composition. The essential amino acid profiles of scented varieties when compared with non-scented, revealed that these varieties exhibited higher values, which ranged from 2.82 to 4.86 gm/100 gm protein for lysine, 1.92 to 3.13 for methionine, 1.67 to 4.23 for tyrosine, 3.65 to 4.91 for phenylalanine, 5.50 to 8.95 for leucine, 2.25 to 3.40 for isoleucine, 2.84 to 3.46 for threonine, and 3.36 to 5.33 for valine. When these values were compared to FAO recommended standards, it was observed that most of the scented varieties had comparable or superior values, while varieties such as, 'Type 3', 'Basmati sufaid 100', 'Likitimachi', 'Randhunipagalu' and 'Basmati 370' showed superior lysine, phenylalanine, leucine and methionine content. These observations suggest that the scented varieties possess better amino acid profiles and exhibit superior nutritional qualities, which could be utilised in breeding varieties with improved amino acid composition.

The free amino acid (FAA) concentration of rice is becoming an increasingly important grain quality factor because of its apparent influence on the organoleptic acceptability of cooked rice. To determine the variability of this character among rice cultivars the FAA profiles of 49 cultivars were determined using Pico-Tag method. Among these 13 cultivars were selected to determine variation in FAA accumulation pattern after 24-hour germination treatment. The results show significant variation in the concentrations of total as well as individual FAAs among cultivars. There were also significant differences between *indica*- and *japonica*-type cultivars in the concentrations of some FAAs. The ratio of the total concentration of aspartate-derived to glutamate-derived FAAs (A/G ratio) evaluated for the *japonica* group (0.68) was significantly lower than that for the *indica* group (1.07). This suggests that typically, *japonica*-type rice grains tend to accumulate more Glu-derived than Asp-derived FAAs. Other results show a decline in the A/G ratios of both groups in response to germination treatment, with the *indica* group exhibiting a more rapid response. These results appear to suggest key differences in the FAA accumulation patterns of *japonica*- and *indica*-type rice grains especially with respect to the contents of aspartate-derived and glutamate-derived amino acids.

Five Malaysian rice (*Oryza sativa* L.) varieties, MR33, MR52, MR211, MR219 and MR232, were tested in pot culture under different salinity regimes for biochemical response, physiological activity, and grain yield. Three different levels of salt stresses, namely, 4, 8 and 12 dS m^{-1}, were used in a randomized complete block design with four replications under glass house conditions. The results revealed that the chlorophyll content, proline, sugar content, soluble protein, free amino acid, and yield per plant of all the genotypes were influenced by different salinity levels. The chlorophyll content was observed to decrease with salinity level but the proline increased with salinity levels in all varieties. Reducing sugar and total sugar increased up to 8 dS m^{-1} and decreased up to 12 dS m^{-1}. Nonreducing sugar decreased with increasing the salinity levels in all varieties. Soluble protein and free amino acid also decreased with increasing salinity levels. Cortical cells of MR211 and MR232 did not show cell collapse up to 8 dS m^{-1} salinity levels compared to susceptible checks (IR20 and BRRI dhan29). Therefore, considering all parameters, MR211 and MR232 showed better salinity tolerance among the tested varieties. Both cluster and principal component analyses depict the similar results (Hakim et al. 2014).

The content of free amino acid in rice leaves of eight rice varieties significantly decreased with increasing of salinity levels. The application of different levels of salinity decreased the accumulation of free amino acid in leaves of all rice varieties and the reduction of free amino acid was prominent in salt-sensitive varieties (IR20 and BRRI dhan 29).

The results showed that the content of free amino acid in the eight rice varieties varied significantly due to the mean effect of salinity levels. The highest amount of free amino acid (16.31 mg g^{-1} fw) was obtained in the leaves of MR33, which was statistically identical with MR211 while the lowest was (7.87 mg g^{-1} fw) in IR20.

The different salinity levels had significant effect on free amino acid content in leaves of eight rice varieties. At 4 dS m^{-1}, the highest amount of free amino acids (19.20 mg g^{-1} fw) was observed in MR33 followed by MR211 (19.09 mg g^{-1} fw). The lowest amount (9.81 mg g^{-1} fw) was found in the IR20. At 8 dS m^{-1} level of salinity, the highest amount of free amino acids was produced in MR232 (15.98 mg g^{-1} fw) and the lowest was recorded in BRRI dhan29 (5.90 mg g^{-1} fw). It was observed that the highest value in Pokkali followed by MR211 and MR232 at higher salinity level. Though the leaves of rice variety BRRI dhan29 contained 2nd highest amount of free amino acids at control treatment but it reduced dramatically with increasing the salinity level .

A study was done to identify simple sequence repeat (SSR) markers associated with the amino acid content of rice (*Oryza sativa* L.). SSR markers were selected by prescreening for the relationship to amino acid content. Eighty-four rice landrace accessions from Korea were evaluated for 16 kinds of amino acids in brown rice and genotyped with 25 SSR markers. Analysis of population structure revealed four subgroups in the population. Linkage disequilibrium (LD) patterns and distributions are of fundamental importance for genome-wide mapping associations. The mean r^2 value for all intrachromosomal loci pairs was 0.033. LD between linked markers decreased with distance. Marker-trait associations were investigated using the unified mixed-model approach, considering both population structure (Q) and kinship (K). A total of 42 marker-trait associations with amino acids ($P < 0.05$) were identified using 15 different SSR markers covering three chromosomes and explaining more than 40% of the total variation. These results suggest that association analysis in rice is a viable alternative to quantitative trait loci mapping and should help rice breeders develop strategies for improving rice quality (Zhao et al. 2008).

Rice phloem sap was obtained through severed stylets of brown planthoppers and its chemical composition was determined. Sucrose, the only sugar detected, was present at 17–25% (w/v) in the sap. Amino acids were present at a total of 3–8% (w/v) and were found to be mostly in a free form. The amount of bound amino acids was estimated to be small, if any were present. Among the free amino acids, asparagine (17–33% on a molar basis), glutamate (6–14%), serine (10–13%), glutamine (7–15%), threonine (5–6%) and valine (6–7%) were dominant, while cystine and methionine (0–0.2%) were present in minor amounts and γ-aminobutyric acid was not detected. The sap had a slightly alkaline pH (ca. 8.0). The inorganic constituents detected by electron probe x-ray microanalyzer were Na, S, P and K, with the K content being the highest. The osmotic pressure was estimated to be 13–15 atm. The amino acid composition of the plant parts was determined and the differences in the case of phloem loading among amino acids were compared.

D-Alanylglycine was detected in the leaf blades of axenic rice seedlings grown under light/dark regime and its content appeared to increase with age. D-Alanylglycine was not detected in rice seedlings grown in the dark, but it was formed with light irradiation of the tissues.

In a study, a total of 85 *AAT* genes were identified in rice genome and were classified into eleven distinct subfamilies based upon their sequence composition and phylogenetic relationship. A large number of *OsAAT* genes were expanded via gene duplication, 23 and 24 *OsAAT* genes were tandemly and segmentally duplicated, respectively. Comprehensive analyses were performed to investigate the expression profiles of *OsAAT* genes in various stages of vegetative and reproductive development by using data from EST, Microarrays, MPSS and Real-time PCR. Many *OsAAT* genes exhibited abundant and tissue-specific expression patterns. Moreover, 21 *OsAAT* genes were found to be differentially expressed under the treatments of abiotic stresses. Comparative analysis indicates that 26 *AAT* genes with close evolutionary relationships between rice and *Arabidopsis* exhibited similar expression patterns (Zhao et al. 2012).

7.3. 4-Aminobutyrate

The accumulation of the amino acid 4-aminobutyrate (gamma-aminobutyrate) (GABA) is markedly stimulated under anaerobic conditions (Streeter and Thompson, 1972; Stewart and Larher, 1980; Aurisano et al. 1995; Ratcliffe, 1995).

In part this may be a consequence of cytoplasmic acidification, stimulating the activity of glutamate decarboxylase, which has an acid pH optimum (~5.8) (Patterson and Graham, 1987).

Glutamate decarboxylase [EC 4.1.1.15]

$$\text{Glutamate} + H^+ \rightarrow \text{GABA} + CO_2$$

Because this reaction is proton consuming, it could represent an adaptive response contributing to regulation of cytoplasmic pH (Patterson and Graham, 1987; Crawford et al. 1994; Ratcliffe, 1995).

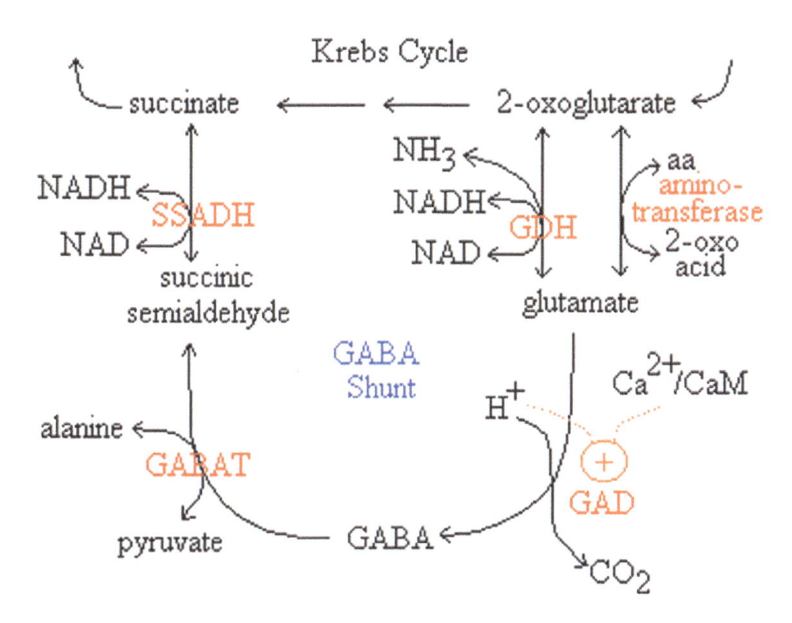

FIGURE 7.3 4-aminobutyric acid synthesis.

Crawford et al. (1994) have shown that weak acids causing cytoplasmic acidification also induce GABA accumulation, and they conclude that this response is consistent with a role for GABA synthesis in active pH regulation.

GABA can be transaminated with pyruvate to yield alanine (or with 2-oxoglutarate to yield glutamate), generating succinic semialdehyde, which can then be metabolized to succinate via the action of succinate semialdehyde dehydrogenase (Patterson and Graham, 1987; Shelp et al. 1995).

The latter enzymes are mitochondrial (Hearl and Churchich, 1984; Breitkreuz and Shelp, 1995), and have alkaline pH optima (Patterson and Graham, 1987). GABA accumulation promoted by cytosolic acidification may result in part from inhibition of GABA-transaminases.

The conversion of glutamate to succinate via the action of glutamate decarboxylase, GABA transaminase [EC 2.6.1.19], and succinic semialdehyde dehydrogenase [EC 1.2.1.16 or 1.2.1.24] is known as the "GABA shunt" (Patterson and Graham, 1987; Breitkreuz and Shelp, 1995; Shelp et al. 1995) affording an alternative pathway for glutamate entry into the TCA (Krebs) cycle.

Although the glutamate decarboxylase pathway of GABA synthesis is thought to be the predominant pathway of GABA synthesis in plants, it should be noted that GABA can also be derived from putrescine (via gamma-aminobutyraldehyde) through the reactions catalyzed by diamine oxidase [EC 1.4.3.6] and gamma-aminobutyraldehyde dehydrogenase [EC 1.2.1.19] (Flores et al. 1989) (see Polyamines). The latter enzyme may be the same as betaine aldehyde dehydrogenase (BADH) [EC 1.2.1.8] involved in glycinebetaine synthesis (Trossat et al. 1997).

GABA accumulation is induced in response to a sudden decrease in temperature (Wallace et al. 1984; Patterson and Graham, 1987), in response to heat shock (Mayer et al. 1990), mechanical manipulation (Wallace et al. 1984), and water stress (Rhodes et al. 1986). Heat shock is known to induce rapid, transient changes in cytoplasmic calcium (Gong et al. 1998).

Rapid GABA accumulation in response to wounding may play a role in plant defense against insects (Ramputh and Brown, 1996).

Glutamate decarboxylase is a cytosol localized enzyme (Breitkreuz and Shelp, 1995) and has recently been shown to be a calmodulin-binding protein that is Ca^{2+}/calmodulin activated (Ling et al. 1994; Baum et al. 1996; Arazi et al. 1995; Snedden et al. 1995). Crawford et al. (1994) noted that reduced cytosolic pH values increase Ca^{2+} levels, and rapid and transient increases in Ca^{2+} levels are known to occur in response to mechanical stress and cold stress; conditions which elicit rapid GABA accumulation (Wallace et al. 1984). Mitochondria contribute to the anoxic Ca^{2+} signal in maize suspension-cultured cells (Subbaiah et al. 1998).

The pretreatment of rice roots for 1 h in aerobic conditions with the Ca^{2+}-channel blockers ruthenium red (RR) and verapamil and the calmodulin (CaM) antagonists *N*-(6-aminohexyl)-5-chloro-1-naphtylenesulfonamide (W-7) and trifluoperazine, induced during 3 h of anoxia: (i) inhibition of amino acid and γ-aminobutyric acid (Gaba) accumulation; (ii) a decline in the protein content; and (iii) a release of amino acids and K^+ into the growth media. The calcium ionophore A23187 reversed these effects in RR-treated roots. Moreover, the aerobic pretreatment of rice roots with A23187 alone or $CaCl_2$ increased the accumulation of amino acids and Gaba. These data indicate that the Ca^{2+}/CaM complex is involved in the transduction of an anaerobic signal by inhibiting proteolysis and solute release, and activating the Ca^{2+}/CaM-dependent glutamate decarboxylase (Aurisano et al. 1995).

The Ca^{2+}/calmodulin activation of glutamate decarboxylase provides a link between intermediary amino acid metabolism and perturbations of cytosolic Ca^{2+} (Ling et al. 1994; Baum and Fridmann., 1996; Arazi et al. 1995; Snedden et al. 1995) known to regulate a host of other metabolic activities (Allan and Trewavas, 1987; Bush, 1995; Sanders et al. 1999). It has recently been shown that GABA stimulates ethylene production in sunflower, apparently by causing increases in ACC synthase [EC 4.4.1.14], mRNA accumulation, ACC levels, ACC oxidase mRNA levels and ACC oxidase activity, suggesting that GABA may play a role in signaling (Kathiresan et al. 1997).

7.4. Glutamate

The only other enzyme of glutamate metabolism known to be stimulated by Ca^{2+} in plants is glutamate dehydrogenase (GDH), a mitochondrial enzyme (Turano et al. 1997).

Glutamate dehydrogenase [EC 1.4.1.2]

$$NH_3 + \text{2-oxoglutarate} + NADH \leftrightarrow \text{Glutamate} + NAD^+$$

In both maize and *Arabidopsis*, GDH is a hexameric enzyme whose subunits are encoded by two separate genes, *gdh1* and *gdh2* (Pryor, 1990; Magalhaes et al. 1990; Turano et al. 1997). In wildtype plants this results in seven isoenzymes whose relative abundance is largely determined by the relative abundance of mRNA transcript abundance of the two genes (Turano et al. 1997). In *gdh1* null mutants of maize a single GDH isozyme is detected, corresponding to the hexamer of the *gdh2* gene product (Magalhaes et al. 1990).

A calcium binding domain has been identified in the B-subunit of GDH encoded by *GDH2*, but not in the A-subunit of GDH encoded by *GDH1* in *Arabidopsis*, suggesting that the different isoforms of GDH composed of different combinations of subunits, may be differentially regulated by Ca^{2+} (Turano et al. 1997).

The Ca^{2+} stimulation of the aminating activity of the B-subunit of GDH encoded by *GDH2* could possibly serve to provide the glutamate substrate required for GABA synthesis and accumulation in response to environmental stress-induced perturbations of intracellular Ca^{2+}.

7.5. Alanine

The magnitude of the difference between alanine levels in the resistant and susceptible varieties, and the fact that several other of the amino acids in these varieties vary in concentration, suggest that it would be worthwhile to pursue quantitative determination of amino acids in other varieties of plants resistant and susceptible to this plant pathogen. Since nitrogen fertilization apparently influenced the alanine level and since nitrogen utilization was reported to influence severity of smut, the usual relationship of these factors to disease development must be studied further and under more controlled environmental conditions.

A cDNA clone encoding alanine aminotransferase (AlaAT) has isolated from randomly sequenced clones derived from a cDNA library of maturing rice seeds by comparison to previously identified genes. The deduced amino acid sequence was 88% and 91% homologous to those of the enzymes from barley and broomcorn millet (*Panicum miliaceum*), respectively. Using this cDNA as a probe, it was isolated and sequenced the corresponding genomic clone. Comparison of the sequences of the cDNA and the genomic gene revealed that the coding region of the gene was interrupted by 14 introns 66 to 1547 bp long. Northern and western blotting analyses showed that the gene was expressed at high levels in developing seeds.

When the 5'-flanking region between −930 and +85 from the site of initiation of transcription was fused to a reporter gene for β-glucuronidase (GUS) and then introduced into the rice genome, histochemical staining revealed strong GUS activity in the inner endosperm tissue of developing seeds and weak activity in root tips. Similar tissue-specific expression was also detected by in situ hybridization. These results suggest that AlaAT is involved in nitrogen metabolism during the maturation of rice seed.

Crop plants require nitrogen for key macromolecules, such as DNA, proteins and metabolites, yet they are generally inefficient at acquiring nitrogen from the soil. Crop producers compensate for this low nitrogen utilization efficiency by applying nitrogen fertilizers. However, much of this nitrogen is unavailable to the plants as a result of microbial uptake and environmental loss of nitrogen, causing air, water and soil pollution. It was noted that engineered rice over-expressing alanine aminotransferase (AlaAT) under the control of a tissue-specific promoter that showed a strong nitrogen use efficiency phenotype. In a study, it was examined the transcriptome response in roots and shoots to the over-expression of AlaAT to provide insights into the nitrogen-use-efficient phenotype of these plants. Transgenic and control rice plants were grown hydroponically and the root and shoot gene expression profiles were analysed using Affymetrix Rice GeneChip microarrays. Transcriptome analysis revealed that there was little impact on the transgenic transcriptome compared with controls, with 0.11% and 0.07% differentially regulated genes in roots and shoots, respectively. The most up-regulated transcripts, a glycine-rich cell wall (GRP) gene and a gene encoding a hypothetical protein (Os8823), were expressed in roots. Another transgenic root-specific up-regulated gene was leucine rich repeat (LRR). Genes induced in the transgenic shoots included GRP, LRR, acireductone dioxygenase (OsARD), SNF2 ATP-translocase and a putative leucine zipper transcription factor (Beatty et al. 2009). This study provides a genome-wide view of the response to AlaAT over-expression, and elucidates some of the genes that may play a role in the nitrogen-use-efficient phenotype.

Bacilio-Jimenez et al. (2003) studied the amino acid composition of root exudates from rice at different stages of growth in an axenic hydroponic system and observed that histidine, proline and valine come out in significantly high amount.

7.6. Proline

Proline as an amino acid is essential for primary metabolism in plants during salt and drought stresses, showing a molecular chaperone role due to its stabilizing action either as a buffer to maintain the pH of the cytosolic redox status of the cell (Verbruggen and Hermans, 2008; Kido et al. 2013) or as antioxidant through its involvement in the scavenging of free highly reactive radicals (Smirnoff and Cumbes, 1989) or acting as a singlet oxygen quencher (Matysik et al. 2002). In higher plants, proline biosynthes is may proceed either via glutamate, by successive reductions catalyzed by pyrroline-5-carboxylate synthase (P5CS) and pyrroline-5-carboxylate reductase (P5CR) or by ornithine pathway and ornithine

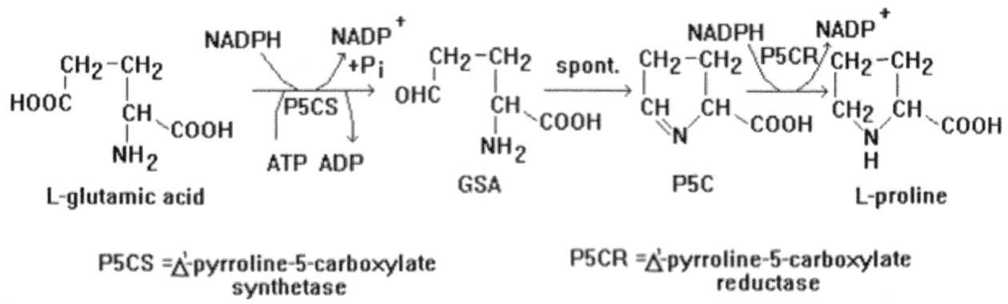

FIGURE 7.4 Pathway of proline synthesis in higher plants. Delauney, A.J. and D.P.S. Verma 1993.

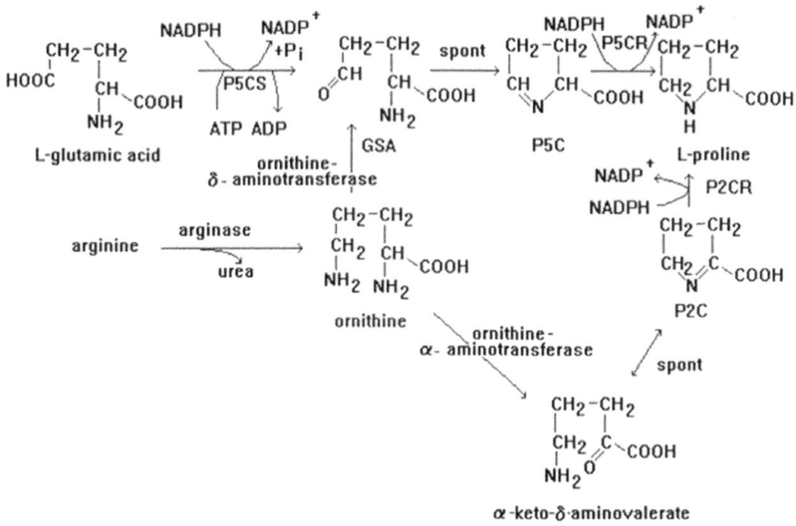

FIGURE 7.5 Alternative pathways of proline synthesis in higher plants. Delauney, A.J. and D.P.S. Verma 1993.

d-aminotransferase (OAT), representing generally the first activated osmoprotectant after stress perception (Savouré, 1995; Parida et al. 2008). Proline accumulation in transgenic rice plants with P5CS cDNA was reported and proved stress-induced overproduction of the P5CS enzyme under salinity stress (Zhu et al. 1998). A cDNA clone encoding P5CS was later isolated from rice and characterized. The expression of P5CS and the accumulation of proline in salt tolerant cultivar are much higher than in salt sensitive lines (Igarashi et al. 1997). When P5CS gene was overexpressed in the transgenic tobacco plants, an increased production of proline coupled with salinity tolerance was noted (Kishor et al. 1995). Thus, P5CS may not be the rate-limiting step in proline accumulation (Delauney and Verma, 1993).

7.7. Glycine Betaine

Betaines are amino acid derivatives in which the nitrogen atom is fully methylated such as that of quaternary ammonium compounds. Among the many quaternary ammonium compounds known in plants, glycine betaine (GB) occurs most abundantly in response to dehydration stress where it reduces lipid peroxidation, thereby helps in maintaining the osmotic status of the cell to ameliorate the abiotic stress effect (Chinnasamy, 2005). In higher plants, glycine betaine is synthesized in the chloroplast from serine via choline by the action of choline monooxygenase (CMO) and betaine aldehyde dehydrogenase (BADH) enzymes (Ashraf and Foolad, 2007).

Genes involved in osmoprotectant biosynthesis are upregulated under salt stress, and the concentrations of accumulated osmoprotectants correlates with osmotic stress tolerance (Chen and Murata, 2002; Zhu, 2002).

Choline dehydrogenase gene (codA) from *Arthrobacter globiformis*aids in improving the salinity tolerance in rice (Vinocur and Altman, 2005). Tolerant genotypes normally accumulate more glycine betaine than sensitive genotypes in response to stress. This relationship, however, is not universal.

The osmolyte that plays a major role in osmotic adjustment is species dependent. Some plant species such as rice (*Oryza sativa*),mustard (*Brassica*spp.), Arabidopsis (*Arabidopsis thaliana*) and tobacco (*Nicotiana tabacum*) naturally do not produce glycine betaine under stress or nonstress conditions (Rhodes and Hanson, 1993). In these species,transgenic plants with overexpressing glycine betaine synthesizing genes exhibited abundant production of glycine betaine, which leads plants to tolerate stresses, including salinity stress (Rhodes and Hanson, 1993). The limitation in production of glycine betaine in high quantities in transgenic plants is reportedly due to either low availability of substrate choline or reduced transport of choline into the chloroplast where glycine betaine is naturally synthesized (Huang et al. 2000).

Thus, to engineer plants for overproduction of osmolytes such as glycine betaine, other factors such as substrate availability and metabolic flux must also be considered. In rice *indica* plant (cv. IR36), deficiency of glycine betaine has been attributed to the absence of the two enzymes, choline monooxygenase and betaine aldehyde, in the biosynthetic pathway (Rathinasabapathi et al. 1993). However, the enzymatically active BADH is detectable in Japonica variety of rice (cv. Nipponbare) (Nakamura et al. 1997). This apparent discrepancy is yet subject for further investigation. Exogenous foliar application of glycine betaine to *Oryzasativa* (Harinasut et al. 1996) resulted in improved growth of plants under salinity stress condition. Further, a decrease in Na^+ and an increase in K^+ concentrations in shoots were observed in GB-treated plants under salinity. This indicates the possible role of glycine betaine in signal transduction and ion homeostasis as well.

Proline accumulates in many plant species in response to environmental stress. Although much is now known about proline metabolism, some aspects of its biological functions are still unclear. The compartmentalization of proline biosynthesis, accumulation and degradation in the cytosol, chloroplast and mitochondria is interesting. It had also been noted the role of proline in cellular homeostasis, including redox balance and energy status. Proline can act as a signaling molecule to modulate mitochondrial functions, influence cell proliferation or cell death and trigger specific gene expression, which can be essential for plant recovery from stress. Although the regulation and function of proline accumulation are not yet completely understood, the engineering of proline metabolism could lead to new opportunities to improve plant tolerance of environmental stresses.

7.8. Proline Accumulation in Plants

Proline is a proteinogenic amino acid with an exceptional conformational rigidity, and is essential for primary metabolism. Since the first report on proline accumulation in wilting perennial rye grass (*Loliumperenne*) (Kemble and MacPherson, 1954), numerous studies have shown that the proline content in higher plants increases under different environmental stresses. Proline accumulation has been reported during conditions of drought (Choudhary et al. 2005) high salinity (Yoshida et al. 1995) high light and UV irradiation (Saradhi et al. 1995), heavy metals (Schat et al. 1997), oxidative stress (Yang et al. 2007) and in response to biotic stresses (Fabro et al. 2004; Haudecoeur et al.2009). An osmoprotective function of proline was discovered first in bacteria, where a causal relationship between proline accumulation and salt tolerance has long been demonstrated (Csonka et al. 1988; Csonka and Hanson, 1991). Such data led to the assumption that proline accumulation in stressed plants has a protective function, which has been emphasized in numerous reviews (Verbruggen and Hermanns, 2008). However, the correlation between proline accumulation and abiotic stress tolerance in plants is not always apparent. For example, high proline levels can be characteristic of salt and cold-hypersensitive Arabidopsis (*Arabidopsis thaliana*) mutants (Liu and Zhu, 1997; Jin and Brows, 1998). Proline content is also high in drought-tolerant rice varieties, but is not correlated with salt tolerance in barley (*Hordeum vulgare*) (Chen et al. 2007; Widodo et al. 2009). Nevertheless, several comprehensive studies using transgenic plants or mutants demonstrate that proline metabolism has a complex effect on development and stress responses, and that proline accumulation is important for the tolerance of certain adverse environmental conditions (Mattioli et al. 2008; Szekely et al. 2008; Miller et al. 2009)

7.9. Compartmentalization of Proline Metabolism in Plants

In plants, proline is synthesized mainly from glutamate, which is reduced to glutamate-semialdehyde (GSA) by the pyrroline-5-carboxylate synthetase (P5CS) enzyme, and spontaneously converted to pyrroline-5-carboxylate (P5C) (Strizhov et al. 1997; Armengaud, 2004). P5C reductase (P5CR) further reduces the P5C intermediate to proline (Szoke et al. 1992; Verbruggen et al. 1993). In most plant species, P5CS is encoded by two genes and P5CR is encoded by one (Armengaud, 2004). Proline catabolism occurs in mitochondria via the sequential action of proline dehydrogenase or proline oxidase (PDH or POX) producing P5C from proline, and P5C dehydrogenase (P5CDH), which converts P5C to glutamate. PDH is encoded by two genes, whereas a single P5CDH gene has been identified in Arabidopsis and tobacco (*Nicotiana tabacum*) (Ribarits et al. 2007). As an alternative pathway, proline can be synthesized from ornithine, which is transaminated first by ornithine-delta-aminotransferase (OAT) producing GSA and P5C, which is then converted to proline. Intracellular proline levels are determined by biosynthesis, catabolism and transport between cells and different cellular compartments.

Computer predictions suggest a mainly cytosolic localization of the biosynthetic enzymes (P5CS1, P5CS2 and P5CR), whereas a mitochondrial localization is predicted for the enzymes involved in proline catabolism, such as PDH1/ERD5, PDH2, P5CDH and OAT. Although signal peptides could not be identified within the primary structure of P5CS1, P5CS2 and P5CR enzymes, the PDH1, P5CDH and OAT proteins have well recognizable mitochondrial targeting signals. P5CS1-GFP is normally localized in the cytosol of leaf mesophyll cells, but in embryonic cells and roots it is associated with organelles that are similar to fusiform bodies. When cells are exposed to salt or osmotic stress, P5CS1-GFP, but not P5CS2-GFP, accumulates in the chloroplasts. GFP-labeled Arabidopsis P5CS2 has been shown to be predominantly localized in the cytosol (Szekley et al. 2008).The P5CR protein and activity has been detected in the cytosol and plastid fraction of leaf, root and nodule cells of soybean (*Glycine max*) (Szoke et al. 1992; Kohl et al. 1988). In pea (*Pisum sativum*) mesophyll protoplasts, P5CR activity was localized in chloroplasts, suggesting that P5CR accumulates in plastids under high osmotic conditions. Proline biosynthesis probably occurs in the cytosol and, in Arabidopsis, it is controlled by the P5CS2 gene (Szekely et al. 2008).

During osmotic stress, proline biosynthesis is augmented in the chloroplasts, which is controlled by the stress induced P5CS1 gene in Arabidopsis. Therefore, proline can be synthesized in different subcellular compartments, depending on the environmental conditions.

Salt-stress effects on osmotic adjustment, ion and proline concentrations as well as proline metabolizing enzyme activities were studied in two rice (*Oryza sativa* L.) cultivars differing in salinity resistance: I Kong Pao (IKP; salt-sensitive) and Nona Bokra (salt-resistant). The salt-sensitive cultivar exposed to 50 and 100 mM NaCl in nutritive solution for 3 and 10 days accumulated higher levels of sodium and proline than the salt-resistant cultivar and displayed lower levels of osmotic adjustment. Proline accumulation was not related to proteolysis and could not be explained by stress-induced modifications in Δ^1-pyrroline-5-carboxylate reductase (P5CR; EC 1.5.1.2) or proline dehydrogenase (PDH; EC 1.5.1.2) activities recorded *in vitro*. The extracted ornithine Δ-aminotransferase (OAT; EC 2.6.1.13) activity was increased by salt stress in the salt-sensitive cultivar only. In both genotypes, salt stress induced an increase in the aminating activity of root glutamate dehydrogenase (GDH; EC 1.4.1.2) while deaminating activity was reduced in the leaves of the salt-sensitive cultivar.

The total extracted glutamine synthetase activity (GS; EC 6.3.1.2) was reduced in response to salinity but NaCl had contrasting effects on GS1 and GS2 isoforms in salt-sensitive IKP. Salinity increased the activity of ferredoxin-dependent glutamate synthase (Fd-GOGAT; EC 1.4.7.1) extracted from leaves of both genotypes and increased the activity of NADH-dependent glutamate synthase (NADH-GOGAT; EC 1.4.1.14) in the salt-sensitive cultivar. It is suggested that proline accumulation is a symptom of salt-stress injury in rice and that its accumulation in salt-sensitive plants results from an increase in OAT activity and an increase in the endogenous pool of its precursor glutamate. The physiological significance of the recorded changes are analyzed in relation to the functions of these enzymes in plant metabolism.

TABLE 7.5 Effects of salt stress on the content of proline and total amino acids in the 2nd leaf blade of NERICA rice seedlings

Varieties	Proline (μmol/g DW)		Total amino acids (μmol/g DW)	
	control	salt stressed	control	salt stressed
NERICA 1	0.42 ± 0.01b	3.29 ± 0.08CD***	52.67 ± 3.15a	80.94 ± 5.91B*
NERICA 2	0.51 ± 0.01ab	11.47 ± 0.71A**	58.72 ± 7.19a	163.20 ± 8.47A***
NERICA 3	0.43 ± 0.00b	5.73 ± 0.44BC**	43.49 ± 4.78a	110.46 ± 4.52B***
NERICA 4	0.43 ± 0.01b	5.68 ± 0.88BC*	55.01 ± 6.74a	97.14 ± 2.53B**
NERICA 5	0.45 ± 0.00b	3.75 ± 0.51BCD*	58.40 ± 8.24a	121.73 ± 6.88AB**
NERICA 6	0.46 ± 0.01b	6.30 ± 1.05BC*	52.38 ± 2.80a	109.81 ± 17.36B*
NERICA 7	0.44 ± 0.03b	2.45 ± 0.25D*	49.04 ± 0.12a	97.64 ± 7.40B*
Nipponbare	0.59 ± 0.04a	6.40 ± 0.36B**	57.96 ± 3.92a	164.38 ± 8.54A***

*** Significant at 1% level, ** Significant at 5% level, * Significant at 10% level.

Isotopic methods show that a substantial portion of protein-bound aspartic acid in tobacco is derived from an aplerotic synthesis via phosphoenolpyruvate (PEP) carboxylase. Similar studies in soybean (*Glycine max* L.) and spinach (*Spinacia oleracea* L.) showed a similar pattern, and this pattern persists with age because of slow protein turnover. A more quantitative analysis indicates that about 40% of protein-bound aspartate is derived in this manner. Analyses of free aspartic and malic acids show that contribution of PEP carboxylase to the synthesis of these acids decreases with increasing age. The C_4 plant *Zea mays* L. did not show this pattern.

Plants, fungi, yeasts and most of the bacteria usually synthesize all the 20 amino acids incorporated in a protein, while monogastric animals can only synthesize 11 of them. The nine remaining amino acids, which are termed essential, need to be provided by the diet.

Proline accumulates in many plant species in response to environmental stress. Although much is now known about proline metabolism, some aspects of its biological functions are still unclear. The compartmentalization of proline biosynthesis, accumulation and degradation in the cytosol, chloroplast and mitochondria is interesting. It had also been noted the role of proline incellular homeostasis, including redox balance and energy status. Proline can act as a signaling molecule to modulate mitochondrial functions, influence cell proliferation or cell death and trigger specific gene expression, which can be essential for plant recovery from stress. Although the regulation and function of proline accumulation are not yet completely understood, the engineering of proline metabolism could lead to new opportunities to improve plant tolerance of environmental stresses.

7.10. *Aspartate Family*

In Asparagine synthesis aspartic acid is converted in to asparagines either by glutamine dependent asparagine synthetic pathway catalyzed by asparagine synthetase in presence of ATP or similarly where ammonium being the donor of second amino group.

Plants, fungi, yeasts and most of the bacteria, usually synthesize all the 20 amino acids incorporated in a protein, while monogastric animals can only synthesize 11 of them. The remaining nine amino acids, which are deemed essential, need to come from diet.

The syntheses of the essential amino acids lysine, threonine, methionine and isoleucine are carried out in a complex and strongly regulated metabolic pathway, which has aspartic acid as a precursor with several enzymes being regulated by feedback inhibition. Since cereal seeds constitute the main source of proteins in plants and are deficient in lysine and threonine, extensive studies of this pathway have been carried out, with special attention to the potential. Such studies allowed the identification of important regulatory mechanisms, showing that many enzymes are positively or negatively regulated by the end-products of the pathway

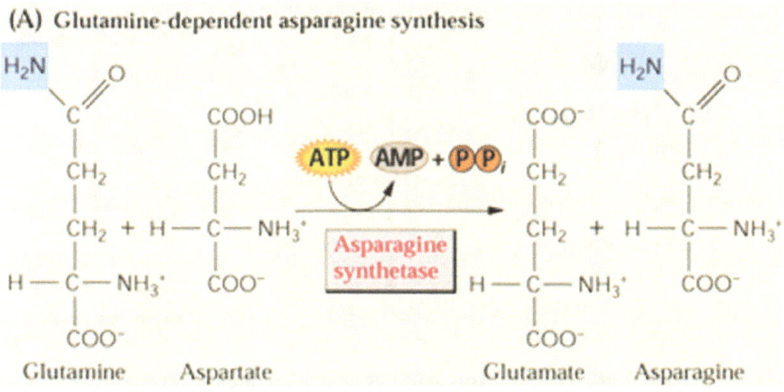

(A) Glutamine-dependent asparagine synthesis

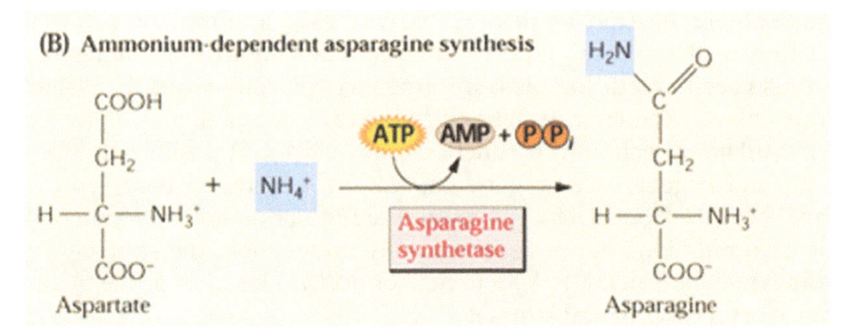

FIGURE 7.6 Asparagine synthesis.

(feedback) or their analogues (Heremans and Jacobs, 1994; Azevedo et al. 1997; Feller et al. 1999).

Aspartate kinase (AK), the first enzyme of the aspartic acid pathway, has been extracted, partially purified and well characterized in several higher plants (Azevedo et al. 1997; Dey and Guha-Mukherjee, 1999). A bifunctional polypeptide with both threonine-sensitive AK and homoserine dehydrogenase (HSDH) activities has been observed in some plant species (Tecxeira et al. 1998). The activity of the enzyme dihydrodipicolinate

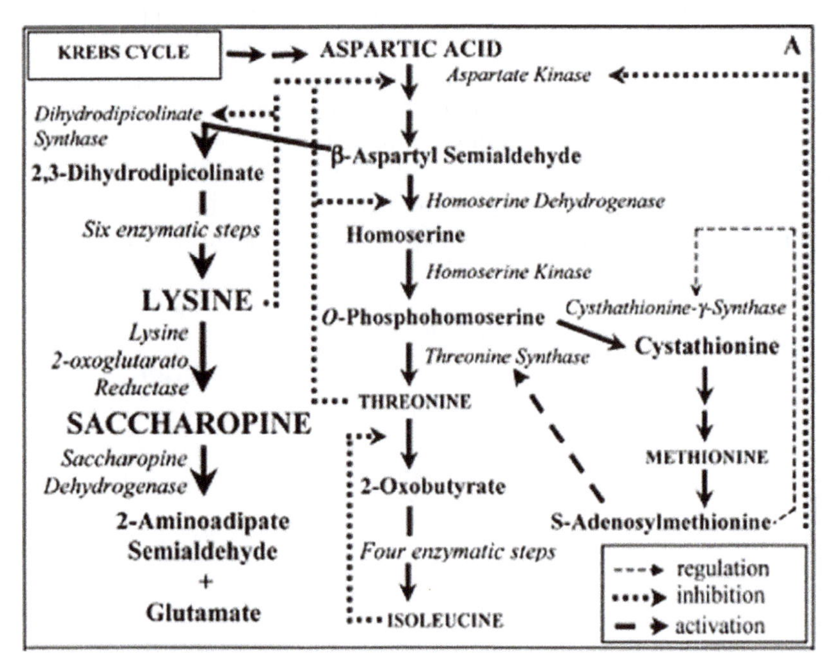

FIGURE 7.7 Amino acids synthesis from aspartic acid.

synthase (DHDPS), which is strongly inhibited by low concentration of lysine, has been shown to be a key regulatory point of the lysine biosynthesis branch of the pathway (Azevedo et al. 1997). Although a great deal of information is available about the lysine biosynthetic enzymes and their regulation, little is known about lysine degradation. Success in obtaining high-lysine plants for human or animal consumption relies on the full understanding of the metabolism of lysine as well as for subsequent genetic manipulation (Azevedo, 2002).

7.11. Lysine Catabolism in Plants

The lysine catabolism in plants was initially studied in wheat, maize and barley in experiments using ^{14}C-lysine, with the radioactivity being incorporated into a-aminoadipic acid and glutamate, indicating that this amino acid is oxidatively degraded through saccharopine (Brandt, 1975; Sodek and Wilson, 1970).

Recent studies have indicated that the lysine catabolism plays an important role for lysine accumulation in plants and the control of the lysine content, particularly in seeds (Arruda et al. 2000). Initial studies with enzymes involved in the lysine degradation reinforced the main role of catabolism in the control of the soluble lysine concentration in the maize endosperm (Arruda et al. 1982; Arruda and da Silva, 1983). The amount of lysine that was shown to be translocated from other tissues to the developing endosperm, for the synthesis of storage proteins, was 2 to 3-fold higher than what would be necessary (Arruda and da Silva, 1983). Due to excess lysine, an accumulation of lysine in the soluble form would be expected; however this does not occur, since the soluble lysine concentration was maintained at a low level during the development of the endosperm. These low lysine levels could contribute by preventing the inhibition of AK activity and facilitate the subsequent biosynthesis of methionine. These results indicated that the soluble lysine concentration is mainly controlled by the catabolic rate, instead of by the feedback inhibition of its synthesis.

7.12. Purification and Characterization of LOR and SDH Enzymes

Two main enzymes have been shown to be part of the lysine catabolic pathway in animals, micro-organisms and recently in plants. Lysine-2-oxoglutarate reductase (LOR) improvement of the nutritious quality. Lysine 2-oxoglutarate reductase (LOR; EC 1.5.1.8), (also known as lysine α-ketoglutarate reductase, LKR) is the first enzyme of the pathway, which condenses lysine and 2-oxoglutarate to form saccharopine, which is then hydrolyzed to a-amino adipic acid and glutamic acid in a reaction catalyzed by the enzyme saccharopine dehydrogenase (SDH; EC 1.5.1.9). The net result of these two reactions resembles a transaminase reaction in which the a-amino group of lysine is transferred to 2-oxoglutarate to form glutamate (Azevedo et al. 1997; Azevedo and Lea, 2001).

Kinetic studies have shown distinct results between maize and rice enzymes, for instance. In maize, saccharopine is a competitive inhibitor of lysine and non-competitive of 2-oxoglutarate, suggesting a mechanism where lysine first interacts with the enzyme, and afterwards with 2-oxoglutarate and NADPH, releasing $NADP^+$ and saccharopine (Brochetto-Braga et al. 1992). The opposite was observed for the rice enzymes (Gaziola et al. 1997). Both enzymes have been well-characterized in mammals and were shown to be part of one single bifunctional polypeptide (Markovitz and Chuang, 1987). The human bifunctional enzyme is a tetramer with a molecular mass of 460 kDa, with 115 kDa subunits (Fjellstedt and Robinson, 1975; Markovitz and Chuang, 1987). In fungi and yeast, the structures of LOR and SDH are comprised of monomers of 49 kDa and 73 kDa, encoded by the *Lys1* and *Lys9* genes, respectively (Ramos et al. 1988; Feller et al. 1999). It was only recently that these enzymes have received more attention in plants, being isolated and characterized in such species as maize, rice, soybean, *Phaseolus*, *Arabidopsis*, canola and coix.

In maize, rice and coix the bifunctional enzyme LOR-SDH was shown to be endosperm-specific (Gonçalves-Butruille et al. 1996; Gaziola et al. 1997; Lugli et al. 2002; Azevedo and Lea, 2001). In maize (Gonçalves-Butruille et al. 1996), rice (Gaziola et al. 1997), soybean (Miron et al. 2000), *Phaseolus vulgaris* (Cunha-Lima et al. 2003) and coix (Lugli et al. 2002) the activities of LOR and SDH reside in the same bifunctional polypeptide, similar to what has been observed in mammals (Fjellstedt and Robinson, 1975; Markovitz and Chuang, 1987). Recently, the presence of one additional monofunctional SDH enzyme was demonstrated in *Arabidopsis* (Tang et al. 1997) and canola (Zhu et al. 2000) which is interesting since both have already been reported as having a bifunctional LOR-SDH enzyme (Tang et al. 1997; Zhu et al. 2000). Furthermore, a new monofunctional LOR has now been detected in *Arabidopsis* (Galili et al. 2001) and in cotton.

The molecular mass of the LOR-SDH enzyme exhibits some variation among plant species (Azevedo and Lea, 2001). In maize, the polypeptide presents a monomeric form of 125 kDa (Gonçalves-Butruille et al. 1996) or 140 kDa (Brochetto-Braga et al. 1992), when determined by SDS-PAGE or native PAGE, respectively, or 260 kDa when determined by gel filtration, in a dimmer structure, with two 117 kDa subunits, which constitutes the native form of the enzyme (Gonçalves-Butruille et al. 1996; Kemper et al. 1999). These subunits could be cleaved by elastase digestion into five bands ranging from 35 kDa to 65 kDa (Kemper et al. 1998). The separation of the five bands during the course of proteolyses could be associated with LOR and SDH activities, and the predominant 65 and 57 kDa bands contained the functional domains of LOR and SDH activities, respectively (Kemper et al. 1998).

In rice, the LOR-SDH protein exhibited a molecular mass of approximately 203 kDa when determined by PAGE and gel filtration, with the presence of multimeric forms, probably dimmeric or tetrameric states 396 kDa (Gaziola et al. 1997). In *Arabidopsis*, a monomeric form of 116 kDa was observed (Tang et al. 1997). In *Phaseolus vulgaris*, the activities of LOR-SDH also reside in a bifunctional protein and depending on the purification procedure, may elute as a monomer of 94 kDa with SDH activity only, or a dimmer of 190 kDa with both enzyme activities (Cunha-Lima et al. 2003). In soybean, monomeric forms of 100 and 123 kDa, and a 256 kDa dimmeric form were identified (Miron et al. 2000).

7.13. Regulation of LOR and SDH Enzymes in Plants

Recent studies with different plant species have demonstrated that lysine may autoregulate its own catabolism *in vivo*, with the enzymes differentially modulated by a intracellular signaling cascade, involving mainly Ca^{2+}, protein phosphorylation-dephosphorylation and ionic strength (Karchi et al. 1995; Miron et al. 1997; Kemper et al. 1998; Gaziola et al. 2000). Karchi et al. (1995) working with tobacco seeds observed that the activity of LOR could be stimulated by exogenous lysine and this stimulatory effect was significantly reduced when the seeds were treated with the Ca^{2+} chelator EGTA, an inhibitory effect that could be overwhelmed with addition of Ca^{2+}. In maize, Ca^{2+} was also shown to modulate LOR activity, whereas SDH activity was not. The Ca^{2+}-dependent LOR activity increase was also tested for inhibition by two structurally different calmodulin inhibitors, which almost completely inhibited the activity of LOR (Kemper et al. 1998). In rice, the results pertaining to the regulation were similar to those observed in maize, for both enzymes (Gaziola et al. 2000). Kemper et al. (1999) reported evidence for a Ca^{2+} effect on the oligomerization state of LOR-SDH from maize. Ca^{2+} stimulated LOR activity through the dimerization of only the LOR domain, and had no effect on the SDH activity.

In addition, LOR modulation has been demonstrated in maize with ionic strength, whereas the SDH activity remained unaltered (Kemper et al. 1998). Organic solvents at concentrations that lowered the water activity increased LOR activity (Kemper et al. 1998). In tobacco and soybean, the LOR activity was modulated with bifunctional polypeptide phosphorylation, but SDH activity was not modulated. The phosphorylation-dephosphorylation with kinase-casein II and alkaline phosphatase respectively,

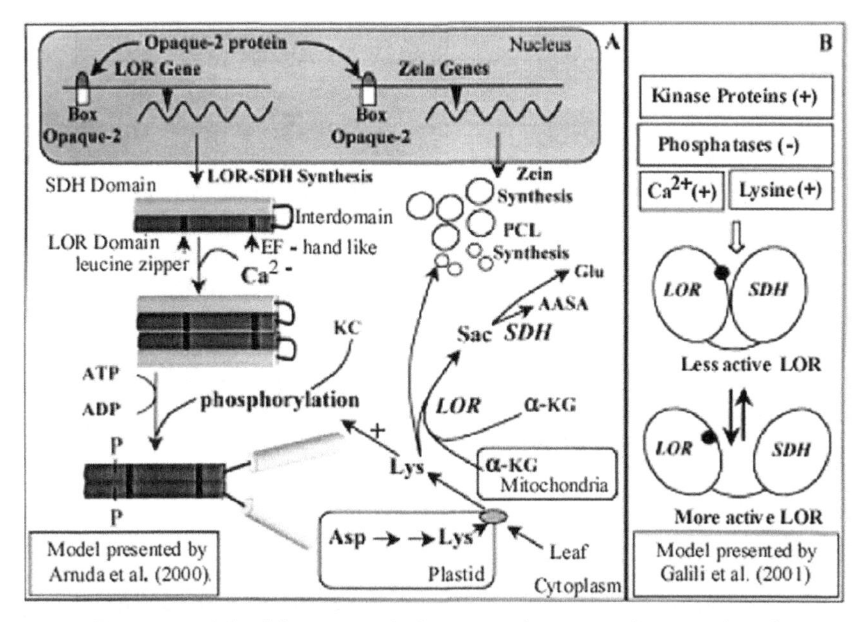

FIGURE 7.8 Model of lysine catabolism regulation in the cereal endosperm cells (A). Synthesis in plastids via the aspartate pathway and lysine translocation from vegetative tissues to the endosperm. Part of the lysine is incorporated into proteins containing lysine (PCL). However the largest storage protein fraction is the prolamins, which are deficient in lysine. In maize, the transcripional activator opaque-2 controls teh expression of genes the cncode zeins and the bifunctional LOR-SDH enzyme, which is regulated by Ca^{2+}, and is involved in enzyme dimerization, and phosphorylation by casein kinase (KC) in a lysinedependent manner. As soon as the lysine concentration increases, the activity of LOR increases due to lysine-dependent phosphorylation. The phosphorylation of the LOR domain could inhibit the enzyme, which would be inhibited by the SDH domain and/or interdomain region. The lysine catabolic process leads to an increase in glutamic acid and amino adipic semialdehyde (AASA). Hypothesis suggests a conformational modulation of LOR-SDH, where the two states may to be found. Calcium, proteins, kinases and phosphatases gegulate the alteration betweeen the two forms (B).

indicated that active LOR is a phosphoprotein with the activity being modulated by the opposite actions of the kinase and phosphatase proteins (Karchi et al. 1995; Miron et al. 1997).

In recent reports, the effects on LOR-SDH activity caused by aminoethyl-l-cysteine (AEC), a lysine analogue, and Sadenosylmethionine (SAM) have also been tested (Gaziola et al. 2000; Lugli et al. 2002). In rice, AEC was shown to be able to substitute for lysine as a substrate for LOR, but less efficiently, whereas SAM did not produce any significant changes (Gaziola et al. 1999; Gaziola et al. 2000). On the other hand, in maize, AEC was not able to substitute for lysine as a LOR substrate (Brochetto-Braga et al. 1992).

Although LOR-SDH from animals, yeast and plants have some different properties, others are common, such as optimum pH, which are neutral for LOR (7.0 to 7.5) and basic for SDH (8.0 to 9.0) (Gonçalves-Butruille et al. 1996; Gaziola et al. 2000).

Regulatory mechanisms controlling lysine metabolism are still not fully understood and some hypothesis have been suggested. Arruda et al. (2000) and Galili et al. (2001) reported alternative hypothesis, which considers the linkage between LOR and SDH domains that may be responsible for LOR activity modulation through protein intramolecular interactions. If such an *in vivo* mechanism really occurs, theoretically it would be possible to minimize the linkage through the alteration of the ionic strength in enzyme assays *in vitro*. Reinforcing this idea, a low LOR activity was detected in buffers without the addition of NaCl, when compared to the LOR activity levels obtained in buffer containing 100 mM NaCl. In addition, in *Arabidopsis* transformed with a construction, in which the interdomain and SDH domains were deleted, the LOR activity was not affected by salt concentrations (Galili et al. 2001). These results suggest that the interdomain region, as well as the SDH domain, may play a role in an interdomain interaction that affects LOR activity.

Peptides derived from the SDH domain or the interdomain were shown, *in vitro*, to be able to inhibit the activity of peptides derived from the LOR domain (Kemper et al. 1998). This fact suggests the existence of *in vivo* inhibition of the monofunctional LOR by the monofunctional SDH, although this inhibition shows less efficiency than that which occurs in the case of the bifunctional LOR. On the other hand, the SDH activity does not appear to be affected by this linkage (Zhu 2000). In a recent report, Zhu (2002) showed that the functional interaction between the LOR and SDH domains is mediated by the linker region and not by specific affinities between these domains.

7.14. *Metabolic Flow*

In some plant tissues, such as seeds, in which the bifunctional LOR-SDH protein apparently corresponds to the majority of the total LOR activity, the lysine catabolic flow is regulated by LOR modulation via the linkage with SDH. This may contribute to the control of lysine homeostasis through lysine-dependent stimulation of LOR activity (Karchi et al. 1994, 1995).

Dominant induction of the monofunctional LOR and SDH proteins during the abscission process and under stress conditions may maintain a high and temporary catabolic flow, leading to glutamate production. Such a flow would probably be temporary, otherwise it could lead to the depletion of the soluble-lysine pool (Galili et al. 2001). Important information concerning such an aspect has been provided by studies of canola, which demonstrated an increased SDH activity, including the monofunctional isoenzyme, under osmotic stress conditions and followed the increase in LOR activity, when the stress was more severe (Moulin et al. 2000). It has also been suggested that this linkage may influence the metabolic flow, allowing the LOR product, saccharopine, to be sent directly to the catalytic site of SDH (Gonçalves-Butruille et al. 1996). This hypothesis may be questionable, based on the results obtained by Falco et al. (1995), who reported saccharopine accumulation in lysine overproducing transgenic soybean seeds, maintaining some SDH activity.

Considering the regulatory mechanisms described above with the different LOR and SDH K_m values for their substrates (Gonçalves-Butruille et al. 1996; Gaziola et al. 1997; Miron et al. 2000) and the fact that plant LOR-SDH isoenzymes have to be located in the cytosol, the SDH domain may act in a physiological non-optimal pH for its activity (Kemper et al. 1999), suggesting that both LOR and SDH activities represent a rate-limiting step in the lysine catabolism (Miron et al. 2000) and that the metabolic flow through the saccharopine pathway is different among plant species and thus subjected to various regulatory points. This observation can be further supported by the work of Falco et al. (1995), who reported saccharopine accumulation in transgenic soybean, whereas canola exhibited accumulation of α-amino adipate semialdehyde, another intermediate of the lysine catabolic pathway. Moreover, the identification of monofunctional SDH enzyme in a limited number of plant species (*Arabidopsis,* canola and cotton) and the studies of its properties, which are similar to those determined for the bifunctional enzyme, suggests that the presence of a monofunctional SDH may provide an increase in metabolic flow, compensating for the limitation generated by the physiological cellular pH.

The metabolic flow through the degradation pathway may be subjected to diverse regulatory mechanisms that have been studied, with special attention to the possible roles of lysine catabolism in distinct metabolic processes, such as growth, development and response to environmental changes or stress (Arruda et al, 2000; Galili et al. 2001). It may also provide further insights into the role of glutamate, which also originates from lysine catabolism. Glutamate may be utilized as a primary precursor of the metabolite stress-related compounds such as proline, arginine, and γ-aminobutyric acid (GABA), which constitute stress-related signaling (Galili et al. 2001). Reinforcing the different metabolic roles it has recently been demonstrated in plants the existence of animal homologues of glutamate receptors, which appear to regulate different physiological processes (Lam et al. 1998). In transgenic *Arabidopsis* plants, the over-expression of these glutamate receptors changed the Ca^{2+} balance, leading to hypersensitivity to ionic stress.

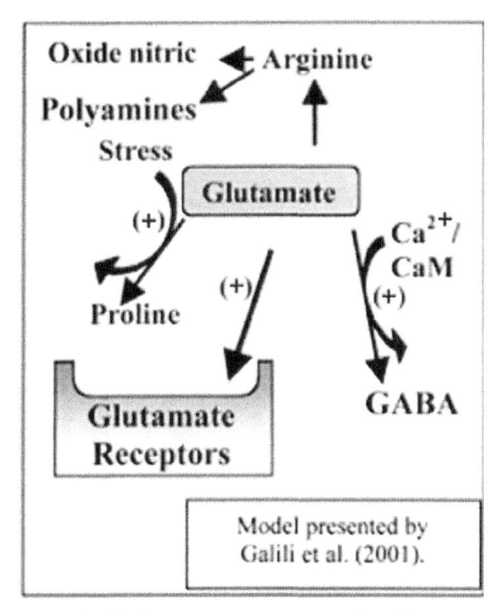

FIGURE 7.9 Conversion of glutamate to stress related compounds. The conversion of glutamate to a strong osmolite, proline, by D-pyrroline-5-carboxylate synthase; to GABA via glutamate decarboxylase calcium/calmodulin modulation; to nitric oxide (signaling molecule) via arginine. Glutamate is also a stimulator of glutamate receptors. Model presented by Galili et al. (2001).

7.15. Expression and Characteristics of LOR and SDH Genes

The LOR-SDH gene is abundantly expressed in floral tissues and seeds in development. The *in situ* mRNA hybridization suggests that the *Arabidopsis* LOR-SDH gene is up-regulated in ovarian tissues, embryos in development and in the outer layers of the endosperm (Tang et al. 1997). Kemper et al. (1999) have demonstrated by *in situ* analysis of SDH activity that the bifunctional enzyme is located in the outer layer of maize developing seeds, whereas in embryos the activity was only slightly detectable, contradicting other studies, which showed an over production of lysine in embryos and subsequent lysine catabolic products (Mazur et al. 1999). These results suggest the possibility of a putative lysine transport mechanism from embryos to the outer layers of the endosperm where lysine is then degraded (Galili et al. 2001). In developing maize seeds, the LOR-SDH gene expression is mediated by the opaque-2 transcription factor, which also controls the expression of genes that encode zein storage proteins (Kemper et al. 1999). cDNA studies of maize (Kemper et al. 1999) and

Arabidopsis (Tang et al. 1997) have shown the expression of the LOR-SDH bifunctional enzyme. One distinct and short mRNA sequence is translated from the same LOR-SDH gene that encodes the monofunctional enzyme in *Arabidopsis* (Tang et al. 1997). Short maize mRNA sequences have also been observed; however, they do not appear to be translated (Kemper et al. 1999).

Sequencing analysis has revealed that maize and *Arabidopsis* LOR-SDH genes contain the CCAAT and TATA box sequences in the promoter and in an internal region of the same gene, possibly controlling the transcripts of the bifunctional LOR-SDH and the monofunctional SDH (Arruda et al. 2000). In addition, GCN4-like sequences, which are involved in the transcription of genes related to nitrogen metabolism in yeast (Hinnebusch, 1988) and plants (Muller and Kanudsen, 1993), have been found in both the upstream and internal promoters in maize and in the internal promoters in *Arabidopsis*. Furthermore, sites for the linkage of opaque-2 have also been found in the upstream and internal promoters of *Arabidopsis,* but only in the upstream promoter in maize. The absence of this site in the promoter of the maize LOR-SDH gene could be an explanation for the presence of the monofunctional SDH in this plant species (Arruda et al. 2000).

The LOR-SDH gene expression is not restricted to reproductive tissues, since mRNAs have been observed in canola leaves when submitted to osmotic stress. Expression analyses of sequences (ESTs) related to the LOR-SDH gene in several plants suggest an abundant expression in the cell division process in various tissues, as well as in cells in the abscission zone and in tissues treated with biotic elicitors (Arruda et al. 2000).

The LOR-SDH locus, as already mentioned, is not restricted to the encoding of the bifunctional LOR-SDH and monofunctional SDH. This locus can also encode for a new monofunctional LOR as in cotton and *Arabidopsis* (Galili et al. 2001). In cotton, monofunctional LOR cDNAs have an identical DNA sequence to the LOR domain, suggesting that this monofunctional LOR is encoded by the same composite locus. The EST database of the abscission zone of cotton contains 1800 sequenced ESTs and presents a relatively high frequency of the monofunctional LOR.

7.16. Biochemical Mutants and Transgenic Plants for Lysine

Mainly in the last four decades, research groups have focused attention on understanding the biochemical and genetic controls of the aspartate pathway (Azevedo et al. 1997). The data allows researchers to induce and select for lysine and threonine overproducing plants through genetic manipulation of key points of the pathway such as catalysis by the enzymes; AK, HSDH, DHDPS, threonine synthase (TS), LOR and SDH (Heremans and Jacobs, 1994, 1997; Ravanel et al. 1998; Laber et al. 1999; Azevedo and Lea, 2001).

The development of plant tissue culture and *in vitro* regeneration technologies have facilitated the selection of biochemical mutants. Such mutants can be selected in cell cultures treated with mutagenic agents and selected on solid or in liquid medium amended with selective agents, such as amino acids or their analogues. The cells that eventually grow in such conditions may be mutants containing enzymes with altered regulatory characteristics (Azevedo, 2002). A similar system can also be used for embryos of seeds submitted to mutagenesis (Lea et al. 1992). Independent of the procedure utilized, the selected plants need to be genetically evaluated and biochemically characterized, as well as submitted to a complete agronomic analysis (Lea et al. 1992).

Specifically in the case of the aspartate pathway, several mutants were selected in a large number of plant species, with the aim of obtaining cereal plants with the accumulation of lysine in seeds, which exhibited altered enzymes (Azevedo et al. 1997; Molina et al. 2001). Mutants were obtained with isoenzymes of AK that were insensitive to lysine plus threonine feedback inhibition Muehlbauer et al. 1994a; Heremans and Jacobs, 1997), which exhibited an overproduction and accumulation of threonine in the leaves and in the seeds but no significant changes in the soluble lysine concentration in the seeds. These results indicated a major role of DHDPS in lysine biosynthesis, since the mutants were still sensitive to lysine feedback inhibition of the DHDPS step of the pathway, therefore driving carbon molecules to threonine biosynthesis (Azevedo and Lea, 2001). Hesse et al. (2002) suggested that after lysine biosynthesis, methionine would be considered the main route for the carbons entering the pathway, instead of threonine biosynthesis. Chiba et al. (1999) showed in *Arabidopsis* that the Cystathionine γ-synthase (CgS) is not feedback-inhibited by end products, but its expression is regulated by methionine at the level of mRNA stability in a process that is activated by methionine or one of its catabolites. This result suggests a central role for CgS in methionine biosynthesis indicating an important flux into the aspartate pathway.

Based on the information provided by the work with the biochemical mutants and on newly developed transformation techniques, a similar strategy has been used to obtain plants that accumulate lysine in the seeds. Transgenic tobacco plants expressing a lysine-insensitive AK from *E. coli* exhibited similar results to those observed for the biochemical mutants, with threonine accumulation, but without changes in the soluble lysine content of the seeds. Other transgenic plants produced with altered enzyme regulation did not result in accumulation of lysine in seeds (Falco et al. 1995; BrinchPedersen et al. 1996).

Soluble lysine accumulation was obtained when a lysine-insensitive DHDPS from *Coryne bacterium* was expressed in transgenic maize embryos (Falco, 2001). Moreover, the knockout of LOR-SDH by T-DNA insertion resulted in a loss of lysine and its catabolism products, but the combination of these transgenic maize plants resulted in a soluble lysine content in the seeds of about 2- to 3-fold higher than the DHDPS transgenic maize plant (Falco, 2001). Transgenic rice plants have also been obtained in order to improve the nutritional value of the seed, by elevating the lysine concentration (Lee et al. 2001). A constitutive and seed-specific expression of feedback-insensitive maize DHDPS lead to a higher content of soluble lysine in the seeds. The higher rate of lysine biosynthesis obtained with the introduction of the altered DHDPS encoding gene also resulted in an increased rate of lysine catabolism. Even so, the over-expression of the mutant gene of DHDPS in a constitutive manner appears to overcome the lysine catabolism, thus maintaining higher lysine concentrations in the mature seeds (Lee et al. 2001).

Azevedo and Lea (2001), in a recent review, suggested that lysine overproduction and accumulation in cereal seeds might be obtained by combining the genetic manipulation of the biosynthesis and lysine degradation mechanisms. Such a suggestion was supported mainly by the fact that the manipulation of enzymes involved in lysine biosynthesis did not produce lysine accumulation in cereal seeds. This could be explained by the fact that vegetables and the maize opaque-2 mutants, which exhibit higher concentration of soluble lysine in the seeds, exhibited a drastic reduction in the lysine catabolic rate in the endosperm, allowing excess lysine to be incorporated into storage proteins, as well as the accumulation in the soluble form (Azevedo and Lea, 2001; Molina et al. 2001). The maize opaque-2 mutant has been extensively studied (Gaziola et al. 1999). This mutation is characterized by an opaque phenotype with a farinaceous endosperm. The high lysine concentration observed in the endosperm is related to an increase in the concentration of soluble lysine and storage proteins with the simultaneous reduction of the prolamin fraction, which has only trace amounts of lysine (Lefèvre et al. 2002). The introduction of the opaque phenotype modifier genes allowed the production of opaque-2 maize lines with good grain productivity, that also exhibit characteristics of high lysine and tryptophan contents, but with a translucent phenotype, which have been denominated as quality protein maize (QPM) (Vasal, 1994; Gaziola et al. 1999). QPM inbred lines have been included in breeding programs with several hybrid of QPM been produced and agronomically tested that are now commercially available (Gaziola et al. 1999).

Through transcriptome and proteome approaches, the regulatory role of the opaque-2 gene has been confirmed, since a 3′ restriction site was shown to be associated with LOR-SDH mRNA abundance (Lefèvre et al. 2002). The use of such techniques certainly will contribute significantly in the future. Azevedo et al. (1997) suggested that cereal cultivars with high lysine content seeds would probably be available in a short period of time. In a similar manner, Hesse et al. (2001) suggested possible traits to increase methionine synthesis in plants. Seed companies and research institutions have already confirmed such a possibility. Even so, additional studies will still be necessary to completely understand the regulatory aspects of lysine, threonine and methionine metabolism and how these mechanisms can be controlled.

Several informations can be obtained by the investigation of protein concentrations of the opaque and floury maize mutants, and of similar mutants of barley, sorghum and other cereal crops. It is surprising that based on the available information, and to the best of our knowledge, other cereals with high lysine mutants, similar to the opaque-2 mutants of maize, have not been utilized in research programs to study the aspartate metabolic pathway, which could further increase our understanding of lysine metabolism (Azevedo, 2002).

7.17. Arginine and Ornithine

Arabidopsis genes encoding enzymes for each of the eight steps in L-arginine (Arg) synthesis were identified, based upon sequence homologies with orthologs from other organisms. Except for *N*-acetylglutamate synthase (NAGS; EC 2.3.1.1), which is encoded by two genes, all remaining enzymes are encoded by single genes. Targeting predictions for these enzymes, based upon their deduced sequences, and subcellular fractionation studies, suggest that most enzymes of Arg synthesis reside within the plastid. Synthesis of the L-ornthine (Orn) intermediate in this pathway from L-glutamate occurs as a series of acetylated intermediates, as in most other organisms. An *N*-acetylornithine: glutamate acetyltransferase (NAOGAcT; EC 2.3.1.35) facilitates recycling of the acetyl moiety during Orn formation (cyclic pathway). A putative *N*-acetylornithine deacetylase (NAOD; EC 3.5.1.16), which participates in the "linear" pathway for Orn synthesis in some organisms, was also identified. Previous biochemical studies have indicated that allosteric regulation of the first and, especially, the second steps in Orn synthesis (NAGS; *N*-acetylglutamate kinase (NAGK), EC 2.7.2.8) by the Arg end-product are the major sites of metabolic control of the pathway in organisms using the cyclic pathway. Gene expression profiling for pathway enzymes further suggests that NAGS, NAGK, NAOGAcT and NAOD are coordinately regulated in response to changes in Arg demand during plant growth and development. Synthesis of Arg from Orn is further coordinated with pyrimidine nucleotide synthesis, at the level of allocation of the common carbamoyl-P intermediate.

Arabidopsis genes encoding enzymes for each of the eight steps in l-arginine (Arg) synthesis were identified, based upon sequence homologies with orthologs from other organisms. Except for *N*-acetylglutamate synthase (NAGS; EC 2.3.1.1), which is encoded by two genes, all remaining enzymes are encoded by single genes. Targeting predictions for these enzymes, based upon their deduced sequences, and subcellular fractionation studies, suggest that most enzymes of Arg synthesis reside within the plastid. Synthesis of the l-ornithine (Orn) intermediate in this pathway from l-glutamate occurs as a series of acetylated intermediates, as in most other organisms. An *N*-acetylornithine: glutamate acetyltransferase (NAOGAcT; EC 2.3.1.35) facilitates recycling of the acetyl moiety during Orn formation (cyclic pathway). A putative *N*-acetylornithine deacetylase (NAOD; EC 3.5.1.16), which participates in the "linear" pathway for Orn synthesis in some organisms, was also identified. Previous biochemical studies have indicated that allosteric regulation of the first and, especially, the second steps in Orn synthesis (NAGS; *N*-acetylglutamate kinase (NAGK), EC 2.7.2.8) by the Arg end-product are the major sites of metabolic control of the pathway in organisms using the cyclic pathway. Gene expression profiling for pathway enzymes further suggests that NAGS, NAGK, NAOGAcT and NAOD are coordinately regulated in response to changes in Arg demand during plant growth and development. Synthesis of Arg from Orn is further coordinated with pyrimidine nucleotide synthesis, at the level of allocation of the common carbamoyl-P intermediate.

7.18. Ornithine Cycle and Arginine Synthesis

The L-partate family amino acids (AFAAs), L-threonine, L-lysine, L-methionine and L-isoleucine have recently been of much interest due to their wide spectrum of applications including food additives, components of cosmetics and therapeutic agents, and animal feed additives. Among them, L-threonine, L-lysine and L-methionine are three major amino acids produced currently throughout the world. Recent advances in systems metabolic engineering, which combine various high-throughput omics technologies and computational analysis, are now facilitating development of microbial strains efficiently producing AFAAs. Thus, a thorough understanding of the metabolic and regulatory mechanisms of the biosynthesis of these amino acids is urgently needed for designing system-wide metabolic engineering strategies. Here we review the details of AFAA biosynthetic pathways, regulations involved, and export and transport systems, and provide general strategies for successful metabolic engineering along with relevant examples. Finally, perspectives of systems metabolic engineering for developing AFAA overproducers are suggested with selected exemplary stud. All of the amino acids apart from lysine undergo transamination; some transaminases use ketoglutarate as the acceptor keto-acid, forming glutamate which is then a substrate for glutamate dehydrogenase. Many others use oxaloacetate as the acceptor keto-acid, forming aspartate,

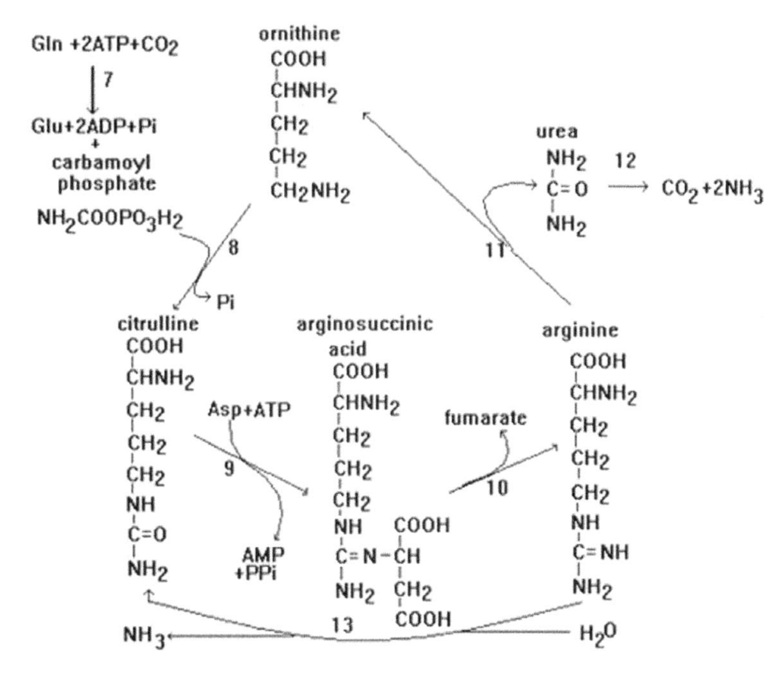

FIGURE 7.10 The Ornithine Cycle.

which then transfers its amino group to ketoglutarate. Thus, by linking a series of transaminases and glutamate dehydrogenase there is a pathway for the overall deamination of most amino acids.

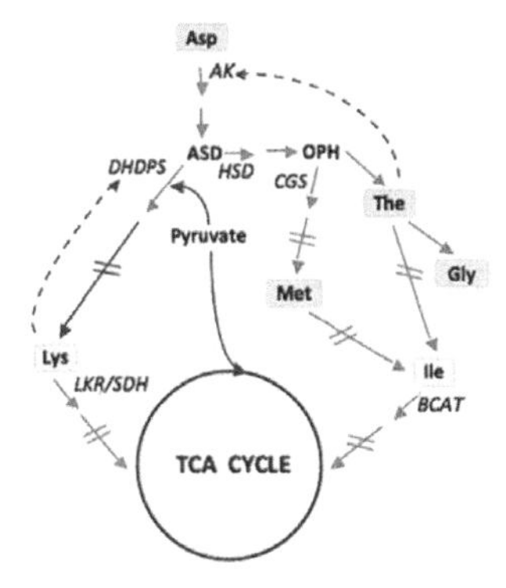

FIGURE 7.11 Schematic diagram of the Asp-family pathway of plants, leading to the synthesis of Lys, Thr, Met, Ile and Gly. Arrows represent individual enzymatic steps. AK, aspartate kinase; ASD, aspartic-semi-aldehyde; BCAT, branch chain aminotransferase; CGS, cystathionine γ-synthase; DHDPS, dihydrodipicolinate synthase; HSD, homoserine dehydrogenase; LKR/SDH, bifunctional Lys-ketoglutarate reductase/saccharopine dehydrogenase; OPH, O-phosphohomoserine.

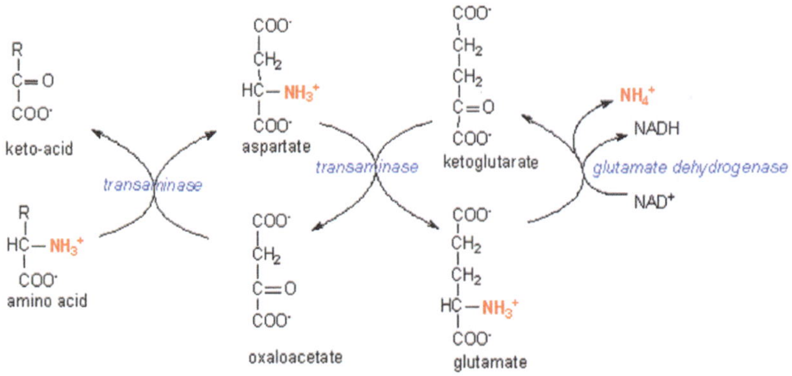

FIGURE 7.12 Transamination process.

Many transaminases use pyruvate as the acceptor keto-acid, forming alanine, and there is an active alanine-glyoxylate transaminase in peroxisomes that transfers the amino group onto glyoxylate, forming glycine, which is then a substrate for glycine oxidase. This provides another pathway for the overall deamination of a variety of amino acids.

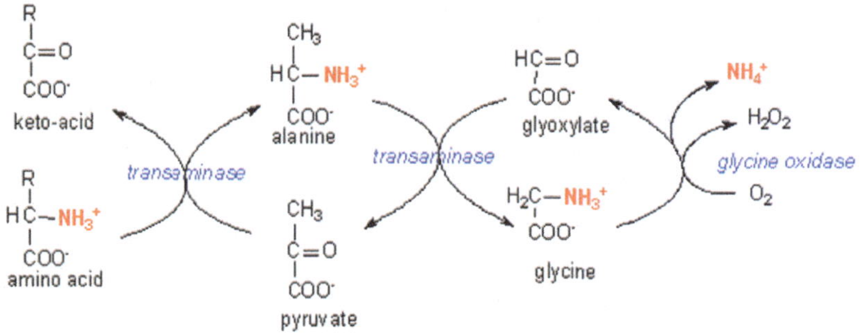

FIGURE 7.13 Alamine and glycine formation.

Genetic defects in alanine-glyoxylate transaminase (either low activity or, rarely, a mutation that leads to the enzyme being in mitochondria rather than peroxisomes) result in hyperoxaluria. The glyoxylate formed by glycine oxidase cannot be recycled to glycine by transamination, but accumulates, and is a substrate for oxidation catalysed by lactate dehydrogenase, forming oxalate.

Auxin is a pivotal plant hormone that regulates many aspects of plant growth and development. Auxin signaling is also known to promote plant disease caused by plant pathogens. However, the mechanism by which this hormone confers susceptibility to pathogens is not well understood. Here, we present evidence that fungal and bacterial plant pathogens hijack the host auxin metabolism in *Arabidopsis thaliana,* leading to the accumulation of a conjugated form of the hormone, indole-3-acetic acid (IAA)-Asp, to promote disease development. We also show that IAA-Asp increases pathogen progression in the plant by regulating the transcription of virulence genes.

The incorporation of C^{14}-amino acids (aspartic acid, glutamic acid, threonine and proline) and C^{14}-nucleic acid bases (adenine, guanine, cytosine and uracil) into the seedling, reproductive stage and young ear portion of rice plant was investigated. It was found that C^{14}-aspartic acid was incorporated into the rice seedling more rapidly than C^{14}-threonine or C^{14}-proline; on the other hand C^{14}-proline was found to be more rapidly incorporated than C^{14}-aspartic acid into reproductive stage plant and young ear portion. Similarly C^{14}-adenine was incorporated into the rice seedling more rapidly than other C^{14}-labelled bases. On the other hand C^{14}-uracil was preferentially incorporated to C^{14}-adenine or C^{14}-guanine into the reproductive stage plant and young ear portion.

It is suggested from the results obtained that proline is polymerized into polypeptide or protein in the rice plant more rapidly at the reproductive stage than at the seedling stage and that a higher proportion of pyrimidine bases might be involved into the metabolic process at the reproductive stage of rice plant. Lys is considered as the first limiting essential amino acid for the synthesis of protein in rice, and its content is marked as the main indicator to judge the rice nutritional quality. Thr and Ile, the other two important essential amino acids in Asp-family, are considered as the second and third limiting amino acids for the synthesis of protein in rice.

Aspartate aminotransaminase (AAT), one of the most active aminotransferases, catalyzes the reaction of producing Asp which is the precursor of Asp-family amino acids in higher plants (Jander and Joshi, 2010). Global warming causes the exacerbation of rice growing environment, which seriously affects rice growth and reproduction, and finally results in the decrease of rice yield and quality. The activities of aspartate metabolism enzymes in grains, and the contents of Aspartate-family amino acids and protein components were investigated to further understand the effects of high temperature (HT) on rice nutritional quality during rice grain filling. Under HT, the average activities of aspartate aminotransferase (AAT) and aspartokinase (AK) in grains significantly increased, the amino acid contents of aspartate (Asp), lysine (Lys), threonine (Thr), methionine (Met) and isoleucine (Ile) and the protein contents of albumin, globulin, prolamin and glutelin also significantly increased.

The results indicated that HT enhanced Asp metabolism during rice grain filling and the enhancement of Asp metabolism might play an important role in the increase of Asp-family amino acids and protein components in grains. In case of the partial appraisal of the change of

Asp-family amino acids and protein components under HT, it had been introduced eight indicators (amino acid or protein content, ratio of amino acid or protein, amino acid or protein content per grain and amino acid or protein content per panicle) to estimate the effects of HT. It is suggested that HT during rice grain filling was helpful for the accumulation of Asp-family amino acids and protein components. Combined with the improvement of Asp-family amino acid ratio in grains under HT, it is suggested that HT during grain filling may improve the rice nutritional quality. However, the yields of parts of Asp-family amino acids and protein components were decreased under HT during rice grain filling.

Protein and amino acids, especially the content and ratio of essential amino acids, are the most important characters of rice nutritional quality (Kawakatsu and Takaiwa, 2010). Although higher plants can produce all protein amino acids from the existing available precursors, they often lack some essential amino acids. For example, methionine (Met) is low in beans and lysine (Lys) is low in cereal grains. Aspartate (Asp) family consists of Asp, Lys, threonine (Thr) and increased amino acid and protein contents in rice seeds, which indicated the important role of AAT in regulating amino acid synthesis in rice. Aspartokinase (AK) is the key enzyme to catalyze Asp catabolism, which regulates the rate of Lys, Thr and Ile synthesis (Galili, 1995; Azevedo et al. 2006; Jander and Joshi, 2010).

Lysine (Lys) is the first limiting essential amino acid in rice, a staple food for half of the world population. Efforts, including genetic engineering, have not achieved a desirable level of Lys in rice. Genetically engineered rice to increase Lys levels by expressing bacterial lysine feedback-insensitive aspartate kinase (AK) and dihydrodipicolinate synthase (DHPS) to enhance Lys biosynthesis; through RNA interference of rice lysine ketoglutaric acid reductase/saccharopine dehydropine dehydrogenase (LKR/SDH) to down-regulate its catabolism; and by combined expression of AK and DHPS and interference of LKR/SDH to achieve both metabolic effects. In these transgenic plants, free Lys levels increased up to ~12-fold in leaves and ~60-fold in seeds, substantially greater than the 2.5-fold increase in transgenic rice seeds reported by the only previous related study. To better understand the metabolic regulation of Lys accumulation in rice, metabolomic methods were employed to analyse the changes in metabolites of the Lys biosynthesis and catabolism pathways in leaves and seeds at different stages. Free Lys accumulation was mainly regulated by its biosynthesis in leaves and to a greater extent by catabolism in seeds. The transgenic plants did not show observable changes in plant growth and seed germination nor large changes in levels of asparagine (Asn) and glutamine (Gln) in leaves, which are the major amino acids transported into seeds. Although Lys was highly accumulated in leaves of certain transgenic lines, a corresponding higher Lys accumulation was not observed in seeds, suggesting that free Lys transport from leaves into seeds did not occur.

Lysine is the most deficient essential amino acid in cereal grains. A bifunctional lysine-degrading enzyme, lysineketoglutarate reductase/saccharopine dehydrogenase (LKR/SDH), is one of the key regulators determining free lysine content in plants. In rice (*Oryza sativa*. L), a bifunctional OsLKR/SDH is predominantly present in seeds. Here, we show that *OsLKR/SDH* is directly regulated by major transcriptional regulators of seed storage protein (SSP) genes: the basic leucine zipper (bZIP) transcription factor (TF), RISBZ1, and the DNA-binding with one finger (DOF) transcription factor, RPBF. *OsLKR/SDH* was highly expressed in the aleurone and subaleurone layers of the endosperm. Mutation analyses in planta, trans-activation reporter assays *in vivo* and electrophoretic mobility shift assays *in vitro* showed that the RPBF recognizing prolamin box (AAAG)and the RISBZ1-recognizing GCN4 motif (TGAG/CTCA) act as important cis-elements for proper expression of *OsLKR/SDH* like SSP genes. However, mutation of the GCN4 motif within *ProOsLKR/SDH* did not alter the spatial expression pattern, whereas mutation of the GCN4 motif within *ProGluB-1* did alter spatial expression. Reducing either RISBZ1 or RPBF decreased OsLKR/SDH levels, resulting in an increase in free lysine content in rice grain. This result was in contrast to the fact that a significant reduction of SSP was observed only when these transcription factors were simultaneously reduced, suggesting that RISBZ1 and RPBF regulate SSP genes and *OsLKR/SDH* with high and limited redundancy, respectively. The same combinations of TF and *cis*-elements are involved in the regulation of *OsLKR/SDH* and SSP genes, but there is a distinct difference in their regulation mechanisms.

Nutritional and physical quality of rice grains is a complex agronomic trait and is determined by a number of factors, such as the amount of storage compounds, texture, appearance and size. The three major storage compounds in the endosperm arestarch, seed storage protein (SSP) and lipid. In cereals, lysine and tryptophanare the major limiting essential amino acids. In cereal seeds, most amino acids are usually incorporated into proteins, especially SSP, in contrast to free amino acids that are not incorporated into proteins (Kawakatsu and Takaiwa, (2010). Sinceenzymes involved in amino acid metabolism and their regulation mechanisms have been elucidated, it is now possible to manipulate amino acid metabolism by gene engineering. Free forms of deficient essential amino acids have been enhanced by manipulating the genes encoding key steps in amino acid synthetic pathways through up-regulation of feedback-insensitive enzymes or suppression of catabolism (Ufaz and Galili, 2008). OPAQUE2 (O2) is a basic leucine zipper (bZIP) transcription factor (TF) regulating the zein genes, ribosome-inactivating factor, and improving the deficient essential amino acids in cereal seeds is a major target of plant breeding.

For the purpose of improving the nutritional value of rice endosperm, we need to increase the content of lysine and threonine at the amino acid level. Schaeffer and Sharpe (1987) reported the high lysine mutants selected from cultured cells by cell selection. The N-methyl-N-nitrosourea (MNU) induction method developed by Satoh and Omura (1979) induces many kinds of mutants in high frequencies. Therefore, it may be possible to induce mutants for altered amino acid content by MNU treatment. It was searched for amino acid mutants in mutagenised materials.

A screening of about 360 endosperm mutants lines (EM lines), which were mutagenised by treating fertilized egg cells in rice cultivars, Kinmaze and Taichung 65 were undertaken. The total amino acids extracted from the rice grain powder of each line were analyzed by the high performance liquid chromatography system using the amino acid analysis column.

TABLE 7.6 Lysine and histidine contents in high lysine lines and original strains

Line	Amino Acid Content (%)
	Lysine Histidine
EM 109	5.42(0.04) 3.77(0.85)
EM 137	5.58(1.86) 3.94(1.07)
EM 139	5.44(0.60) 3.91(1.84)
EM 143	5.76(1.06) 4.04(0.66)
EM 246	5.42(0.52) 3.94(0.57)
EM 253	5.14(0.17) 4.10(0.76)
EM 279	5.19(0.29) 4.19(0.69)
EM 280	5.65(0.29) 3.93(1.25)
EM 317	6.38(1.44) 4.32(1.97)
EM 392	5.10(0.27) 4.38(0.95)
Kinmaze	4.23(0.44) 1.03(1.22)
EM 579	5.67(0.08) 4.00(0.67)
EM 596	5.24(0.39) 3.17(0.10)
Taichung 65	4.01(0.35) 1.21(0.13)

The values shown in parentheses stand for the standard values. The original variety of EM 579 and EM 596 is Taichung 65, while that of the others is Kinmaze.

As the result of screening, 12 high lysine mutant lines could be selected. The lysine contents of the original varieties and the high lysine lines are shown in Table 7.6. The lysine content in the high lysine mutants was 5.10-6.8%, while those of Kinmaze and Taichung 65, were 4.23% and 4.10%, respectively. The histidine content in high lysine lines was 3.17-4.30%, while it was 1.30% and 1.21% in Kinmaze and Taichung 65, respectively (Table 7.6). The resolution of the relationship between high lysine and high histidine needs to be investigated.

The endosperms of all the high-lysine mutant lines were mostly of floury appearance. The genetic relationship between lysine content and floury endosperm was investigated in the cross between Kinmaze and EM-317. The F_2 and parental seeds were analyzed for amino acid content on a single grain basis.

In F_2, the normal and floury seeds were 62 and 21, respectively. The mode of segregation gave a good fit to the 3:1 ratio. Figure 7.14 shows the results of the amino acid analysis. In normal seeds, the lysine content varied from 5.60% to 7.40% with a mean of 6.29%. In floury seeds, the lysine content varied from 7.10% to 9.32% with a mean of 8.24%. The lysine content of Kinmaze and EM-317 were 5.56% and 7.99%, respectively. These results suggest that the high lysine content and floury endosperm in EM-317 are the result of pleiotropic gene action.

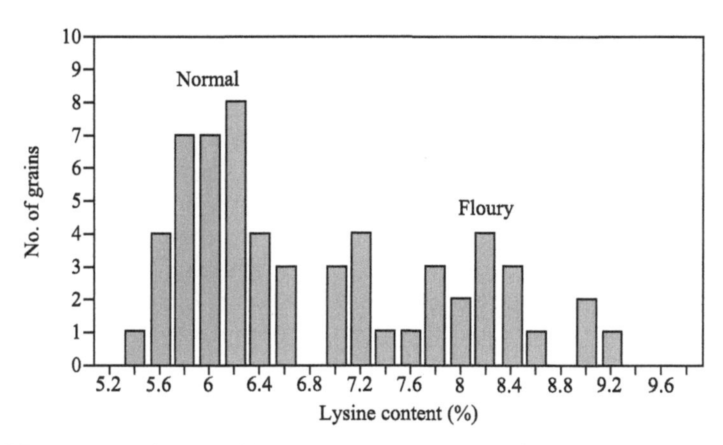

FIGURE 7.14 Distribution of lysine contents in the F_2 seeds from Kinmaze × EM-317.

With the aim of increasing the cysteine level in rice (*Oryza sativa* L.) and thus improving its nutritional quality, transgenic rice plants were generated expressing an *Escherichia coli* serine acetyltransferase isoform (*EcSAT*), the enzyme synthesizing *O*-acetylserine, the precursor of cysteine. The gene was fused to the transit peptide of the *Arabidopsis* Rubisco and driven by a ubiquitin promoter to target the enzyme to plastids. Twenty-two transgenic plants were examined for transgene protein expression, and five lines with a high expression level and enzymatic activity, respectively, were selected for further analysis. In these lines, the contents of cysteine and glutathione increased 2.4-fold and 2-fold, respectively. More important is the increase in free methionine and methionine incorporated into the water-soluble protein fraction in seeds. Free methionine increased in leaves up to 2.7-fold, in seeds up to 1.4-fold, and bound to seed proteins up to 4.8-fold, respectively, while the bound methionine level remained constant or even decreased in leaves. Notably, the transgenic lines exhibited higher isoleucine, leucine, and valine contents (each up to 2-fold depending on tissue, free, or bound), indicating a potential conversion of methionine via methionine γ-lyase to isoleucine. As the transgenic rice plants overexpressing *EcSAT* had significantly higher levels of both soluble and protein-bound methionine, isoleucine, cysteine and glutathione in rice they may represent a model and target system for improving the nutritional value.

Human beings are only capable of synthesizing ten of twenty naturally occurring amino acids. The other ***essential amino acids*** are obtained from the diet. Cereal grains are often limiting for lysine, tryptophan and threonine, while the legume seeds have an adequate level of lysine but are limiting for the sulphur containing amino acids, methionine and cysteine. Animals can convert methionine into cysteine, but not the reverse. Low level of methionine in plants diminishes their value as a source of dietary protein for human and animals. There are several attempts to improve the methionine level in plants. This study gives an overview of various technology for enhancement of methionine level in plants, including traditional plant breeding methods and selection of mutant; synthesis of an artificial gene rich in methionine and cysteine residue; genetic modification to increase methionine storage in protein; genetic modification to increase methionine biosynthesis and co-expressing methionine-rich storage proteins with enzymes that lead to high soluble methionine level, with minimal interference on plant growth, phenotype and productivity. The studies have resulted in the identification of steps important for the regulation of flux through the pathways and for the production of transgenic plants having increased free and protein bound methionine. The goal of increasing methionine content, and therefore nutritive value, of plant protein is presently being achieved and will no doubt continue to progress in the near future. Seeds are major sources of dietary protein for large vegetarian populations around the world and intensively farmed animals. Though, the protein in seeds can have a skewed amino acid composition due to the high abundance of a limited number of individual seed storage proteins. Deficiency of certain essential amino acids can be a cause of malnutrition in countries that are dependent on a diet of lowdiversity and can limit the efficiency of animal production.

Recent developments in Recombinant DNA Technology, plant tissue culture and *in vitro* regeneration are proposing new ways of increasing the level of essential amino acids, including methionine, by manipulating existing genes and/or introducing foreign genes into plants. Three important metabolic functions of methionine are: (*1*) trans-methylation to form a primary methyl donor, *S*-adenosylmethionine (SAM), which methylates compounds to form products such as creatine and phosphatidylcholine, follow-on in the product of methylation homocysteine being produced; SAM can also form decarboxylated SAM, which then provides an aminopropyl group for polyamine synthesis after which methionine will be reproduced; SAM also influence DNA synthesis and repairs the expression of genes; its deficiency has been associated with DNA fragmentation and strand breaks; (2) transsulfuration to form cysteine, which in turn is incorporated into glutathione or catabolized to taurine; and (3) protein synthesis. Methionine can also go intothe body pool by re-methylation of homocysteine or protein breakdown.

In developing countries in which plant-derived foods are predominant, low level of methionine can lead to non-specific signs of protein deficiencies in humans, such as decreased blood proteins, lowered resistance to disease and methylation retarded disorders, such as fatty liver, atherosclerosis, neurological disorders and tumorigenesis, retarded mental and physical development in young children. In animals, methionine depletion lowers the thresholdof chemical-induced toxicity, suggesting that this may be significant in carcinogenesis processes. A prominent role of methionine is its powerful antioxidant action against free radicals produced in the natural metabolic processes of the body. Methionine is also an excellent source for the essential mineral sulphur, which quickly inactivates free radicals produced in the body. Patients of Gilbert's syndrome, which results in an abnormality of liver functioning, are also benefited by supplements of the amino acid methionine. It is also required during the synthesis of collagen, nucleic acids and different proteins found in almost every cell of the human body as well as it is a constituent of many enzymes and proteins found in different parts of the body. The amino acid methionine reduces the level of histamine present in the body which is very useful for people affected by schizophrenia and related conditions, in which the levels of histamine are generally higher than those found in normal healthy adults.

It also promotes the excretion of estrogen from the body of women. In the body, methionine can be converted into the amino acid cysteine, which itself is a precursor of the vital compound called glutathione. Glutathione is a vital neutralizer of toxins present in the liver; the chemical thus protects the liver from the damaging effects of toxic compounds produced as a result of general metabolism. Glutathione is thus afforded a level of protection by methionine, as levels of methionine inhibit the depletion of glutathione when the body becomes overloaded with accumulated toxins and chemicals. It is also believed that glutathione carries nutrients to lymphocytes and phagocytes, important immune system cells. The levels of the neurotransmitting substances such as dopamine, nor-epinephrine and epinephrine are increased by methionine. Methionine is also used to bring relief from chronic pain, controlling hypertension, lower the potency of allergic symptoms, as an aid to reduce all kinds of inflammation, to lower cholesterol level and to protect the person from the bad effects of aspirin and related chemicals.

Food sources which are abundant in methionine include foods such as beans, various lentils, eggs, fish, meat, onions, garlic, soybeans, seeds and yogurt. Methionine is used by the body to synthesize a particular molecular choline–a brain food. Diets must be supplemented either with choline or lecithin (another compound high in choline) so as to ensure an adequate supply of methionine at all times. Daily amino acid requirements of a person may be determined by their body weight and this requirement spans a range of values for different body types. About 800-1000 mg of methionine is required by an average sized adult per day; this is an amount of the amino acid that is exceeded by the total methionine intake found in the majority of diets in the western world.

7.19. Methionine Biosynthesis and Metabolism in Plants

In plants and micro-organisms, methionine is synthesized via a pathway that uses both aspartic acid (figure below) and cysteine. Methionine is derived from cysteine by the sequential action of three enzymes, the first of which, cystathionine gamma-synthase (CGS), combines *O* phosphohomoserine from the aspartate amino acid pathway and cysteine. *O*-phosphohomoserine (OPHS), which is a common substrate for both threonine synthase (TS) and cystathionineg-synthase (CgS). OPHS is directly converted to threonine by TS, while methionine is synthesised in three steps. Condensation of cysteine and OPHS is catalysed by CgS resulting in cystathionine, which is subsequently converted to homocysteine by cystathione b-lyase, and methionine by methionine synthase. SMM is synthesized from Methionine by Met *S*-methyltransferase (MMT) and is recycled back Methionine by homo-Cystein *S*-methyltransferase (HMT).

Virtually all eukaryotic a-tubulins harbour a C-terminal tyrosine that can be reversibly removed and relegated, catalyzed by a specific tubulinï-tyrosine carboxypeptidase (TTC) and a specific tubulinï-tyrosine ligase (TTL), respectively. The biological function of this post-translational modification has remained enigmatic. 3-nitro-L-tyrosine (nitrotyrosine, NO2Tyr), can be incorporated into detyrosinated a-tubulin instead of tyrosine, producing irreversibly nitrotyrosinated a-tubulin. To gain insight into the possible function of detyrosination, the effect of NO2Tyr has been assessed in two plant model organisms (rice and tobacco). NO2Tyr causes a specific, sensitive, and dose-dependent inhibition of cell division that becomes detectable from 1 h after treatment and which is not observed with non-nitrosylated tyrosine. These effects are most pronounced in cycling tobacco BY-2 cells, where the inhibition of cell division is accompanied by a stimulation of cell length, and a misorientation of cross walls. NO2Tyr reduces the abundance of the detyrosinated form of a-tubulin whereas the tyrosinated a-tubulin is not affected. These findings are discussed with respect to a model where NO2Tyr is accepted as substrate by TTL and subsequently blocks TTC activity. The irreversibly tyrosinated a-tubulin impairs microtubular functions that are relevant to cell division in general, and cell wall deposition in particular.

Tobacco, rice, carrot and tomato tissue cultures were grown in liquid media containing L-phenylalanine or L-tyrosine, or both together. The addition of these amino acids increased their respective cellular levels (4–20 fold), but did not lower the level of chorismate mutase, an enzyme in the biosynthetic pathway of phenylalanine and tyrosine. These results indicate that the biosynthesis of phenylalanine and tyrosine in cultured plant cells is not regulated by repression of the synthesis of chorismate mutase by phenylalanine or tyrosine.

The aromatic amino acids phenylalanine, tyrosine and tryptophan in plants are not only essential components of protein synthesis, but also serve as precursors for a wide range of secondary metabolites that are important for plant growth as well as for human nutrition and health. The aromatic amino acids are synthesized via the shikimate pathway followed by the branched aromatic amino acids biosynthesis pathway, with chorismate serving as a major intermediate branch point metabolite. Yet, the regulation and coordination of synthesis of these amino acids are still far from being understood. Recent studies on these pathways identified a number of alternative cross-regulated biosynthesis routes with unique evolutionary origins. Although the major route of Phe and Tyr biosynthesis in plants occurs via the intermediate metabolite arogenate, recent studies suggest that plants can also synthesize phenylalanine via the intermediate metabolite phenylpyruvate (PPY), similarly to many microorganisms. Recent studies also identified a number of transcription factors regulating the expression of genes encoding enzymes of the shikimate and aromatic amino acids pathways as well as of multiple secondary metabolites derived from them in *Arabidopsis* and in other plant species.

Metabolic manipulation of plants to improve their nutritional quality is an important goal of plant biotechnology. Expression in rice (*Oryza sativa* L.) of a transgene (OASA1D) encoding a feedback-insensitive subunit of rice anthranilate synthase results in theaccumulation of tryptophan (Trp) in calli and leaves. It is shown that the amount of free Trp in the seeds of such plants is increased by about two orders of magnitude compared with that in the seeds of wild-type plants. The total Trp content in the seeds of the transgenic plants was also increased. Two homozygous lines, HW1 and HW5, of OASA1D transgenic rice were generated for characterization of agronomic traits and aromatic metabolite profiling of seeds. The marked overproduction of Trp was stable in these lines under field conditions, although spikelet fertility and yield, as well as seed germination ability, were reduced compared with the wild type. These differences in agronomic traits were small, however, in HW5. In spite of the high Trp content in the seeds of the HW lines, metabolic profiling revealed no substantial changes in the amounts of other phenolic compounds. The amount of indole acetic acid was increased about 2-fold in the seeds of the transgenic lines. The establishment and characterization of these OASA1D transgenic lines have thus demonstrated the feasibility of increasing the Trp content in the seeds of rice (or of other crops) as a means of improving its nutritional value for human consumption or animal feed (Wakasa et al. 2006).

The ability to increase the level of Trp in food crops by commonly synthesizing tryptophan from shikimic acid or anthranilate. The latter condenses with phosphoribosylpyrophosphate (PRPP), generating pyrophosphate as a by-product. After ring opening of the ribose moiety and following reductive decarboxylation, indole-3-glycerinephosphate is produced, which in turn is transformed into indole. In the last step, tryptophan synthase catalyzes the formation of tryptophan from indole and the amino acid serine.

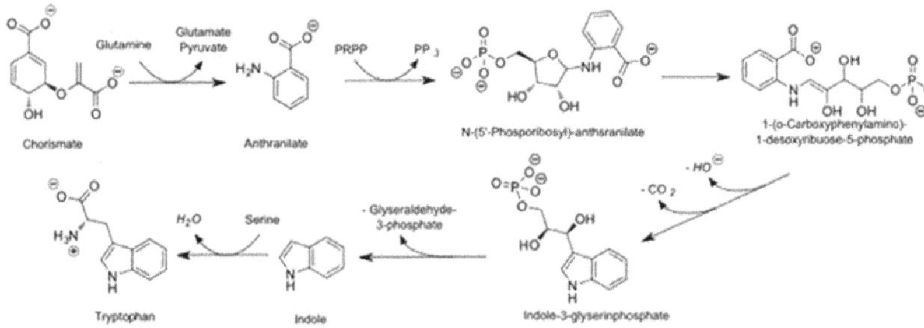

FIGURE 7.15 Tryptophan biosynthesis.

Main pathway to *de novo* biosynthesis of serine starts with the glycolytic intermediate 3-phosphoglycerate. An NADH-linked dehydrogenase converts 3-phosphoglycerate into a keto acid, 3-phosphopyruvate, suitable for subsequent transamination. Aminotransferase activity with glutamate as a donor produces 3-phosphoserine, which is converted to serine by phosphoserine phosphatase.

As indicated in figure 7.16, serine can be derived from glycine (and *vice versa*) by a single step reaction that involves serine hydroxymethyltransferase and tetrahydrofolate (THF).

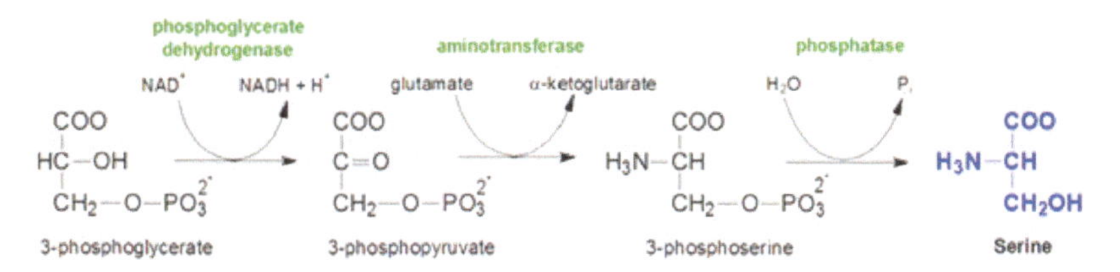

FIGURE 7.16 Serine biosynthesis.

The branched-chain amino acids (BCAAs) valine, leucine and isoleucine are essential amino acids that play critical roles in animal growth and development. Animals cannot synthesize these amino acids and must obtain them from their diet. Plants are the ultimate source of these essential nutrients, and they synthesize BCAAs through a conserved pathway that is inhibited by its end products. This feedback inhibition has prevented scientists from engineering plants that accumulate high levels of BCAAs by simply over-expressing the respective biosynthetic genes. To identify components critical for this feedback regulation, genetic screening had been carried out for Arabidopsis mutants that exhibit enhanced resistance to BCAAs. Multiple dominant allelic mutations in the VALINE-TOLERANT 1 (VAT1) gene were identified that conferred plant resistance to valine inhibition. Map-based cloning revealed that VAT1 encodes a regulatory subunit of acetohydroxy acid synthase (AHAS), the first committed enzyme in the BCAA biosynthesis pathway. The VAT1 gene is highly expressed in young, rapidly growing tissues. When reconstituted with the catalytic subunit *in vitro*, the *vat1* mutant containing AHAS holoenzyme exhibits increased resistance to valine. Importantly, transgenic plants expressing the mutated vat1 gene exhibit valine tolerance and accumulate higher levels of BCAAs. Studies not only uncovered regulatory characteristics of plant AHAS, but also identified a method to enhance BCAA accumulation in crop plants that will significantly enhance the nutritional value of food and feed(Chen et al. 2010).

7.20. Branched Chain Amino Acids

Branched-chain amino acids (BCAAs) are essential amino acids that play important roles in protein anabolism and neurotransmitter biosynthesis in animals. Because animals lack the ability to synthesize BCAAs, they must acquire them from their diet. Plants are the ultimate source of these essential amino acids for animals. ALS is the first enzyme in the pathway for the biosynthesis of BCAAs; ALS catalyzes the condensation of two pyruvate molecules to form acetolactate (a precursor of valine and leucine), and the condensation of pyruvate and α-ketobutyrate to yield acetohydroxybutyrate (a precursor of isoleucine). Plant ALS is inhibited by all three BCAAs, with valine and leucine being particularly potent. As the key enzyme in BCAA biosynthesis and a primary target site of action for at least four structurally distinct classes of herbicides (sulfonylureas, imidazolinones, triazolopyrimidine sulfonamides and pyrimidinyl carboxy herbicides), ALS has been well characterized in certain organisms. To date, most ALS enzymes that have been characterized have both a catalytic subunit (65 kDa) and a smaller regulatory subunit, which varies in size between 9 and 54 kDa, depending on the species of origin. The plant ALS regulatory subunit has been suggested to be able to stabilize and enhance catalytic subunit. The ALS regulatory subunit also mediates end-product inhibition by BCAAs through a complex domain interaction.

A major advantage of ALS-inhibiting herbicide compounds is that they are non-toxic to animals, highly selective, and very potent, thereby requiring only low ap- plication rates. Thus, ALS-inhibiting herbicides are an essential part of the multibillion-dollar weed-control market. Because ALS-inhibiting herbicides control a broad spectrum of grass and broadleaf weeds, including weeds that are closely related to the crop itself and some key parasitic weeds, several herbicide-resistant plant mutants have been screened and herbicide-resistant mutations are well studied. Because herbicide-resistance and BACC accumulation are favorable traits for cultivation and nutrition, respectively, modification of ALS is currently a hot topic in crop molecular breeding, and detailed analysis of ALS-modified crops is important.

In a previous study, we succeeded in creating herbicide-resistant rice plants in which point mutations—a tryptophan (TGG) to leucine (TTG) change at amino acid 548 (W548L), and a serine (AGT) to isoleucine (ATT) change at amino acid 627 (S627I)—were introduced into the endogenous ALS catalytic subunit using a homolo- gous recombination-dependent gene targeting system. These double mutations in the rice ALS gene represented a novel combination of spontaneous mutations. Although each individual amino acid change in ALS resulted in a phenotype tolerant to the sulfonylurea herbicide bispyripac-sodium, conversion of both amino acids conferred increased tolerance to bispyripac-sodium when expressed in *Escherichia coli* . To demonstrate the effects of W548L and S627I mutations on BCAA synthesis and feedback regulation by BCAAs in rice, we investigated the activity of the ALS holoenzyme in the presence of BCAAs and measured amino acid content in ALS modified rice plants.

To understand the effect of the W548L/S627I double mutation on feedback regulation of ALS by BCAAs, it was analyzed ALS enzymatic activity in the presence of valine and leucine. Protein extracts from wild-type and ALS mutants [W548L/S627I (m)], in which W548L and S627I double mutations were introduced into the ALS catalytic subunit, as well as extracts of plants overexpressing ALS harboring the double mutation [W548L/S627I (ox)] were examined for their sensitivities to valine and leucine. The

ALS activity of wild-type was inhibited by 65% in the presence of 1 mM valine and leucine, compared with the activity apparent in the absence of these amino acids. In contrast, the inhibition rates of ALS activity in W548L/S627I (m) and W548L/S627I (ox) were 22% and 53%, respectively. In W548L/S627I (ox) plants, the mutated ALS gene is driven by a strong 35S promoter, thus the transcriptional level of this mutated ALS gene is much higher than that in wild-type or W548L/S627I (m) (approximately 20-fold higher than wild-type, data not shown). Regardless of the fact that the proportion of wild-type ALS transcript may be less than 5% that in W548L/S627I (ox) plants,

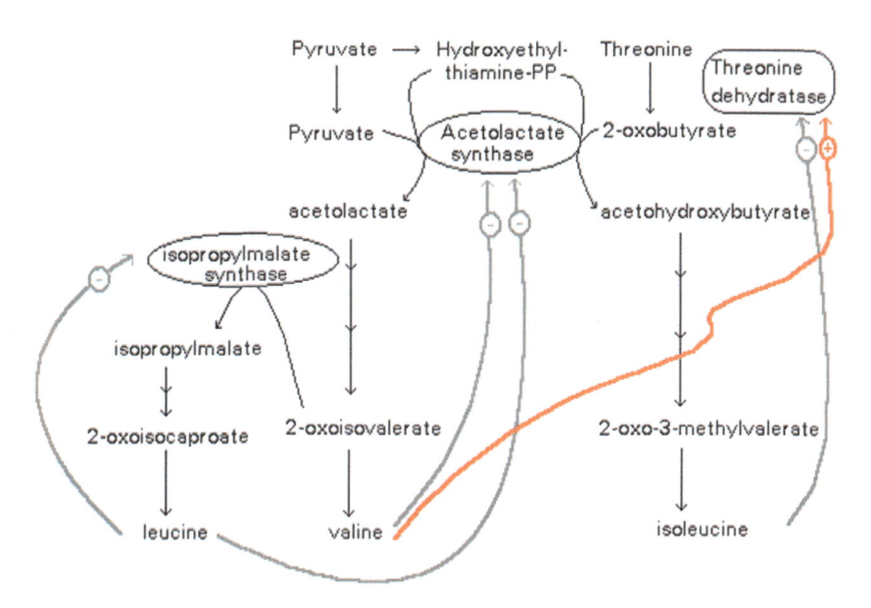

FIGURE 7.17 Feedback inhibition of valine, leucine and isoleucine biosynthesis Feedback control of he branched chain amino acid biosynthesis.

In plants and microorganisms, it is synthesized via several steps, starting from pyruvic acid and alpha-ketoglutarate. Enzymes involved in this biosynthesis include:

1. Acetolactate synthase (also known as acetohydroxy acid synthase)
2. Acetohydroxy acid isomeroreductase
3. Dihydroxyacid dehydratase
4. Valine aminotransferase
5. The inhibitory effects of isoleucine, leucine and valine single and in combination on the activity of acetohydroxyacid synthetase has been determined using extracts of a range of higher plants. In all

cases leucine and valine were significantly inhibitory on their own, but more inhibitory when supplied together. There is evidence that all the plants tested contain an acetohydroxyacid synthetase that is regulated by co-operative feedback by leucine and valine in a manner similar to that previously reported for barley. The possibility that the branched-chain amino acids could repress the synthesis of the enzyme was also tested, but no evidence for repression was found.

Valine and isoleucine biosynthesis

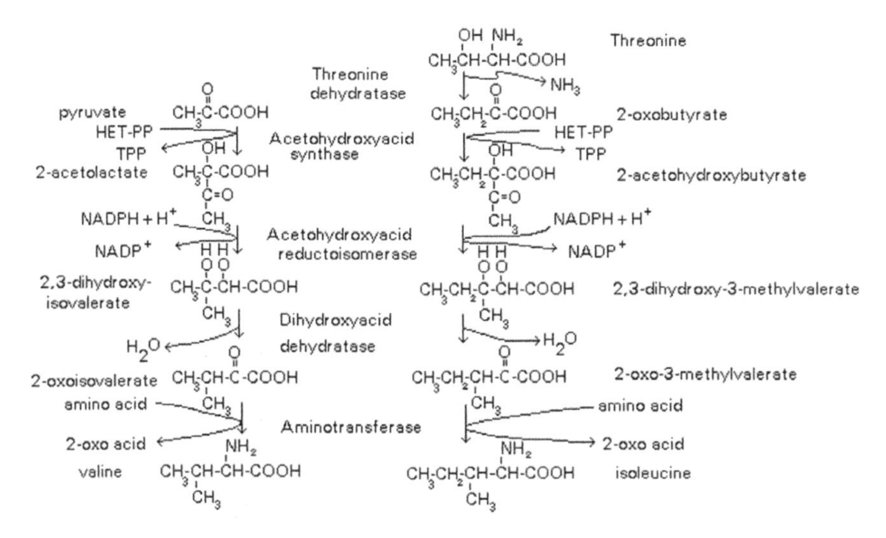

FIGURE 7.18 Valine and Isoleucine Biosynthesis.

Acetolactate synthase (ALS) is a thiamine diphosphate-dependent enzyme in the biosynthetic pathway leading to isoleucine, valine and leucine in plants. ALS is the target of several classes of herbicides that are effective to protect a broad range of crops. In this study, the functional analysis of a gene encoding for ALS from rice (*OsALS*) was described. Sequence analysis of an EST from rice revealed that it harbors a full-length open reading frame for *OsALS* encoding a protein of approximately 69.4 kDa and the N-terminal of *OsALS* containsa feature of chloroplast transit peptide. The predicted amino acid sequence of *OsALS* is highly homologous to those of weed ALSs among plant ALSs. The *OsALS* expression showed that the gene was functionally capable of complementing the two *ilvH* mutant strains of *Escherichia coli*. These results indicate that the *OsALS* encodes for an enzyme in acetolactate synthase in rice.

Rice (*Oryza sativa* L.) is one of the most widely grown crops and a main staple food for about half the world's population. The nutritional quality of rice grain is important to all rice consumers, especially where it is the population's main staple. The contents of protein and amino acids are the major factors of nutritional quality, and their regulation has increasingly become a major breeding objective. It was reported that rice protein content (PC) varied from 4.9% to 19.3% in indica and from 5.9% to 16.5% in japonica. The genetic variation in PC provides a basis for breeding PC. However, manipulating this trait in traditional breeding is difficult because such substantial variation is quantitatively inherited. With the advent of molecular markers such as restriction fragment length polymorphisms (RFLPs) and microsatellites and their maps, quantitative traitlocus or loci (QTL) that control quality traits can be dissected. Several recent studies have been undertaken to decipher the genetic basis of PC in rice by QTL mapping Hu et al. 1992, providing useful information for improving the nutritional quality of rice. However, some problems still remain. In previous studies, PC was calculated from Kjeldahl nitrogen multiplied by a factor of 5.95, which is based on the nitrogen content (16.8%) of the major rice protein glutelin, so it can only indicate the relative quantity of protein. In fact, the amino acids in total protein of grain can be unbalanced because certain essential amino acids are extremely low. Lysine (Lys) and threonine (Thr) were identified as the first and second limiting essential amino acids omilled rice protein, based on human requirements as estimated by the World Health Organization. Therefore, special attention should be paid to the quality of rice protein, which is improving usually represented by essential amino acid content (AAC) and essential amino acid index (FAO, 1970). Unfortunately, little is known about the genetic basis of the AAC in rice, particularly the genes associated with their metabolism pathways. The present study uses a recombinant inbred line (RIL) population planted in two different years to dissect the genetic basis of the AAC in milled rice as illustrated by each amino acid, the total contents of essential amino acids and all amino acids. The results will help rice breeders develop strategies for improving the quality of rice grain.

During the model simulations, the rice cells were freely allowed to consume Gln and Asn as nitrogen sources for the amino acid biosynthesis (Bewley and Black, 1994). Under both aerobic and anaerobic conditions, the consumed Gln and Asn were completely converted into Glu and Asp, and subsequently into α-K G and O A A, for amino acid synthesis. Furthermore, the amounts of Asn consumed in aerobic conditions were reasonably higher, since an enhanced pyruvate pool facilitated the amino acid biosynthesis. Our simulation results also highlighted the functional ability of rice to synthesize all amino acids via biosynthetic pathways even under anoxia, rather than protein degradation, as often speculated.

Interestingly, the flux analysis also suggested the possibility for GABA, a nonprotein amino acid, to be synthesized under both aerobic and anaerobic conditions. As mentioned previously, although the synthesis of GABA and its subsequent utilization in the GABA shunt did not influence cellular growth under aerobic conditions, it enhanced the growth rate slightly under anaerobic conditions, owing to the crucial role in Gly biosynthesis. Under anoxia, GABA is first synthesized from Glu by Glu decarboxylase (GAD) and then converted into succinate via GABA aminotransferase and succinic semialdehyde dehydrogenase (SSADH). During these conversions, NADH is liberated in the SSADH step and recycled via a series of enzymes including Ser hydroxymethyltransferase (SHM1), producing net amounts of Gly. These observations are in very good agreement with earlier experiments (Shingaki-Wells et al. 2011) that reported the anaerobic accumulation of GABA along with an increase in the expression of SHM1.

References

Allan, E.F. and A.J. Trewavas. 1987. The role of calcium in metabolic control. In : The Biochemistry of Plants (Ed DD Davies) vol. 12 Academic Press, New York pp. 117-149.

Arazi, T.,G. Baum, W.A. Sneddon, B.J. Shelp and H. Fromm 1995. Molecular and biochemical analysis of calmodulin interactions with the calmodulin-binding domain of plant glutamate decarboxylase. Plant physiol. **108:** 551-561.

Armengaud, P., L. Thiery, N. Buhot, G. DeMarch and A. Savoure. 2004. Transcriptional regulation of proline biosynthesis in Medicago turncatulareveals developmental and environmental specific features. *Physiol. Plant,* **120:** 442-450.

Arruda, P., L. Sodek and W.J. Da Silva. 1982. Lysine-Ketoglutarate reductase activity in developing maize endosperm. *Plant Physiol.,* **69:** 988-989.

Arruda, P. and W.J. Da Silva. 1983. Lysine-Ketoglutarate reductase activity in maize – its possible role in lysine metabolism of developing endosperm. *Phytochemistry,* **22:** 206-208.

Arruda, P., E.L. Kemper, F. Papes and A. Leite. 2000. Regulation of lysine catabolism in higher plants. *Trends Plant Sci.,* **5:** 324-330.

Ashraf, M. and M.R. Foolad. 2007. Roles of glycine betaine and proline in improving plant biotic stress resistance. *Environ. Exp.Bot.* **59:** 206-216.

Aurisano, N., A. Bertani and R. Reggiani. 1995. Involvement of calcium and calmodulin in protein and aminoacid metabolism in rice roots under anoxia. *Plant Cell Physiol.* **36:** 1525-1529.

Azevedo, R.A., P. Arruda, W.L. Turner and P.J. Lea. 1997. The biosynthesis and metabolism of the aspartate derived amino acids in higher plants. *Phytochem.* **46:** 395-419

Azevedo, R.A. and P.J. Lea. 2001. Lysine metabolism in higher plants. *Amino Acids,* **20:** 261-279.

Azevedo, R.A. 2002. Analysis of the aspartic acid metabolic pathway using mutant genes. *Amino Acids.,* **22:** 217-230.

Azevedo, R.A., M. Lancien and P.J. Lea. 2006. The aspartic acid metabolic pathway: an exciting and essential pathway in plants. *Amino Acids,* **30:** 143-162.

Bacilio-Jimenez, M., S. Aguilar-Flores, E. Ventura-Zapata, E. Perezcapos, S. Bouquelet and E. Zenteno. 2003. Chemical characterization of root exudates from rice (Oryza sativa) and their effects on the chemotactic response of endophytic bacteria. *Plant and Soil.,* **249:** 271-277.

Baum, G. and Y. Fridmann. 1996. Calmodulin binding to glutamate decarboxylase is required for regulation of glutamate and GABA metabolism and normal development in plants. *EMBO J.*, **15:** 2988-2996.

Beatty, P.H., A.K. Shrawat, R.T. Carroll, T. Zhu and A.G. Good. 2009. Transcriptome analysis of nitrogen efficient rice under over-expressing alanine aminotransferase. *Plant Biochem.J.*, **7:** 562-576.

Brandt, A. 1975. Invivo incorporation of 14C lysine into the endosperm of proteins of wild type and high lysine barley. *FEBS Letters*, **52:** 288-291.

Breitkreuz, K.E. and B.J. Shelp. 1995. Subcellular compartmentation of the 4-aminobutyrate shunt in protoplasts from developing soybean cotyledons. *Plant Physiol.*, **108:** 99-103.

Brinch-Pederson, H., G. Galili, S. Knudsen and P.B. Holm. 1996. Engineering of the aspartate family biosynthetic pathway in barley (Hordeum Vulgare L.) by transformation with heterologomgenes encoding feedback–Insensitive aspartate kinase and dehydrodipicdinate synthase. *Plant Mol.Biol.*, **32:** 611-620.

Brochetto-Braga, M.R., A. Leite and P. Arruda. 1992. Partial purification and characterization of lysine-ketoglutarate reductase in normal and Opaque 2 maize endosperm. *Plant Physiol.*, **98:** 1139-1147.

Bush, D.S. 1995. Calcium regulation in plant cells and its role in signalling. Annu. Rev. *Plant Physiol.* Plant Mol. Biol. **46:** 95-122

Chen, Z., T.A. Cuin, M. Zhou, A. Twomey, B.P. Naidu and S. Shabala. 2007. Compatible solute accumulation and stress mitigating effects in barley genotypes contrasting in their salt tolerance. *J.Exp.Bot.*, **58:** 4245-4255.

Chen, H., K. Saksa, F. Zhao, J. Qiu and L. Xiong. 2010. Genetic analysis of pathway regulation for enhancing branched chain amino acid biosynthesis in plants. *Plant J.*, **63:** 573-583.

Chen, T.H.H. and N. Murata. 2002. Enhancement of tolerance of abiotic stress by metabolic engineering of betaines and other compatible solutes curr. Opin. Plant Biol. **5:** 250-257

Chinnasamy, G. 2005. A proteomics perspective on biocontrol and plant defense mechanism. *In:* Siddiqui, Z.A. (Ed) PGPR: Biocontrol and biofertilization. pringer, Dordrchet, Netherlands. pp. 233-256.

Chiba, Y., M. Ishikawa, F. Kijima, R.H. Tyson, J. Kim, A. Yamamoto, E. Mambra, T. Leustek, R.M. Wallsgrove and S. Naito. 1999. Evidence for autoregulation of cystathionine synthase mRNA stability in Arabiidopsis. *Science*, **285:** 1371-1374.

Choudhary, N.L., R.K. Sairam and A. Tyagi. 2005. Expression of delta 1-pyrroline-5-carboxylate synthetase gene during drought in rice (Oryzasativa L.). *Ind.J. Biochem., Biophys.* **42:** 366-370.

Cunha-Lima, S.T., R.A. Azevedo, L.G. Santoro, S.A. Gaziola and P.J. Lea. 2003. Isolation of the bifunctional enzyme lysine-Ketoglutarate reductase –saccharopine dehydrogenase from phaseolus vulgaris. *Amino Acids*, **24:** 179-186.

Crawford, L.A., A.W. Brown, K.E. Breitkreuz and P.C. Guinel. 1994. The synthesis of gamma–aminobutyric acid in response to treatments reducing cytosolic pH. *Plant Physiol.*, **104:** 865-871.

Creighton, T.H. 1993. Proteins: Structures and molecular properties; WH Freemman, *Sano.Franciso.*

Csonka, L.N., S.B. Gelvin, B.W. Goodner, C.S. Orser, D. Siemieniak and J.L. Slighton. 1988. Nucleotide sequence of a mutation in the proB gene of Escherichia coli that confers proline overexpression and enhanced tolerance to osmotic *stress.Gene*, **64:** 199-205.

Csonka, L.N. and A.D. Hanson. 1991. Prokaryotic osmoregulation: Genetics and physiology. *Annu. Rev. Microbiol.*, **45:** 569-606.

DasGupta, D.K. and P. Basuchaudhuri. 1974. Effect of molybdenum on nitrogen metabolism of rice. *Expt.Agric.*, **10:** 251-255.

Dey, M. and S. Guha-Mukherjee. 1999. Phytochrome activation of aspartate kinase in etiolated chickpea (*Cecer arietinum*) seedlings. *J.Plant Physiol.*, **154:** 454-458.

Delauney, A.J., C. Hu, K. Kishor and D.P.S Verma. 1993. Cloning of ornithine delta-aminotransferase cDNA from Vigna aconitifolia by trans complementation in Escherichia coli and regulation of proline biosynthesis *J.Biol.Chem.*, **268:** 18673-18678.

Delauney, A.J. and D.P.S. Verma. 1993. Proline biosynthesis and osmoregulation in plants. *Plant J.*, **4:** 215-223.

Deuschle, K.,D. Funck, H. Hellmann and K. Daschner. 2001. A nuclear genen-coding mitochondrial delta-pyrroline-5-carboxylate dehydrogenase and its potential role in protection from proline toxicity. *Plant J.*, **27:** 345-356.

Fabro, G., I. Kovacs, V. Pavet, L. Szabados and M.E. Alvarez. 2004. Proline accumulation and At P5CS2 gene activation are induced by plant pathogen incompatible interactions in Arabidopsis. *Mol.Plant Microbe Interact.*, **17:** 343-350.

Falco, S.C. 2001. Increasing lysine in corn. *Amino Acids,* **21:** 57-58.

Falco, S.C., T. Guida, M. Locke, J. Mauvais, C. Sanders, R.T. Ward and P. Webber. 1995. Transgenic canola and soybean seeds with increased lysine. *Biotechnology,* **13:** 577-582.

Feller, A., F. Ramos, A. Pierard and E. Dubois. 1999. In Sacharomyces cerevisae, feedback inhibition of homocitrate synthase isoenzymes by lysine modulates the activation of LYS gene expression by LYS14p. *Eur.J.Biochem,* **261:** 163-170.

Flores, H.E., C.M. Protacio and M.W. Signs. 1989. Primary and secondary metabolism of polyamines in plants. *In:* Plant Nitrogen Metabolism. J.E. Poulton, J.T. Romeo and E.E. Coun (Eds). Plenum Press, New York. pp.329-393.

Fjellstedt, T.A. and J.C. Robinson. 1975. Purification and properties of L-lysine alpha-ketoglutarate reductase from human placenta. *Arch. Biochem.Biophys.* **168:** 536-548.

Galili, G. 1995. Regulation of lysine and threonine synthesis. Plant cell **7:** 899-906.

Galili, G., G. Tang, X. Zhu and B. Gakiere. 2001. Lysine catabolism: a stress and development super regulated metabolic pathway. *Curr.Opin.Plant Biol.* **4:** 261-2661.

Gaziola, S.A., C.M.G. Teixeira, J. Lugli, L. Sodek and R. A. Azevedo. 1997. The enzymology of lysine catabolism in rice seeds. Isolation, characterization and regulatory properties of a lysine 2 oxoglutarate/saccharopine dehydrogenase bifunctional polypeptide. *Eur.J.Biochem.* **247:** 364-371.

Gaziola, S.A., L. Sodek, P. Arruda, P.J. Lea and R.A. Azevedo. 2000. Degradation of lysine in rice seeds : effect of calcium, ionic strength. S-adenosylmethionine and S-2-aminoethyl-1-cysteine on the lysine 2-oxoglutarate reductase-saccharopine dehydrogenase bifunctional enzyme *Physiol. Plant.* **110:** 164-171

Gaziola, S.A., E.S. Alessi, P.E.O. Guimaroes, C.Damerval and R.A. Azevedo. 1999. Quality protein maize : a biochemical study of enzymes involved in metabolism. *J. Agri. Food Chem.* **47:** 1268-1275.

Gong, M., A.H. vander Luit, M.R. Knight and A.J. Trewavas. 1998. Heat shock induced changes in intracellular Ca_{2+} level in tobacco seedlings in relation to thermo tolerance. *Plant Physiol.*, **116:** 429-437.

Goncalves-Butruille, M., P. Szajner, E. Torigoi, A. Leite and P. Arruda. 1996. Purification and characterization of the bifunctional enzyme lysine-keto-glutarate reductase –saccharopine dehydrogenase from maize. *Plant Physiol.* **110:** 765-771.

Gupta, N., A.K. Gupta, V.S. Gaur and A. Kumar. 2012. Relationship of nitrogen use efficiency with the activities of enzymes involved in nitrogen uptake and assimilation of finger millet genotypes grown under different nitrogen inputs. *Sci.World J.,* **10:**

Hakim, M.A., A.S. Juraimi, M.M. Hanafi, M. Ismail, A. Salawat, M.Y. Rafii and M.A. Latif. 2014. Biochemical and anatomical changes and yield reduction in rice (Oryzasativa L.) under varied salinity regimes. *BioMed Res.Int.*

Hare, P. and W. Cress. 1997. Metabolic implications of stress induced proline accumulation in plants. *Plant Growth Regul,* **21:** 79-102.

Haudecoeur, E., S. Planamente, A. Cirou, M. Tannieres, B.J. Shelp, S. Morera and D. Faure. 2009. Proline antagonizes GABA induced quenching of quorum-sensing in Agrobacterium tumefaciens. *Proc.Natl.Acad.Sci.USA.,* **106:** 14587-14592.

Hearl, W.G. and J.E. Churchich. 1984. Interactions between 4-aminobutyrate aminotransferase and succinic semialdehyde dehydrogenase, two mitochondrial enzymes. J.Bid Chem. **259:** 11459-11463.

Heremans, B. and M. Jacobs. 1994. Selection of Arabidopsis thaliana (L) Heynt mutants resistant to aspartate derived amino acids and analogues. *Plant Sci.,* **101:** 151-162.

Heremans, B. and M. Jacobs, 1997. A mutant of Arabidopsis thaliana (L). Heynt with modified control of aspartate kinase by threonine. *Biochem.Genet.* **35:** 139-153.

Hesse, H., O. Kreft, S. Maimann, M. Zeh, L. Willmitzer and R. Hoefgen. 2001. Approaches towards understanding methionine biosynthesis in higher plants. *Amino Acids,* **20:** 281-289.

Hinnebusch, A.G. 1988. Mechanism of gene regulation in the general control of amino acid biosynthesis in Saccharomyces cervisiae. *Microbiol. Rev.,* **52:** 248-273.

Hong, Z., K. Lakkineni, Z. Zhang and D.P.S. Verma. 2000. Removal of feed back inhibition of delta-1-pyrroline-5-carboxylate synthetase results in increased proline accumulation and protection of plants from osmotic stress. *Plant Physiol.,* **122:** 1129-1136.

Hu, C.A., A.J. Delauney and D.P.S. Verma. 1992. A bifunctional enzyme (delta-1-pyrroline-5-carboxylate synthetase) catalyzes the first two steps in proline biosynthesis in plants. *Proc.Natl.Acad. Sci.USA,* **89:** 9354-9358.

Huang, J., R. Hirji, L. Adam, K.L. Rozwadowski, J.K. Hammerlindl, R.A. Keller and G.Selvaraj.2000. Genetic engineering of glycinebetaine production toward enhancing stress tolerance in plants metabolic limitations. *Plant Physiol.* **122:** 747-756.

Igarashi, Y., Y. Yoshiba, Y. Sanada, K. Yamaguchi-Shinozaki, K. Wada and K. Shinozaki. 1997. Characterization of the gene for delta-1-pyrroline-5-carboxylate synthetase and correlation between the expression of the gene and salt tolerance in Oryzasativa. *Plant Mol.Biol:* **33:** 857-865.

Jander, G. and V. Joshi. 2010. Recent progress in deciphering the biosynthesis of aspartate derived amino acids in plants. *Mol.Plant,* **3:** 54-65.

Kawakatsu, T. and F. Takaiwa. 2010. Cereal seed storage protein synthesis: fundamental processes for recombinant protein production incereal grains. *Plant Biotechnol.J.,* **8:** 939-953.

Kathiresan, A., P. Tung, C.C. Chinnappa and D.M. Reid. 1997. Gamma amino butyric acid stimulates ethylene biosynthesis in sunflower. *Plant Physiol.,* **115:** 129-136.
Karchi, H., O. Shaul and G. Galili. 1994. Lysine synthesis and catabolism are coordinately regulated during tobacco seed development. *Proc. Natl. Acad.Sci. USA.,* **91:** 2577-2981.
Karchi, H.,D. Miron, S.Ben-Yaacov and G. Galili. 1995. The lysine-dependent stimulation of lysine catabolism in tobacco seeds requires calcium and protein phosphorylation *Plant cell* **7:** 1963-1973.
Kemble, A.R. and H. T. MacPherson. 1954. Liberation of amino acids in perennial ray grass during wilting. *Biochem J.* **58:** 46-59.
Kemper, E.L., G. Cord-Neto, A.N. Capella, M. Gongalves – Butruitte, R.A. Azevedo and P. Arruda. 1998. Structure and regulation of the bifunctional enzyme lysine-ketoglutarate reductase-saccharopine dehydrogenase in maize. *Eur. J. Biochem.,* **253:** 720-729.
Kemper, E.L., G. Cord-Neto, A.N. Capella, M. Goncalves-Butruitte, R.A. Azevedo and P. Arruda. 1999. The role of Opaque-2 on the control of lysine degrading activities in developing maize endosperm. *Plant Cell.,* **11:** 1981-1994.
Kido, E.A. J.R.C.F. Neto, R.L. Silva, L.C. Belarmino, J.P.B. Neto, N.M. Soares-cavalcanti, V. Pandolfi, M. D silva, A.L. Nepomuceno and A.M. Benko-Iseppon. 2013. Expression dynamics and genome distribution of osmoprotectants in soybean : identifying important components to face abiotic stress. BMC Informatics, **14:** 51-57
Kishor, P.B.K., Z. Hong, G.H. Miao, C.A.A. Hu and D.P.S. Verma. 1995. Over expression of delta-1-pyrroline-5-carboxylate synthetase increases proline production and confers osmotolerance in transgenic plants. *Plant Physiol.,* **1108:** 1387-1394.
Kiyosue, T., Y. Yoshida, K. Yamaguchi-Shinozaki and K. Shinozaki. 1996. A nuclear gene encoding mitochondrial proline dehydrogenase, an enzyme involved in proline metabolism, is upregulated by proline but down regulated by dehydration in Arabidopsis. *Plant cell,* **8:** 1323- 1335.
Kohl, D.H., K.R. Schubert, M.B. Carter, C.H. Hagedom and G. Shearer. 1988. Proline metabolism in N^{2-} fixing root nodules: Energy transfer and regulation of purine synthesis. *Proc.Natl.Acad.Sci/USA.,* **85:** 2036-2040.
Kumamaru, T., H. Sato and H. Satoh. 2012. High lysine mutant of rice. *Gramene,*
Laber, B., W. Maurer, S. Scharf, K. Stepusin and F.S. Schmidt. 1999. Vitamin B biosynthesis: Formation of pyridoxine-5-phosphate from 4 (Phosphohydoxy)-L-threonine and 1-deoxy –D-xylulose-5-phosphate by PdxA and PdxJ protein. *FEBS Letters,* **449:** 45-48.
Lakshmanam, M., Z. Zhang, B. Mohanty, J.Y. Kwou, H.Y. Choi, H.J. Nam, D. Kim and D.Y. Lee. 2013. Elucidating rice cell metabolism under flooding and drought stresses using flux based modeling and analysis. *Plant Physiol.,* **162:** 2140-2150.
Lam, H.M., J. Chiu, M.H. Hsieh, L. Meisel, I.C. Oliveira, M. Shin and G. Coruzzi. 1998. Glutamate receptor genes in plants. *Nature,* **396:** 125-126.
Lea, P.J., R.D. Blackwell and R.A. Azevedo 1992. Analysis of barley metabolism using mutant genes. *In:* Shewry (Ed) Barley: Genetics, biochemistry, molecular biology and biotechnology. CABI, Wallingford, pp. 181-208.
Lee, S.I., H.U. Kim, Y.H. Lee, S.C. Suh, Y.P. Lim, H.Y. Lee and H.I. Kim. 2001. Constitutive and seed specific expression of a maize lysine feedback–insensitive dihydrodipicolinate synthase gene leads to increase free lysine in rice seeds:, *Mol. Breed.,* **8:** 75-84.

Lefevre, A., L. Consoli, S.A. Gaziola, A.P. Pellegrino, R.A. Azevedo and C. Damerval. 2002. Dissecting the Opaque-2 regulatory network using transcriptome and proteome approaches along with enzyme activity measurements. *Scientia Agricola,* **59:** 407-414.

Lefevre, A., L. Consoli, S.A. Gaziola, R.J. Smith, P.J. Lea and R.A. Azevedo. 2002. Enzymes of lysine metabolism from Coix Lacryma-jobi seeds. *Plant Physiol. Biochedm.* **40:** 25-32.

Ling, V., W.A. Snedden, B.J. Shelp and S.M. Assman 1994. Analysis of a soluble calmodulin binding protein from fava bean roots: identification of glutamate decarboxylase as a calmodulin activated enzyme. Plant Cell, **6:** 1135-1143

Liu, J. and J.K. Zhu. 1997. Proline accumulation and salt-stress induced gene expression in a salt-hypersensitive mutant of Arabidopsis. *Plant Physiol.* **114:** 591-596.

Lugli, J., A. Cambell, S.A. Gaziola, R.J. Smith, P.J. Lea and R.A. Azevedo. 2002. Enzymes of lysine metabolism from coix lacryma-jobi seeds. *Plant Physiol. Biochem.,* **40:** 25-32.

Magee, T. and M.C. Seabra. 2005. Fatly acylation and prenylation of proteins: What's not in fat. *Curr.Opin.Cell Biol.,* **17:** 190-196.

Magalhaes, J.R., G.C. Ju, P.J. Rich and D. Rhodes 1990. Kinetics of $15NH_4$ assimilation in Zea mays: Preliminary studies with a glutamate dehygenase (GDH1) null mutant. *Plant Physiol.,* **94:** 647-656.

Mayer, R.R., J.H. Cherry and D. Rhodes 1990. Effects of heat shock on amino acid metabolism of cow pea cells *Plant Physiol.,* **94:** 796-810.

Markovitz, P.J. and D.T. Chuang. 1987. The bifunctional aminoadipic semialdehyde synthase in lysine degradation. *J. Biol.Chem.,* **262:** 9353-9358.

Mattioli, R., P. Costantino and M. Trovato 2008. Modulation of intracellular proline levels affects flowering time and infloresence architecture in Arabidopsis. *Plant Mol.Biol.,* **66:** 277-288.

Matysik, J., Alia, B. Bhalu and P. Mohanty. 2002. Molecular mechanisms of quenching of reactive oxygen species by protine under stress in plants. Curr. Sci. **82:** 525-532.

Mazur, B., E. Krebbers and S. Tingey. 1999. Gene discovery and product development for grain quality traits. *Science,* **285:** 372-375.

Miller, G., A. Honigs, H. Stein, N. Suzuki, R. Mittlerand A. Zilberstein. 2009. Unraveling delta-1-pyrroline-5-carboxylate proline cycle in plants by uncoupled expression of proline oxidation enzymes. *J. Biol. Chem.,* **284:** 26482-26492.

Moulin, M., C. Deleu and F. Larher. 2000. L-lysine catabolism is osmo regulated at the level of lysine-Ketoglutarate reductase and sacchoropine dehydrogenase in rape seed leaf discs. *Plant Physiol. Biochem,* **38:** 1577-1585.

Miron, D.,S. Ben-Yaacov, H. Karchi and G. Galili 1997. *In vitro* dephosphorylation inhibits the activity of soybean lysine-keto glutarate reductase in a lysine regulated manner. *Plant J.* **12:** 1453-1458

Miron, D., S. Ben-Yaacov, D. Reches, A. Schupper and G. Galili. 2000. Purification and characterization of bifunctional lysine-ketoglutarate/saccharopine dehydrogenase from developing soybean seeds. *Plant Physiol:* **123:** 665-668.

Molina, S.M.G., S.A. Gaziola, P.J. Lea and R.A. Azevedo. 2001. Manipulating cereal crops for high lysine accumulation in seeds. *Scientia Agricola,* **58:** 205-211.

Muehlbaner, G.J., B.G. Gengenbach, D.A. Somers and C.M. Donovan. 1994. Genetic and amino acid analysis of two maize threonine over producing lysine insensitive aspartate kinase mutants. *Threo.Apple.Genet.,* **89:** 767-774.

Muller, M. and S. Kundsen. 1993. The nitrogen response of a barley C-hordein promoter is controlled by positive and negative regulation of the GCN4 and endosperm box. *Plant J.*, **4:** 343-355.

Nakamura, T., S. Yokota, Y. Muramoto, K. Tsu-Sui, Y. Oguri, K. Fukui and T. Takabe. 1997. Expression of a betaine aldehyde dehydrogenase gene in rice, a glycine-betaine nonaccumulator and possible localization of its protein in peroxisomes. *Plant J.*, **11:** III5-II20.

Nquven, H.C., R. Hoefgen and H. Hesse 2012. Improving the nutritive value of rice seeds: Elevation of cysteine and methionine contents in rice plants by ectopic expression of a bacterial serineacetyl transferase. *J. Exp.Bot.*, **63:** 5991-6001.

Patterson, B.D. and D. Graham. 1987. Temperature and metabolism. *In:* Biochemistry of plants. Vol. 12 D.D. Davies (Ed) Academic Press, New York. pp.153-199.

Parida, A.K., V.K. Dagaonkar, M.S. Phalak and L.P. Aurangabadkar. 2008. Differential responses of the enzymes involved in proline biosynthesis and degradation in drought tolerant and sensitive cotton genotypes during drought stress and recovery. *Acta Physiol. Planta*, **30:** 619-627.

Pilobello, K.T. and L.K. Mahal 2007. Deciphering the glycocode: The complexity and analytical challenge of glycines. *Curr. Opin.Chem.Biol.*, **11:** 300-305.

Pryor, A. 1990. A maize glutamate dehydrogenase null mutant is cold temperature sensitive. *Maydica*, **35:** 367-372

Ramputh, A.E. and A.W. Brown. 1996. Rapid gama-aminobutyric acid synthesis and the inhibition of the growth and development of oblique banded leaf roller larvae. *Plant Physiol.*, **111:** 1349-1352.

Ramos. F., E. Dubois and A. Pierard. 1988. Control of enzyme synthesis in the lysine biosynthetic pathway of saccharomyces cerevisiae. *FEBS. J.* **171:** 171-176

Ratcliffe, R.G. 1995. Metabolic aspects of the anoxic response in plant tissue. *In:* Environment and Plant Metabolism: Flexibility and Acclimation. N. Snumoff (Ed) Bios Scientific, Oxford., pp. 111-127.

Rhodes, D., S. Handa and R.A. Bressau. 1986. Metabolic changes associated with adaptation of plant cells to water stress. *Plant Physiol.*, **82:** 890-903.

Rhodes, D. and A.D. Hanson. 1993. Quaternary ammonium and tertiary sulfonium compounds in higher plants. *Annu.Rev.Plant Physiol. Plant Mol.Biol.*, **44:** 357-384.

Rodnina, M.V., M. Bringer and W. Wintermeyer 2007. How ribosomes make peptide bonds. *Trends Biochem.Sci.*, **32:** 20-26.

Ribarits, A., A. Abdullaev, A. Tashpulatov, A. Richter, E. Heberle-Bors and A. Touraev 2007. Two tobacco proline dehydrogenases are differentially regulated and play a role in early plant development. *Planta.*, **225:** 1313-1324.

Roosens, N.H., T.T. Thu, H.M. Iskandar and M. Jacobs 1998. Isolation of the ornithine-delta-aminotransferase c DNA and effect of salt stress on its expression in Arabidopsis thaliana. *Plant Physiol.*, **117:** 263-271.

Ravanel, S., B. Gakiere, D. Job and R. Douce. 1998. The specific features of methionine biosynthesis and metabolism in plants. *Proc.Natl.Acid.Sci.USA.*, **95:** 7805-7812.

Rathinasabapathi, B., D.A. Gage, D.J. Mackill and A.W. Hanson. 1993. Cultivated and wild rices do not accumulate glycinebetaine due to deficiencies in two biosynthetic steps. *Crop Science*, **33:** 534-538.

Saradhi, P.P., Alia, S. Arora and K.V. Prasad. 1995. Proline accumulates in plants exposed to UV radiation and protects them against UV induced peroxidation. *Biochem. Biophys.Res.Commu.*, **209:** 1-5.

Satoh, H. and T. Omura. 1979. Induction of mutation by the treatment of fertilized egg cell with N-methyl-N-nitrosourea in rice. *J.Facul.Agri.Kyushu Univ.,* **24:** 165-174.
Sanders, D.C., Braonlee and J.F. Harper. 1999. Communicating with calcium. *Plant cell,* **11:** 691-706.
Savoure, A., S. Jaoua, X. Hua, W.A. Diaz, M.V. Montagu and N. Verbruggen. 1995. Isolation, characterization, and chromosomal location of a gene encoding the delta-1-pyrroline-5-carboxylate synthetase in Aratsidopsis thaliana. *FEBS Letters,* **372:** 13-19.
Schaeffer, G.W. and F.T. Sharpe. 1987. Increased lysine and seed storage protein in rice plants recovered from calli selected with inhibitory levels of lysine plus threonine and S-(aminoethyl) cystein. *Plant Physiol.,* **84:** 509-515.
Schat, H., S.S. Sharma and R. Vooijs. 1997. Heavy metal induced accumulation of free proline in a metal tolerant anda nontolerant ecotype of Silene vulgaris. *Physiol.Plant.,* **101:** 477-482.
Sekhar, B.P. and G.M. Reddy. 1982. Amino acid profiles in some scented rice varieties. *Theor. Appl.Genet.,* **62:** 1525-1529.
Shelp, B.J., C.S. Walton, W.A. Snedden, L.G. Tuin, I.J. Oresnik and D.B. Layzell. 1995. GABA shunt in developing soybean seed is associated with hypoxia. *Plant Physiol.,* **94:** 219-228.
Shingaki-Wells, R.N., S. Huang, N. L. Taylor, A.J. Carroll, W.Zhou and A.H. Millar. 2011. Differential molecular responses of rice and wheat coleoptiles to anoxia reveal metabolic adaptations in amino acid metabolism for tissue tolerance. *Plant Physiol.* **156:** 1706 -1724.
Smotrys, J.E. and M.E. Linder. 2004. Palmitoylation of intracellular signaling proteins: Regulation and function. *Annu.Rev.Biochem.,* **73:** 559-587.
Smirnoff, N. and Q.J. Cumbes. 1989. Hydroxyl radical scavenging activity of compatible solutes. *Phytochem* **28:** 1057-1060.
Snedden, W.A., T. Arazi, H. Fromm and B.J. Shelp. 1995 calcium/calmodulin activation of soybean glutamate decarboxylase. *Plant Physiol.* **108:** 543-549.
Stewart G.R. and F. Larher. 1980. Accumulation of amino acids and related compounds in relation to environmental stress. In : The Biochemistry of Plants (Ed. B.J. Miflin) vol. 5, Academic Press, New York, pp. 609-635
Strizhov, M., E. Abraham, L. Okresz, S. Blickling, A. Zilberstein, J. Schell, C. Koncz and L. Szabados. 1997. Differential expression of two PSCS genes controlling proline accumulation during salt stress requires ABA and is regulated by ABA1, ABI1 and AXR2 in Arabidopsis. *Plant J.,* **12:** 557-569.
Streeter, J.G. and J.F. Thompson. 1972. Invitro and invivo studies on gama amino butyric acid metabolism with radish plant (*Raphanus sativus* L.). *Plant Physiol.* **49:** 579-584.
Sodek, L. and C.M. Wilson. 1970. Incorporation of leucine-C14 into protein in the developing normal and Opaque 2 corn. *Arch.Biochem. Biophys.,* **140:** 29-38.
Subbaiah, C.C., D.S. Bush and M.M. Sachs. 1998. Mitochondrial contribution to the anoxic Ca_{2+} signal in maize suspension cultured cells. *Plant Physiol.* **118:** 759-771.
Szekely, G., E. Abraham, A. Cseplo, G. Rigo, L. Zsigomond, J. Csiszar, F. Ayaydin, N. Strizhov, D.J. Jaaasik, E. Schmelzer, C. Koucz and L. Szabados. 2008. Duplicated P5CS genes of Arabidopsis play distinct roles in stress regulation and developmental control of proline biosynthesis. *Plant J.,* **53:** 11-28.
Szoke, A., G.H. Miao, Z. Hong and D.P.S. Verma. 1992. Subcellular location of delta pyrroline-5-carboxylate reductase in root/nodule and leaf of soybean. *Plant Physiol.* **99:** 1642-1649.

Tang. G., D. Miron, J.X. Zhu-Shimoni and G. Galili. 1997. Regulation of lysine catabolism through lysine-keto glutarate reductase and saccharopine dehydrogenase in Arabidopsis *Plant Cell.* **9:** 1305-1316

Teixeira, C.M.G., S.A. Gaziola, J. Lugli and R.A. Azevedo. 1998. Isolation, partial purification and characterization of aspartate kinase isoenzymes from rice seeds. *J.Plant Physiol.,* **153:** 281-289.

Turano, F.J., S.S. Thakkar, T. Fang and J.M. Weisemann. 1997. Characterization and expression of NAD(H)-dependent glutamate dehydrogenase genes in Arabidopsis. *Plant Physiol.,* **113:** 1329-1341.

Trossat, C., B. Rathinasabapathi and A.D. Hanson. 1997. Transgenetically expressed betaine aldehyde dehydrogenase efficiently catalyzes oxidation of dimethyl sulfonio propionaldehyde and w-amino aldehydes. *Plant physiol.,* **113:** 1457-1461.

Urry, D.W. 2004. The change in Gibbs free energy for hydrophobic association: Derivation and evaluation by means of inverse temperature transitions. *Chem. Phys. Letters,* **399:** 177-183.

Vasal, S.K. 1994. High quality protein corn. *In:* Hallawer, A.R (Ed) Speciality corns, CRC Press, Boca Raton, FL. pp.79-120.

Verbruggen, N., X.J. Hua, M. May and M. Montagu. 1996. Environmental and developmental signals modulate proline homeostasis: Evidence for a negative transcriptional regulator. *Proc.Natl.Acad.Sci.USA,* **93:** 8787-8791.

Verbruggen, N., R. Villarroel and M. Montagu. 1993. Osmoregulation of a pyrroline-5-carboxylate reductase gene in Arabidopsis thaliana. *Plant Physiol.,* **103:** 771-781.

Verbruggen, N. and C. Hermanas. 2008. Proline accumulation in plants–a review. *Amino Acids,* **35:** 753-759.

Vinocur, B. and A. Altman. 2005. Recent advances in engineering plant tolerance to abiotic stress: achievements and limitations. *Curr.Opin. Biotech.,* **16:** 123-132.

Wallace, W.J. Secor and L.E. Schrader. 1984. Rapid accumulation of gama amino butyric acid and alanine in soybean leaves in response to abrupt transfer to low temperature, darkness or mechanical manipulation. *Plant Physiol.* **75:** 170-175.

Wang, L., M. Zhong, X. Li, D. Yuan, Y. Xu, H. Liu, Y. He, L. Luo and Q. Zhang. 2008. The QTL controlling amino acid content in grains of rice (Oryza sativa) are colocalized with the regions involved in the amino acid metabolism pathway. *Mol.Breed.,* **21:** 127-137.

Wakasa, K., H. Hasegawa, H. Nemoto, F. Matsuda, H. Miyazawa, Y. Tozawa, K. Morino, A. Komatsu, T. Yamada, T. Terakawa and H. Miyagawa. 2006. High level tryptophan accumulation in seed of transgenic rice and its limited effects on agronomic traits and seed metabolite profile *J.Exp.Bot,* **57:** 3069-3078.

Widodo, J., H. Patterson, E. Newbigin, M. Tester, A. Bacic and V. Roessner. 2009. Metabolic responses to salt stress of barley (Hordeum vulgare L.) cultivars Sahara and Clipper, which differ in salinity tolerance. *J.Exp.Bot.,* **60:** 4089-4103.

Xin, Z. and J. Brows. 1998. Eskimol mutants of Arabidopsis are constitutively freezing tolerant. *Proc.Natl.Acad.Sci USA,* **95:** 7799-7804.

Yang, S.L., S.S. Lam and M. Gong 2009. Hydrogen peroxide induced proline metabolic pathway of its accumulation in maize seedlings. *J. Plant Physiol.* **166:** 1694-1699.

Yoshirba, Y., T. Kiyosue, T. Katagiri, H. Ueda, T. Mizoguchi, K. Yamaguchi-Shinozaki, K. Wada, Y. Harada and K. Shinozaki. 1995. Correlation between the induction of a gene for delta 1-pyrroline-5-carboxylate synthetase and the accumulation of proline in Arabidopsis thaliana under osmotic stress. *Plant J.*, **7:** 751-760.

Zhao, W., E.J. Park, J.W. Chung, Y.J. Park, I.M. Chung, J.K. Ahnand G.H. Kim. 2008. Association analysis of the amino acid contents in rice *J. Integer. Plant Biol.* **51:** 1126-1137

Zhao, H., H. Ma, L. Yu, X. Wang and J. Zhao. 2012. Genome-wide survey and expression analysis of amino acid transporter family in rice (*Oryza sativa* L.). *Plos ONE* doi 10.1371/journal pne 0049210

Zhu, J.K. 2002. Salt and drought stress signal transduction in plants. *Annu.Rev. Plant Biol.,* **53:** 247-273.

Zhu, J.K., J. Liu and L. Xiong. 1998. Genetic analysis of salt tolerance in Arabidopsis thaliana : evidence of a crictical role for potassium nutrition. *Plant Cell.* **10:** 1181-1192.

Zhu, J.K. 2000. Genetic analysis of plant salt tolerance using Arabidopsis. *Plant Physiol.* **124:** 941-948.

chapter eight

Chlorophyll

Chlorophyll (also **chlorophyl**) is a green pigment found in cyanobacteria and the chloroplasts of algae and plants. Its name is derived from the Greek words *chloros* ("green") and *phyllon* ("leaf"). Chlorophyll is an extremely important biomolecule, critical in photosynthesis, which allows plants to absorb energy from light. Chlorophyll absorbs light most strongly in the blue portion of the electromagnetic spectrum, followed by the red portion. Conversely, it is a poor absorber of green and near-green portions of the spectrum, hence the green color of chlorophyll-containing tissues. Chlorophyll was first isolated by Joseph Bienaimé Caventou and Pierre Joseph Pelletier in 1817.

Chlorophyll is a chlorin pigment, which is structurally similar to and produced through the same metabolic pathway as other porphyrin pigments such as heme. At the center of the chlorin ring is a magnesium ion.

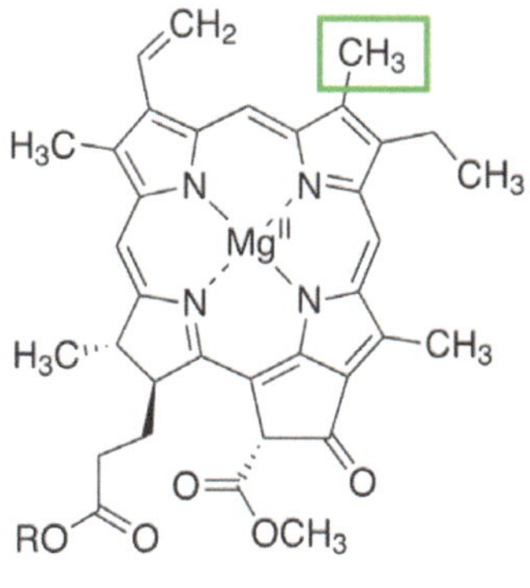

FIGURE 8.1 Chlorophyll.

Chlorophyll *a* is essential for most photosynthetic organisms to release chemical energy but is not the only pigment that can be used for photosynthesis. All oxygenic photosynthetic organisms use chlorophyll *a*, but differ in accessory pigments like chlorophylls *b*. Chlorophyll *a* can also be found in very small quantities in the green sulfur bacteria, an anaerobic photoautotroph. These organisms use bacteriochlorophyll and some chlorophyll *a* but do not produce oxygen. Anoxygenic photosynthesis is the term applied to this process, unlike oxygenic photosynthesis where oxygen is produced during the light reactions of photosynthesis.

8.1. Chlorophyll Biosynthesis

In the first phase of chlorophyll biosynthesis, the amino acid glutamic acid is converted to 5-aminolevulinic acid (ALA). This reaction is unusual in that it involves a covalent intermediate in which the glutamic acid is attached to a transfer RNA molecule. This is one of a very small number of examples in biochemistry in which a tRNA is utilized in a process other than protein synthesis. Two molecules of ALA are then condensed to form porphobilinogen (PBG), which ultimately form the pyrrole rings in chlorophyll. Then is the assembly of a porphyrin structure from four molecules of PBG. This phase consists of six distinct enzymatic steps, ending with the product protoporphyrin IX. The next phase of the chlorophyll biosynthetic pathway is the formation of the fifth ring (ring E) by cyclization of one of the propionic acid side chains to form protochlorophyllide. The pathway involves the reduction of one of the double bonds in ring D, using NADPH. This process is driven by light in angiosperms and is carried out by an enzyme called protochlorophyllide oxidoreductase (POR). Non-oxygen-evolving photosynthetic bacteria carry out this reaction without light, using a completely different set of enzymes. Cyanobacteria, algae, lower plants and gymnosperms contain both the light-dependent POR pathway and the light-independent pathway. Seedlings of angiosperms grown in complete darkness lack chlorophyll, because the POR enzyme requires light. These etiolated plants very rapidly turn green when exposed to light. The final step in the chlorophyll biosynthetic pathway is the attachment of the phytol tail, which is catalyzed by an enzyme called chlorophyll synthetase (Malkin and Nyogi, 2000).

The branched isoprenoid pathway is rather complex and comprises enzymatic steps in at least two compartments. In plants, Phy represents the side chain of Chls, tocopherols (TP) and phylloquinones, and is necessary for their integration into plastid membranes (Soll et al. 1980, 1983; Soll, 1987; Bollivar et al. 1994). In both Chl and TP synthesis, the Phy chain is provided by geranylgeranyl pyrophosphate (GGPP), a plastidial isoprenoid, formed by four molecules of isopentenyl pyrophosphate (IPP), which are derived from the cytosolic and chloroplastidic pathways (Rohmer et al. 1993; Lichtentaler et al. 1997). In Chl synthesis, GGPP can either be reduced to phytyl pyrophosphate (phypp) and esterified with Chlide to generate phytyl Chl (chlphy), or first esterified with Chlide to form geranylgeranylated Chl (chlgg) and then stepwise reduced into chlphy (Soll et al. 1983; Bollivar et al. 1994; Keller et al. 1998; Chew et al. 2008). In the TP pathway, tocopherols are generally believed to arise from the condensation of homogentisic acid and phypp (Schultz et al. 1985; Soll et al. 1980; Collakova and Della Penna, 2001; Savidge et al. 2002; Cahoon et al. 2003).

All the biosynthesis steps up to this point are the same for the synthesis of both chlorophyll and heme. But here the pathway branches, and the

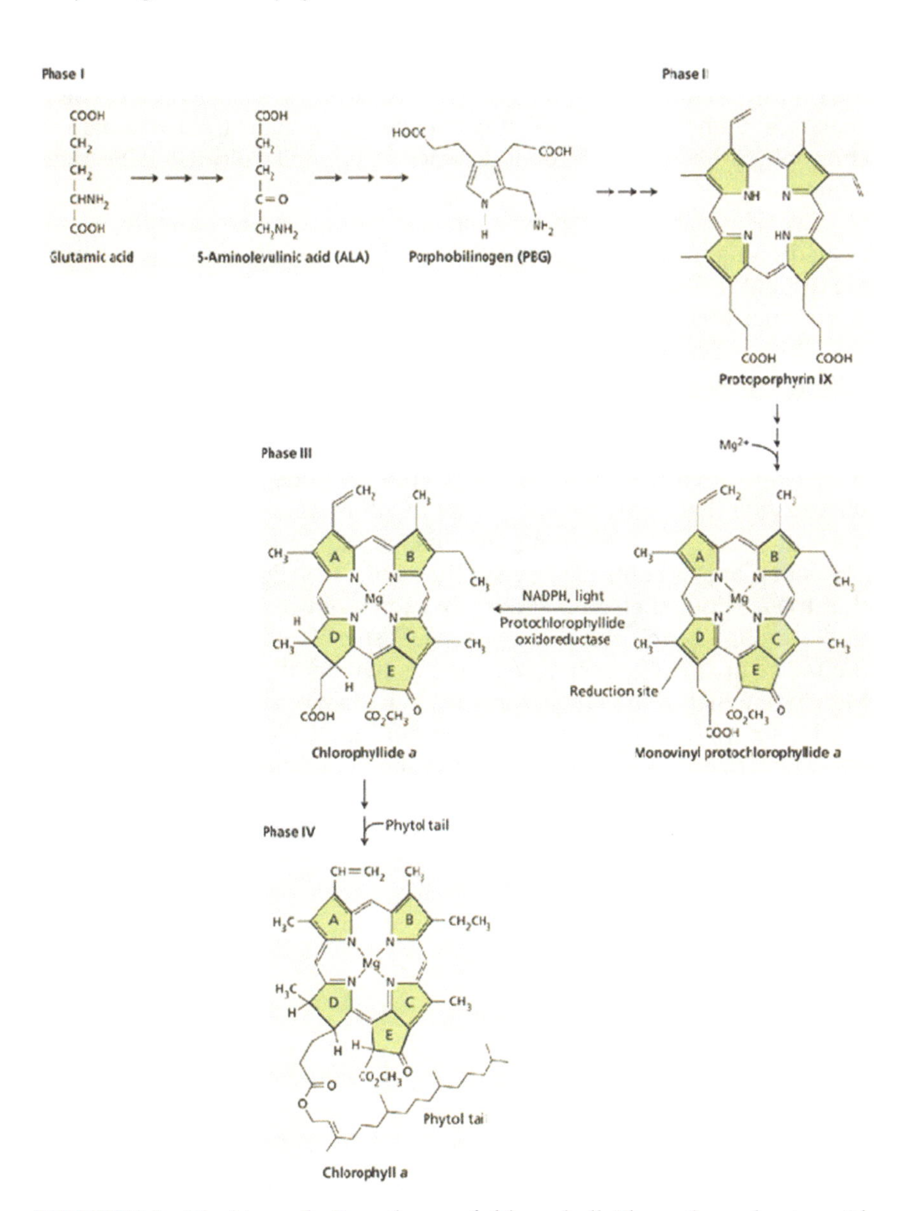

FIGURE 8.2 The biosynthetic pathway of chlorophyll. The pathway begins with glutamic acid, which is converted to 5-aminolevulinic acid (ALA). Two molecules of ALA are condensed to form porphobilinogen (PBG). Four PBG molecules are linked to form protoporphyrin IX. The magnesium (Mg) is then inserted, and the light-dependent cyclization of ring E, the reduction of ring D, and the attachment of the phytol tail complete the process. Many steps in the process are omitted in this figure.

fate of the molecule depends on which metal is inserted into the center of the porphyrin. If magnesium is inserted by an enzyme called magnesium chelatase, then the additional steps needed to convert the molecule into chlorophyll take place; if iron is inserted, the species ultimately becomes heme.

The elucidation of the biosynthetic pathways of chlorophylls and related pigments is a difficult task, in part because many of the enzymes are present in low abundance. Recently, genetic analysis has been used to clarify many aspects of; these processes (Armstrong and Apel, 1998).

Chl *b* is derived from Chl *a* by oxygenation of the 7-methyl group to a formyl group. An extensive amount of evidence has been gathered on the requirement of Chl *b* for accumulation of stable LHCs. For example, Chl *b*-less strains lack most, if not all, of the major LHC apoproteins. In contrast, over-expression of chlorophyllide (Chlide) *a* oxygenase (CAO), the enzyme that catalyzes synthesis of Chlide *b*, in *Arabidopsis* increased the size of the light-harvesting antenna, an indication that the amount of LHCs in plants is controlled by synthesis of Chl *b*. Eggink et al. proposed that the additional electronegative oxygen atom in Chl *b* causes further redistribution of the chlorin π electron system towards the periphery of the molecule, thereby increasing the positive point charge on the central Mg atom, which results in an increase in its Lewis acid strength. Stability of LHCs may thus result from strengthening of the coordination bonds between Chl *b* and electronegative, oxygen-containing ligands because of additional Coulombic attraction. Several of the Chl *b* molecules are also hydrogen-bonded to amino acid side chains, which further strengthens the interaction with the protein.

Chlorophyll b

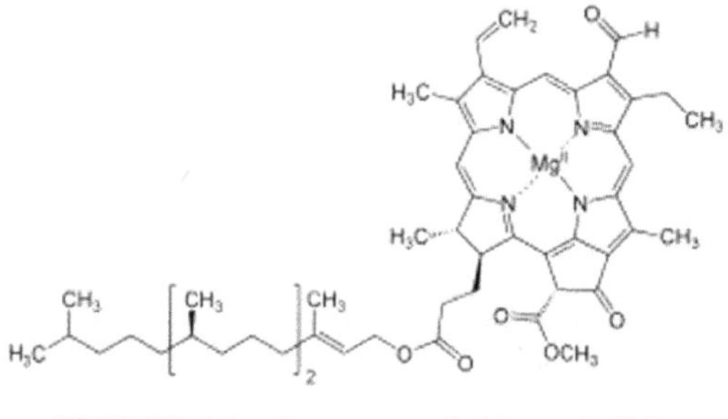

FIGURE 8.3 Structure of chlorophyll b.

Chlorophyll *b* is synthesized from chlorophyll *a* by chlorophyll *a* oxygenase. It was identified two genes (OsCAO1 and OsCAO2) from the rice genome that are highly homologous to previously studied chlorophyll *a* oxygenase (CAO) genes. They are positioned in tandem, probably resulting from recent gene duplications. The proteins they encode contain two conserved functional motifs – the Rieske Fe-sulfur coordinating center and a non-heme mononuclear Fe-binding site. OsCAO1 is induced by light and is preferentially expressed in photosynthetic tissues. Its mRNA level decreases when plants are grown in the dark. In contrast, OsCAO2 mRNA levels are higher under dark conditions, and its expression is down-regulated by exposure to light. To elucidate the physiological function of the CAO genes, it had been isolated knockout mutant lines tagged by T-DNA or Tos17. Mutant plants containing a T-DNA insertion in the first intron of the OsCAO1 gene have pale green leaves, indicating chlorophyll *b* deficiency. It was also isolated a pale green mutant with a Tos17 insertion in that OsCAO1 gene. In contrast, OsCAO2 knockout mutant leaves do not differ significantly from the wild type. These results suggest that OsCAO1 plays a major role in chlorophyll *b* biosynthesis, and that OsCAO2.

Gabaculine is a potent inhibitor of glutamate-1-semialdehyde aminotransferase [EC 5.4.3.8] involved in the synthesis of chlorophyll in plants.

The effects of three allelopathic phenolics, *o*-hydroxyphenyl acetic, ferulic and *p*-coumaric acids, on the chlorophyllase activity of rice leaf (*Oryza sativa* cv. TN67) were investigated. Ten-day-old green seedlings of rice were cultured in greenhouse for 16 d in Kimura's culture solution, which was changed every four days, with or without 50, 100 or 200 ppm of the phenolic compounds. Just before changing the culture solution, leaves were harvested to determine their chlorophyll (Chl) and chlorophyllide (Chlide) contents, and their chlorophyllase *a* and *b* activities. While the Chl and Chlide contents decreased and increased, respectively, causing the molar ratio of Chlide/Chl to increase, as the phenolic concentrations increased, the chlorophyllase *a* and *b* activities drastically increased. This suggests that the consumption-orientation of Chl was significantly stimulated by the exogenously applied phenolics. The order of inhibition of growth of the rice seedlings is: ferulic acid>*p*-coumaric acid>*o*-hydroxyphenylacetic acid. The order of inhibition effect on Chl accumulation is: *p*-coumaric acid>*o*-hydroxyphenylacetic>ferulic acid. The order of stimulation effect on chlorophyllase *a* activity is: *o*-hydroxyphenylacetic acid>ferulic acid>*p*-coumaric acid. The order of promotion effect on chlorophyllase *b* is: ferulic acid>*o*-hydroxyphenylacetic acid>*p*-coumaric acid. The different responses of chlorophyllase *a* and *b* activities to the same concentrations of allelochemical phenolics suggest that they may be two different enzymes. It is apparent that the three phenolics may enhance the activities of enzymes, such as chlorophyllase and Mg-dechelatase, responsible for the Chl degradative pathway. A combination of the present and the preceding data strongly suggest that the three allelopathic phenolics may comprehensively affect the biosynthetic and degradative pathways of Chl.

8.2. Influence of Nitrogen

From all metabolic elements which plants use from soil, nitrogen is needed largest amounts (Tucker, 2004). Nitrogen exists in organic and inorganic forms and the greatest nitrogen content is in seeds, leaves, shoots and roots. Deficiency of nitrogen leads to loss of green color in the leaves, decrease leaf area and intensity of photosynthesis. Understanding the processes that govern N uptake and distribution in crops is of major importance with respect to both environmental concerns and the quality crop products. Nitrogen uptake and accumulation in crops represents two major components of the N cycle in the agro system (Gastal and Lamari, 2002). The relationship between N and biomass accumulation in crops relies on the reciprocal regulation of multiple crop physiological process. Therefore, N uptake and distribution in plant and crops involves many aspects of growth and development.

Modern technology of wheat production is mainly based on numerous scientific farming measures as well as application of mineral fertilizers. Mainly one third of applied nutrient wheat plants are able to use during vegetative period. In the field practices it is very important to optimize quantity of fertilizers, decrease expenses of production and improve efficiency of wheat plant of nitrogen absorption, accumulation and reutilization. The nitrogen plays main role in wheat nutrition because of its importance in protein and nucleic acid. Plant species and cultivars to suboptimal supplies of mineral element, including N, are different (Saric and Kovacevic, 1981; Clark, 1983). Wheat properties are mainly caused by effect of genetic factors in interaction with environment (Pepo, 2005; Balogh et al. 2006). The total N content represent indicator of N accumulation in plant (Desai and Bathia, 1978) which indicate root system activity and translocation of organic and inorganic matter to top of plant. Physiological N efficiency in plant indicating activity of top of plant and involve of absorbed N into processes of synthesis. Leaves exhibit a structural and functional acclimation of the photosynthetic apparatus to the light intensity experienced during their growth (Prioul et al. 1980). Nitrogen supply has large effect on leaf growth because it increases the leaf area of plants and, thus, it influences photosynthesis. Photosynthetic proteins represent a large proportion to total leaf N (Evans, 1983). Chlorophyll content is approximately proportional to leaf nitrogen content, too (Evans, 1983).

A study was conducted during Boro and T. Aman seasons of 2002 at Bangladesh Rice Research Institute (BRRI), Gazipur to see the relationship of SPAD (Soil plant analysis development) reading with chlorophyll and N contents of leaves and to determine the critical LCC value for rice crops. Hybrid varieties Sonarbangla-1 and BRRI hybrid dhan01 were used for both rice crops and BRRI dhan29 and BRRI dhan31 were used as checks for Boro and T. Aman crops, respectively. Sonarbangla-1,

BRRI hybrid dhan0l and BRRI dhan29 had similar leaf chlorophyll contents in Boro season. The maximum chlorophyll content (1.6-1.8 mg/g leaf) was observed at 39-42 soil plant analysis development (SPAD) value. In T. Aman season, the inbred BRRI dhan3l showed lower amount of chlorophyll (1.2-1.4 mg/g leaf) at 39-42 SPAD value compared to the hybrids Sonarbangla-1 and BRRI hybrid dhan 0l. Seasonal variation in chlorophyll content between Sonarbangla-1 and BRRI hybrid dhan0l was not large. Relationship between SPAD value and chlorophyll content was very close ($R^2 \geq 0.8$) at panicle initiation and flower initiation stages for all the varieties. Similar relationship was also observed in case of SPAD value and nitrogen content in leaves. The results indicated that the rice leaves showing higher SPAD readings (>35) had higher chlorophyll and nitrogen contents. The adjusted critical LCC values were 3.0 for Boro and 3.5 for T. Aman seasons for all rice varieties (Islam et al. 2009).

The investigation into the influence of different nitrogen (N) applications to some morphological and physiological features of leaf blades, including leaf thickness, chlorophyll content at different leaf ages and chlorophyll *a*/*b* ratios. A paddy field and a cement tank experiments were conducted simultaneously. Rice leaf thickness was measured through a specially developed displacement sensor. Meanwhile, chlorophyll content was estimated using chlorophyll meter (SPAD) and spectrophotometer after ethanol extraction of leaf samples. With the increase of N application, leaf thickness became thinner and chlorophyll *a*/*b* ratios decreased. Moreover, the sensitivity of the SPAD readings of the same leaf at different leaf ages to N rates was assessed through coefficients of variation (CV). CV of SPAD readings increased from 8.8% to 21.6% during leaf lifetime, which indicates that SPAD readings became more and more sensitive to nitrogen rates as leaf aged. Therefore, SPAD readings of the lower leaves, which were physiologically older than the upper ones, were more sensitive to nitrogen rates (Jinwen et al. 2009).

8.3. Chlorophyll Extraction

The methodologies used for chlorophyll extraction in plant materials are almost always based on methods that destructively extract leaf tissue using organic solvents that include acetone (McKinney, 1941; Bruisna, 1961), dimethylsulfoxide (DMSO) (Hiscox and Israelstam, 1979) methanol, N,N-dimethyl formamide and petroleum ether (Moran and Porath, 1980; Lichtenthaler and Wellburn, 1983; Inskeep and Bloom, 1985). During the extraction and dilution, significant pigment losses may occur thus leading to a high variability in the results. Shoaf and Lium (1976) used DMSO to modify the extraction methodology to eliminate the squashing and centrifuging stage. This method allowed longer storage periods for the extracted pigment, so that the spectrophotometer analyses need not to be performed immediately after extraction. Although a high correlation between the chlorophyll content and photosynthesis rate was not obtained (Marini, 1986), the assessment of photosynthetic pigments, and consequently their relationships, is an important indicator of senescence (Brown et al. 1991).

Chlorophyll loss is associated with environmental stress and the variation in total chlorophyll/carotenoids ratio may be a good indicator of stress in plants (Hendry and Price, 1993). In addition, measuring gas exchange and chlorophyll content repeatedly on the same leaves in field may provide useful information on the relationship between these parameters (Schaper and Chacko, 1991). The chlorophyll meter (or SPAD meter) is a simple, portable diagnostic tool that measures the greenness or the relative chlorophyll concentration of leaves. The meter makes instantaneous and non-destructive readings on a plant based on the quantification of light intensity (peak wavelength: approximately 650 nm: red LED) absorbed by the tissue sample. A second peak (peak wavelength: approximately 940 nm: infrared LED) is emitted simultaneous with red LED to compensate the thickness of leaf (Minolta Camera Co. Ltd. 1989). Compared with the traditional destructive methods, this equipment might provide a substantial saving in time, space and resources. However, to determine the chlorophyll concentration in a sample, calibration curves between meter readings and the chlorophyll concentration in the tissue sample must be made.

Recent research indicates a close link between leaf chlorophyll concentration and leaf N content, which makes sense because the majority of leaf N is contained in chlorophyll molecules (Peterson et al. 1993). Chlorophyll concentration or leaf greenness is affected by a number of factors, one being N status of the plant. Since the chlorophyll meter has the potential to detect N deficiencies, it also shows promise as a tool for improving N management (Peterson et al. 1993; Smeal and Zhang, 1994; Balasubramanian et al. 2000).

TABLE 8.1 Effect of variable nutrient levels on leaf chlorophyll profile, SPAD value and nitrogen % of traditional rice varieties during rainy season 2003

Treatment	Chlorophyll *a*	Chlorophyll *b*	Total chlorophyll	SPAD value	Nitrogen %
Nitrogen (0 kg/ha)	0.820	0.245	1.064	36.36	3.12
Nitrogen (80 kg/ha)	0.990	0.306	1.296	41.85	3.48
Cowdung (10 t/ha)	0.868	0.249	1.117	38.15	3.26

Minimum rate and timing of application of nitrogen (N) fertilizer are most crucial in achieving high yield in irrigated lowland rice. In order to assess leaf N status, a semi dwarf rice cultivar (Khazar) was grown with different N application treatments (0, 40, 80 and 120 kg N ha^{-1} splited at transplanting, mid tillering and panicle initiation stages) in a sandy soil in Guilan Province, Iran, in 2003. The chlorophyll meter (SPAD 502) readings were recorded and leaf N concentrations were measured on the uppermost fully expanded leaf in rice plants at 10-day intervals from 19 days after transplanting to grain maturity. Regression analysis showed that the SPAD readings predicted only 23% of changes in the leaf N concentration based on pooled data of leaf dry weight (N_{dw}) for all growth stages. However, adjusting the SPAD readings for specific leaf weight (SPAD/SLW) improved the estimation of N_{dw}, up to 88%. Specific leaf weight (SLW), SPAD readings, leaf area and weight as independent variables in a multiple regression analysis predicted 96% of the N_{dw} changes, while SPAD readings independently predicted about 80% of leaf N concentration changes on the basis of leaf area (N_a). It seems that chlorophyll meter provides a simple, rapid and nondestructive method to estimate the leaf N concentration based on leaf area, and could be reliably exploited to predict the exact N fertilizer topdressing in rice.

Nitrogen nutrient management is an important strategy in regulating the rice growth and photosynthetic efficiency. For higher rice production, increased nitrogen fertilizer is applied, which not only reduces farmers' profits, but also results in deteriorated ecological environments (Shen et al. 1997; Zhu, 2000). With higher biomass production, super hybrid rice consumes more fertilizers especially nitrogen (N) fertilizer, and this deteriorates environmental conditions. Therefore, there exists a continuing need to improve nitrogen use efficiency. How to achieve higher yield with reduced nitrogen fertilization, which produces no measurable effects on the normal physiological processes of functional leaves, has become an important challenge for the high-yielding cultivation of super hybrid rice (Long et al. 2007). Studies have shown that slow/controlled release fertilizer is an effective way to increase nitrogen use efficiency (Tremkel, 1997; Zhang et al. 2001).

Slow/controlled release fertilizer is found to promote the nitrogen absorption and improve nitrogen use efficiency for rice, and the related physiological mechanisms have been well understood (Liu et al. 2002; Li et al. 2004; Luo et al. 2007). However, few studies have been conducted on the effect of slow/controlled of release of nitrogen fertilizer on rice leaf absorption, transmission and distribution of light energy, and dissipation of excess excitation energy and on the related mechanisms, especially in super-hybrid rice. Chlorophyll fluorescence technique, a new technology using rapid light curve to measure the photosynthetic capacity of a sample, has the advantages of fast measurement and less impact on samples, etc. (Ralph et al. 1998; Kühl et al. 2005). Rapid light curve reflects not only the tolerance of a sample on strong light, but also the relative change of the antenna pigment in a sample indirectly. It has a unique role in detecting crop photosynthesis and can reflect the 'internal' characteristics of the photosynthetic system (Lazar, 1999). In order to investigate the photosynthetic mechanism of slow-release nitrogen fertilizer on light absorption, use and distribution, and light suppression of super hybrid rice leaves.

To compare the effects of slow-release nitrogen fertilizer at six different levels on the flag leaf chlorophyll fluorescence characteristics of super hybrid rice, a field fertilization experiment was conducted with super hybrid rice Y Liangyou 1 as a test material. The photosynthetic electron transport rate (ETR), effective quantum yield (EQY), photochemical quenching coefficient (*q*P), and non-photochemical quenching coefficient (NPQ) of flag leaves were measured at the initial heading, full heading, 10 d after full heading and 20 d after full heading stages. Results showed that the values of ETR, EQY and *q*P increased with rice development from initial heading to 20 d after full heading, whereas the NPQ decreased. During the measured stages, ETR, EQY and *q*P increased initially and then decreased as nitrogen application amount increased, but they peaked at different nitrogen fertilizer levels. The maximum ETR and EQY values appeared at the treatment of 135 kg/hm_2 N. In conclusion, the optimum nitrogen amount for chlorophyll fluorescence characteristics of super hybrid rice was 135–180 kg/hm_2 (Long et al. 2007).

8.4. Photosynthetic Characteristics

Effects of irradiance on photosynthetic characteristics were examined in senescent leaves of rice (*Oryza sativa* L.). Two irradiance treatments (100 and 20% natural sunlight) were imposed after the full expansion of the 13th leaf through senescence. The photosynthetic rate was measured as a function of intercellular $C0_2$ pressure with a gas-exchange system. The amounts of cytochrome *f*, coupling factor 1, ribulose 1, 5bisphosphatecarboxylase/oxygenase (Rubisco), and chlorophyll were determined. The coupling factor 1 and cytochrome *f* contents decreased rapidly during senescence, and their rates of decrease were much faster from the 20% sunlight treatment than from the full sunlight treatment. These changes were well correlated with those in the photosynthetic rate at CO_2 pressure = 600 microbars, but not with those under the ambient air condition (350 microbars CO_2) and 200 microbars CO_2. This suggested that the amounts of coupling factor 1 and cytochrome *f* from the full sunlight treatment cannot be limiting factors for the photosynthetic rate at ambient air conditions. The Rubisco content also decreased during senescence, but its decrease from the 20% sunlight treatment was appreciably retarded. However, this difference was not reflected in the photosynthetic rates at the ambient and 200 microbars CO_2. This implied that *in vivo* Rubisco activity may be regulated in the senescent leaves from the 20% sunlight treatment. The chlorophyll content decreased most slowly. In the

20% sunlight treatment, it remained apparently constant with a decline in chlorophyll *a*/*b* ratio. These photosynthetic characteristics of the senescent rice leaves under low irradiance were related to acclimation of shade plants(Hidema et al. 1991).

Chlorophyll pigments play an important role in the photosynthetic process as well as biomass production. Genotypes maintaining higher leaf chlorophyll-*a* and chlorophyll-*b* during growth period may be considered potential donor for the ability of producing higher biomass and photosynthetic capacity. Higher photosynthetic rate is supported by leaf chlorophyll content in leaf blades (Miah et al. 1997). The chlorophyll meter or SPAD (Soil plant analysis development) offers a new strategy for synchronizing N application with actual crop demand in rice (Peng et al. 1996; Balasubramanian et al. 2000). The chlorophyll meter indicates the need of a nitrogen top dressing that would result greater agronomic efficiency of nitrogen fertilizer than commonly pre-application of nitrogen (Hussain et al. 2000). It was well established that SPAD based nitrogen management needs considerably lower amount of nitrogen than the standard nitrogen management practices without any yield losses (Ali, 2005; Miah and Ahmed, 2002). Considering the above facts, this study was undertaken to evaluate the traditional rice varieties in respect of chlorophyll content, SPAD value and nitrogen use efficiency.

Field experiments were conducted on the experimental farm of the Taiwan Agricultural Research Institute (TARI) to examine seasonal changes in the amounts of chlorophylls in leaves and whole plants of rice (*Oryza sativa* L. cv. Tainung 67) grown in the first and the second crop seasons in 1998. Chlorophyll (Chl) differences between two crops and their relationships with growth were also investigated. The differences in climate conditions between the two crop seasons were quite phenomenal and led to different rates of plant growth and amounts of chlorophyll produced during the growth periods. Seasonal changes in Chl*a*, Chl*b*, and total Chl were curvilinear in style, with Chl*a* making up the major portion of total chlorophyll. Young plants of rice began with lower levels of chlorophyll, which increased as the plants developed, and decreased as the plants aged. A significant increase in chlorophyll occurred in the early stage of the second crop compared to that of the first crop, and the difference was maintained throughout the growing season. When the unit time transformation technique was used to exclude the difference in growth rate, the curves of the chlorophylls (Chl*a*, Chl*b*, and total Chl) were similar in the two crops, but plants grown in the second crop season had higher levels of chlorophylls at the same time unit scale. Generally, the ratio of Chl*a* to Chl*b* (Chl*a*/Chl*b*) decreased after transplanting, indicating a significant reduction ($p < 0.01$) in the green-to-yellow pigment relationship. Plants grown in the first crop had higher Chl*a*/Chl*b* ratios than those grown in the second crop, especially in leaf blades. Good agreement was found between growth traits and the respective total Chl content from regression analyses, and these growth traits were highly correlated with total Chl (R_2 > 0.8). However, rice plants grown in the first crop showed a higher levels of plant growth and leaf area at the same chlorophyll level than those grown in the second crop. These results suggest that rice plants grown in the first crop may exhibit better growth and production than those grown in the second crop under normal growing conditions (Yang and Lee, 2001).

8.5. *Abiotic Stress*

Water shortage is a major abiotic stress for crop production worldwide, limiting the productivity of crop species, especially in dry-land agricultural areas. This investigation aimed to classify the water-deficit tolerance in mutant rice (*Oryza sativa* L. spp. *indica*) genotypes during the reproductive stage. Proline content in the flag leaf of mutant lines increased when plants were subjected to water deficit. Relative water content (RWC) in the flag leaf of different mutant lines dropped in relation to water deficit stress. A decreased RWC was positively related to chlorophyll *a* degradation. Chlorophyll *a*, chlorophyll *b*, total chlorophyll, total carotenoids, maximum quantum yield of PSII, stomatal conductance, transpiration rate and water use efficiency in mutant lines grown under water deficit conditions declined in comparison to the well-watered, leading to a reduction in net-photosynthetic rate. In addition, when exposed to water deficit, panicle traits, including panicle length and fertile grains were dropped. The biochemical and physiological data were subjected to classify the water deficit tolerance. NSG19 (positive control) and DD14 were identified as water deficit tolerant, and AA11, AA12, AA16, BB13, BB16, CC12, CC15, EE12, FF15, FF17, G11 and IR20 (negative control) as water deficit sensitive, using Ward's method (Cha-um et al. 2012).

Photosynthetic pigments, chlorophyll *a* (Chl*a*), chlorophyll *b* (Chl*b*), total chlorophyll a (TC) and total carotenoids (C_{x+c}) in rice crop were decreased when subjected to water deficit stress. In DD14 mutant lines, Chl*a* content in water-deficit stressed plants was maintained. The degradation percentage of Chl*a* (5.1%) and TC (27.4%) in DD14 plants subjected to water-deficit was lower than in other rice lines (65.1-71.7%). Degradation of Chl*a* pigments was positively correlated to maximum quantum yield of PSII (F_v/F_m), leading to reduced net photosynthetic rate (P_n). Photosynthetic abilities including F_v/F_m, P_n, stomatal conductance (g_s) and transpiration rate (E) drastically decreased when exposed to water deficit conditions. In addition, water use efficiency (WUE) was similarly trended, except in AA12, BB13, BB16, CC12 and CC15 where it was alleviated (16.9, 24.4, 8.3, 27.6 and 11.1% reduction). Panicle length in rice lines grown under water deficit stress showed no differences, except in CC12 (25.4% reduction) and FF15 (35.0% reduction). Number of fertile grains decreased ($p \leq 0.01$).

To understand the impact of water stress on the greening process, water stress was applied to 6-day-old etiolated seedlings of a drought-sensitive cultivar of rice (*Oryza sativa*), Pusa Basmati-1 by immersing their roots in 40 mm polyethylene glycol (PEG) 6000 (–0.69 MPa) or 50 mm PEG 6000 (–1.03 MPa) dissolved in half-strength Murashige and Skoog (MS)-nutrient-solution, 16 h prior to transfer to cool-white-fluorescent + incandescent light. Chlorophyll (Chl) accumulation substantially declined in developing water-stressed seedlings. Reduced Chl synthesis was due to decreased accumulation of chlorophyll biosynthetic intermediates, that is, glutamate-1-semialdehyde (GSA), 5-aminolevulinic acid, Mg-protoporphyrin IX monomethylester and protochlorophyllide. Although 5-aminolevulinic acid synthesis decreased, the gene expression and protein abundance of the enzyme is responsible for its synthesis, GSA aminotransferase, increased, suggesting its crucial role in the greening process in stressful environment. The biochemical activities of Chl biosynthetic enzymes, that is, 5-aminolevulinic acid dehydratase, porphobilinogen deaminase, coproporphyrinogen III oxidase, porphyrinogen IX oxidase, Mg-chelatase and protochlorophyllide oxidoreductase, were down-regulated due to their reduced protein abundance/gene expression in water-stressed seedlings. Down-regulation of protochlorophyllide oxidoreductase resulted in impaired Shibata shift. Results demonstrate that reduced synthesis of early intermediates, that is, GSA and 5-aminolevulinic acid, could modulate the gene expression of later enzymes of Chl biosynthesis pathway (Dalal and Tripathy, 2012).

The relative chlorophyll content of rice showed a variation under different water levels. However, the relative chlorophyll content value was less at week 3 and the difference is significant at week 6 and9 respectively.

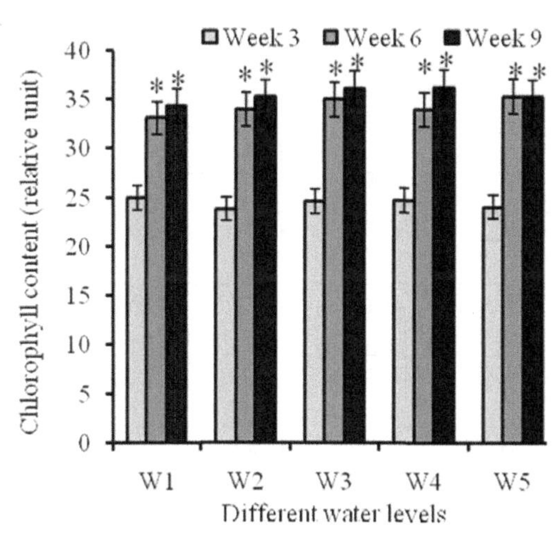

FIGURE 8.4 Variation of chlorophyll content with water level in rice

The chlorophyll *a* (Chl*a*) content in the leaf tissue of RD6 rice grown in salt affected soil decreased, by 5.4, 19.6 and 26.4%, when exposed to 0.3, 1.0 and 2.0% salt levels, respectively. Chlorophyll *b* (Chl*b*), total chlorophyll (TC) and total carotenoids (C_{x+c}) content had a similar trend to that of Chl*a*. The Chl*a* content in salt stressed leaves of RD6 rice was stabilized in the saline soil treated by OM, especially in the 1-2% salt levels. TC pigments in rice plants grown with OM treatment in 2% salt levels were maintained better than in the control (without OM). Photosynthetic pigment contents in the leaves of salt stressed rice plants were decreased, depending on increasing salt levels in the soil. The Chl*a* content in the salt stressed leaves was positively related to maximum quantum yield of PSII (F_v/F_m) ($r^2 = 0.80$). F_v/F_m dropped relating to salt concentration in the soil. Quantum efficiency of PSII (Φ_{PSII}) in the salt stressed leaves was declined by 1.5, 5.3 and 12.7% when exposed to 0.3, 1.0 and 2.0% salt levels, respectively. The Φ_{PSII} in salt stressed leaves of RD6 rice was improved in the saline soil treated by organic matter (OM), especially in the 2.0% salt levels. A positive relationship between F_v/F_m and Φ_{PSII} was displayed ($r^2 = 0.59$) which subsequently affected the net photosynthetic rate (P_n) ($r^2 = 0.46$) (Cha-um and Kirdmanee, 2011).

TABLE 8.2 Chlorophyll concentration (mg/g FW)of different rice genotypes grown under normal and saline environments

Genotype	Chl*a*-control	Chl*a*-EC 8.5 dS/m	Chl*b*-control	Chl*b*-EC 8.5 dS/m	Chl-control	Chl-EC8.5 dS/m
DM63275	0.590	0.389	0.283	0.257	0.822	0.645
NIAB-IRRI-9	0.596	0.440	0.365	0.290	0.938	0.721
KS-282	0.493	0.358	0.288	0.260	0.781	0.618
NIAB-RICE-1	0.479	0.441	0.280	0.256	0.758	0.697
DM25 x NIAB-IRRI-9	0.559	0.412	0.299	0.280	0.858	0.691
Jhona-349	0.523	0.392	0.291	0.278	0.811	0.670
DM-5-89	0.501	0.344	0.301	0.252	0.802	0.601
NIAB-IRRI-9 x DM-25	0.531	0.420	0.279	0.251	0.810	0.671
Jhona-349 x Basmati-370	0.544	0.449	0.320	0.278	0.864	0.727
Basmati-370 x Jhona-349	0.572	0.411	0.317	0.261	0.889	0.671
Basmati-370	0.528	0.366	0.289	0.238	0.817	0.604
DM-59418	0.556	0.405	0.309	0.262	0.866	0.667
Basmati-385 x NIAB-IRRI-9	0.491	0.370	0.300	0.231	0.791	0.601

Super Basmati	0.516	0.376	0.273	0.257	0.789	0.632
DM-38-88	0.560	0.392	0.292	0.255	0.851	0.646
DM-64198	0.602	0.404	0.320	0.261	0.922	0.665
Basmati-370 x NIAB-RICE-1	0.550	0.379	0.301	0.277	0.864	0.434
DM-3-89	0.447	0.449	0.277	0.258	0.799	0.767

Photosynthetic oxygen evolution, chlorophyll contents and chlorophyll *a/b* ratios of 3rd to 6th leaves of rice seedlings were measured to examine whether or not inactivation of photosynthesis during senescence is related to loss of chlorophyll. Photosynthetic activity decreased more rapidly than chlorophyll content during leaf senescence; as a result, the lower the leaf position, the lower was the rate of oxygen evolution determined on the basis of chlorophyll. Chlorophyll *a/b* ratio also decreased with advancing senescence. Electrophoretic analysis revealed that the decline in chlorophyll *a/b* ratio is due to more rapid degradation of the reaction center complexes than light-harvesting chlorophyll a/b proteins of photosystem II and that the photosystem I reaction center disappears in parallel with the inactivation of photosynthesis.

The main strategies enabling rice plants to cope with flash flooding stress require growth regulation during submergence and subsequent rapid growth recovery after de-submergence. The objective of this study was to characterize the response of 56 diverse contrasting rice genotypes to submergence and their recovery following de-submergence. Among these genotypes, nine lines had been developed for anaerobic germination and submergence tolerance (AG + *Sub1*) by IRRI. Fourteen-day-old plants were submerged completely in water for seven days. Subsequently, the plants were kept under normal rice-cultivation conditions as the control for a further period of five days. The tested genotypes were generally classified into three clusters based on shoot elongation rate of submerged to non-submerged treatments (ratio) during submergence period and chlorophyll contents during recovery period using Ward's method. The genotypes in clusters I include most of AG + *Sub1* lines and tolerant genotype FR13A adapted to submergence stress, which get the benefits of quiescence mechanism during submergence coupled with maintenance of higher chlorophyll content during recovery period. In contrast, the cluster III spanned most of intolerant genotypes such as IR42 by enhancing shoot elongation through escape mechanism in response to submergence. This mechanism negatively affected the plant growth recovery due to a great reduction of chlorophyll contents during the recovery period. The genotypes placed in cluster II followed the similar trend as cluster I during the submergence and recovery periods in addition to increases in shoot fresh weight during submergence period. This finding suggests that other mechanisms along with quiescence might be associated with submergence stress in the genotypes placed in cluster II. In conclusion, the contrasting rice genotypes expressed differential growth responses in genotypes with lower shoot elongation ratio using different quiescence strategies during submergence period(El-Hendawy et al. 2012).

Cold stress reduces significantly the concentration of chlorophyll in susceptible rice genotypes (Dai et al., 1990; Aghaee et al., 2011). Chlorophyll content was used as a tool to evaluate the degree of cold tolerance of transgenic plants (Tian et al., 2011), to monitor plant recovery after stress (Kuk et al., 2003), and to compare chilling tolerance between distinct hybrid lines during grain filling (Wang et al., 2006).

Analyses of fluorescence parameters have demonstrated structural and functional alterations in the photosynthetic apparatus of different plant genotypes or transgenic seedlings submitted to low temperature treatments (Saijo et al., 2000; Ji et al., 2003; Hirotsu et al., 2004; Lee et al., 2004a, b; Wang and Guo, 2005; Kim et al., 2009; Lee et al., 2009a, b; Bonnecarrère et al., 2011; Saad et al., 2012). In most cases, the ratio F_V/F_M is used to evaluate cold sensibility or tolerance, since it indicates the maximum photochemical efficiency of PS II. A new and more responsive parameter of fluorescence, photosynthetic performance index (PI) has been used recently in evaluations of cold tolerance in rice plants (Saad et al., 2012). This parameter includes components related to capture, absorption and use of luminous energy (Stirbet and Govindjee, 2011), and complements the evaluations of cold tolerance because it is closely related to the final photosynthetic activity of the plant and, therefore, to survival on stress conditions. However, the N source had little effect on gas exchange, Chl*a* fluorescence parameters, and photosynthetic electron allocation in rice plants, except that NH_4^+- grown plants had a higher O_2-independent alternative electron flux than NO_3^- grown plants. NO_3^- reduction activity was rarely detected in leaves of NH_4^+ grown cucumber plants, but was high in NH_4^+ grown rice plants.

8.6. *Mechanism Associated with Chlorophyll*

The branched isoprenoid pathway is rather complex and comprises enzymatic steps in at least two compartments. In plants, Phy represents the side chain of Chls, tocopherols (TP) and phylloquinones, and is necessary for their integration into plastid membranes (Soll et al. 1980 1983; Soll, 1987; Bollivar et al. 1994). In both Chl and TP synthesis, the Phy chain is provided by geranylgeranyl pyrophosphate (GGPP), a plastidial isoprenoid, formed by four molecules of isopentenyl pyrophosphate (IPP), which are derived from the cytosolic and chloroplastidic pathways (Rohmer et al. 1993; Lichtentaler et al. 1997). In Chl synthesis, GGPP can either be reduced to phytyl pyrophosphate (phypp) and esterified with Chlide to generate phytyl Chl (chlphy), or first esterified with Chlide to form geranylgeranylated Chl (chlgg) and then stepwise reduced into chlphy (Soll et al. 1983; Bollivar et al. 1994; Keller et al. 1998; Chew et al. 2008). In the TP pathway, tocopherols are generally believed to arise from the condensation of homogentisic acid and phypp (Schultz et al. 1985; Soll et al. 1980; Collakova and Dellapenna, 2001; Savidge et al. 2002; Cahoon et al. 2003).

Light amount and quality are powerful regulators of chlorophyll biosynthesis and chloroplast development. Light also establishes circadian and diurnal cycles that provide a constant internal control over gene expression and when in tune with environmental signals, plants display maximum growth (Harmer et al. 2000; McClung, 2006; Nozue et al. 2007). Nitrogen is required for building biological molecules and is therefore also intrinsically linked to both photosynthetic activity and the overall carbon status of the plant (Coruzzi and Bush, 2001; Coruzzi and Zhou, 2001). Nitrogen assimilation in the chloroplast is a prerequisite for chlorophyll biosynthesis, specifically by building up the glutamate pool (Ishizaki et al. 2007; Potel et al. 2009). The Glutamine Synthetase/Glutamate Synthase (GS/GOGAT) pathway is a key point in nitrogen assimilation where ammonium is incorporated into glutamate, providing the precursor for production of all amino acids, nucleic acids and chlorophylls (Potel et al. 2009; Kissen et al. 2010). The subsequent steps involved in chlorophyll biosynthesis are well documented and involve a number of key rate-limiting enzymes (Eckhardt et al. 2004). HEMA1 encodes a Glu-tRNA reductase enzyme that controls flux through the tetrapyrrole biosynthetic pathway and leads to production of 5-aminolevulinic acid (ALA) from which the porphyrin ring system is derived (Kumar and Soll, 2000; Hedtke et al. 2007). Genomes uncoupled 4 (GUN4) subsequently binds protoporphyrin chlorophyll intermediates (Mg-Proto and Mg-ProtoMe), stimulates Mg chelatase activity, and has also been implicated in plastidic retrograde signaling to regulate nuclear gene expression (Adhikari et al. 2009; Peter and Grimm, 2009; Davision et al. 2005). Light-dependent reduction of protochlorophyllide to chlorophyllide is catalyzed by NADPH: protochlorophyllide oxidoreductase (POR) in mature leaves, where the genes PORB and PORC form a redundant system regulating chlorophyll biosynthesis (Paddock et al. 2010). While these genes demonstrate circadian and diurnal patterns of expression (Stephenson et al. 2009), the exact mechanism by which chlorophyll content is adjusted with varying amounts of light and nitrogen is not well documented.

Because nitrogen is a key component of the chlorophyll molecule, the concentration of nitrate available to a plant directly influences chlorophyll biosynthesis and chloroplast development (Peng et al. 2007; Bondada and Syvertsen, 2003). Chlorophyll content is a key indicator of plant health and can be used to optimize nitrogen fertilizer application in order to potentiate larger crop yields with lower environmental load (Hussain et al. 2000;

Scharf et al. 2006). A subset of nitrate responses are mediated by the class of plant hormones known as cytokinins, whose synthesis and transport is linked to the nitrogen status of the plant (Takai et al. 2001; Sakakibara et al. 2006; To and Kieber, 2008). Cytokinin signaling plays a central role in the regulation of cell division, differentiation and various developmental processes including chlorophyll biosynthesis and chloroplast development (Argueso et al. 2010). Cytokinins have also been shown to exert control over the process of N-remobilization and grain development (Muller and Sheen, 2007; Ashikari *et* al. 2005; Bartrina et al. 2001). Understanding the processes involved in fine tuning chloroplast development and grain production with fluctuating light and nitrogen levels is vital for making agricultural improvements in crop plants.

Chloroplast development is an important determinant of plant productivity and is controlled by environmental factors including amounts of light and nitrogen as well as internal phytohormones including cytokinins and gibberellins (GA). The paralog GATA transcription factors GNC and CGA1/GNL up-regulated by light, nitrogen and cytokinin while also being repressed by GA signaling. Modifying the expression of these genes has previously been shown to influence chlorophyll content in Arabidopsis while also altering aspects of germination, elongation growth and flowering time. In this work, we also use transgenic lines to demonstrate that GNC and CGA1 exhibit a partially redundant control over chlorophyll biosynthesis. We provide novel evidence that GNC and CGA1 influence both chloroplast number and leaf starch in proportion to their transcript level. GNC and CGA1 were found to modify the expression of chloroplast localized Glutamate Synthase (GLU1/Fd-GOGAT), which is the primary factor controlling nitrogen assimilation in green tissue. Altering GNC and CGA1 expression was also found to modulate the expression of important chlorophyll biosynthesis genes (GUN4, HEMA1, PORB and PORC). As previously demonstrated, the CGA1 transgenic plants demonstrated significantly altered timing to a number of developmental events including germination, leaf production, flowering time and senescence. In contrast, the GNC transgenic lines we analyzed maintain relatively normal growth phenotypes outside of differences in chloroplast development. Despite some evidence for partial divergence, results indicate that regulation of both GNC and CGA1 by light, nitrogen, cytokinin and GA acts to modulate nitrogen assimilation, chloroplast development and starch production. Understanding the mechanisms controlling these processes is important for agricultural biotechnology.

Chlorophyll (Chl) synthase catalyzes esterification of chlorophyllide to complete the last step of Chl biosynthesis. Although the Chl synthases and the corresponding genes from various organisms have been well characterized, Chl synthase mutants have not yet been reported in higher plants. In this study, a rice (*Oryza Sativa*) Chl-deficient mutant, yellow-green leaf1 (ygl1), was isolated, which showed yellow-green leaves in young plants with decreased Chl synthesis, increased level of tetrapyrrole intermediates, and delayed chloroplast development. Genetic analysis demonstrated that the phenotype of ygl1 was caused by a recessive mutation in a nuclear gene. The ygl1 locus was mapped to chromosome 5 and isolated by map-based cloning. Sequence analysis revealed that it encodes the Chl synthase and its identity was verified by transgenic complementation. A missense mutation was found in a highly conserved residue of YGL1 in the ygl1 mutant, resulting in reduction of the enzymatic activity. YGL1 is constitutively expressed in all tissues, and its expression is not significantly affected in the ygl1 mutant. Interestingly, the mRNA expression of the cab1R gene encoding the Chl *a*/*b*-binding protein was severely suppressed in the ygl1 mutant. Moreover, the expression of some nuclear genes associated with Chl biosynthesis or chloroplast development was also affected in ygl1 seedlings. These results indicate that the expression of nuclear genes encoding various chloroplast proteins might be feedback regulated by the level of Chl or Chl precursors.

References

Adhikari, N.D., R. Orler, J. Chory, J.E. Froehlich and R.M. Larkin. 2009. Porphyrins promote the association of GENOMES UNCOUPLED 4 and a Mg-chelatase subunit with chloroplast membranes. *J.Biol.Chem.*, **284:** 24783-24796.

Aghaee, A., F. Moradi, H. Zare-Mavian, F. ZarinKamar, H. Pour-Irandoost and P. Sharifi. 2011. Physiological responses of two rice (*Oryza sativa* L.) genotypes to chilling stress at seedling stage. *Afr.J.Biotechnol.*, **10:** 7617-7621.

Ali, M.A. 2005. Productivity and resource use efficiency of rice as affected by crop establishment and N management. Ph.D.Thesis, UPLB, Philippines.

Argueso, C.T., T. Raines and J.J. Kiebber. 2010. Cytokinin signaling and transcriptional networks. *Curr. Opin. Plant Biol.*, **13:** 533-539.

Armstrong, G. and K. Apel. 1998. Molecular and genetic analysis of light dependent chlorophyll biosynthesis. *Methods in Enzymol.*, **297:** 237-244.

Ashikari, M., H. Sakakibara, S. Lin, T. Yamamoto, T. Takahashi, A. Nishimura, E.R. Angles,Q. Qian, H. Kitano and M. Matsuoka. 2005. Cytokinin oxidase regulates rice grain production. *Science*, **309:** 741-745.

Balasubramanian, V., A.C. Morales, R.T. Cruz, T.M. Thiyagarajan, R. Nagarajan, M. Babu, S. Abdulrachman and L.H. Hai. 2000. Adaptation of the chlorophyllmeter (SPAD) technology for real time N-management in rice: A review. *Int.Rice Res. Inst.*, 25-26.

Balogh, A., P. Pepo and M. Hornok. 2006. Interaction of crop year, fertilization and variety in winter wheat management. *Cereal Res. Commu.*, **34:** 389-392.

Bartrina, I., E. Otto, M. Strand, T. Werner and T. Schmulling. 2011. Cytokinin regulates the activity of reproductive meristems, flower organ size, ovule formation and thus seed yield in Arabidopsis thaliana. *Plant Cell*, **23:** 69-80.

Bollivar, D.W., S.J. Wang, J.P. Allen and C.E. Bauer. 1994. Molecular genetic analysis of terminal steps in bacteriochlorophylla. Biosynthesis. Characterization of a Rhodobacter capsulatus strain that synthesizes geranyl geraniol esterified bacteriochlorophylla. *Biochemistry*, **33:** 12763-12768.

Bondada, B.R. and J.P. Syvertsen. 2003. Leaf chlorophyll, net gas exchange and chloroplast ultrastructure in citrus leaves of different nitrogen status. *Tree Physiol.*, **23:** 563-569.

Bonnecarrere, V., O. Borsani, P. Diaz, F. Capdevielle, P. Blanco and J. Mouza. 2011. Response to photooxidative stress induced by cold in japonica rice is genotype dependent. *Plant Sci.*, **180:** 726-732.

Brown, S.B., J.D. Houghton and G.A.F. Hendry. 1991. Chlorophyll break down, pp.465-489. *In:* Scheer, H. (Ed). Chlorophylls. CRC Press, Boca Raton.

Bruisna, J. 1961. A comment on the spectrophotometric determination of chlorophyll. *Biochem. Biophys. Acta,* **52:** 576-578.

Cahoon, E.B., S.E. Hall, K.G. Ripp, T.S. Ganzke, W.D. Hitz and S.J. Coughlan. 2003. Metabolic redesign of vitamin E biosynthesis in Arabidopsis thaliana and implications for the evaluation of Prochlorococcus species. *Plant Cell,* **17:** 233-240.

Cha-um, S., S. Yooyongwech and S. Supaibulwatana. 2012. Water deficit tolerant classification in mutant lines of indica rice. *Sci.Agric.*, **69:** 135-141.

Cha-um, S. and C. Kirdmanee. 2011. Remediation of salt affected soil by the addition of organic matter An investigation into improving glutinous rice productivity. *Sci.Agric,* **68:** 406-410.

Chew, A.G.M., N.V. Frigaard and D.A. Bryant. 2008. Identification of the bch P gene, encoding geranyl geranyl reductase in chlorobaculum tepidum. *J. Bacteriol.*, **190:** 747-749.

Clark, R.B. 1983. Plant genotype differences in uptake, translocation, accumulation and use of mineral elements required for plant growth. *In:* Genetic aspects of plant nutrition. Martinus Nijhoff, The Hague.

Collakova, E. and D. Della Penna. 2001. Isolation and functional analysis of homogentisate phytyl transferase from Synechocystis sp. Pcc6803 and Arabiodopsis. *Plant Physiol.*, **127:** 1113-1124.

Coruzzi, G.M. and D.R. Bush. 2001. Nitrogen and carbon nutrient and metabolite signaling in plants. *Plant Physiol.*, **125:** 61-64.

Coruzzi, G.M. and L. Zhou. 2001. Carbon and nitrogen sensing and signaling in plants: Emerging matrix effects. *Curr.Opin. Plant Biol.*, **4:** 247-253.

Dai, Q, S.V. Benito and M.V. Romeo. 1990. Amelioration of cold injury in rice (Oryza sativa L.) improving root oxidizing activity by plant growth regulators. *Philipp. J.Crop Sci.*, **15:** 49-54.

Dalal, V.K. and B.C. Tripathy. 2012. Modulation of chlorophyll biosynthesis by water stress in rice seedings during chloroplast biogenesis. *Plant Cell Environ.*, **35:** 1685-1703.

Davision, P.A., H.L. Schubert, J.D. Reid, C.D. Iorg, A. Heroux, C.P. Hill and C.N. Hunter. 2005. Structural and biochemical characterization of Gun 4 suggests a mechanism for its role in chlorophyll biosynthesis. *Biochemistry,* **44:** 7603-7612.

Eckhardt, U., B. Grimm and S. Hortensteiner. 2004. Recent advances in chlorophyll biosynthesis and breakdown in higher plants. *Plant Mol.Biol.*, **56:** 1-14.

El-Hendawy, S., C. Sone, O. Ito and J.I. Sakagami. 2012. Differential growth response of rice genotypes based on quiescence mechanism under flash flooding stress. *Aus.J.Crop Sci.*, **6:** 1587-1597.

Evans, J.R. 1983. Nitrogen and photosynthesis in the flag leaf of wheat (*Triticum aestivum*). *Plant Physiol.* **72:** 297-302.

Gastal, F. and G. Lemaire. 2002. N uptake and distribution in crops: An agronomical and ecophysiological perspective. *J.Exp.Bot.*, **53:** 789-799.

Harmer, S.L., J.B. Hogenesch, M. Straume, H.S. Chang, B. Han, T. Zhu, X-Wang, J.A. Kreps and S.A. Kay. 2000. Orchestrated transcription of key pathways in Arabidopsis by circadian clock. *Science,* **290:** 2110-2113.

Hedtke, B., A. Alawady, S. Chen, F. Bornke and B. Grimm. 2007. HEMA RNAi silencing reveals a control mechanism of ALA biosynthesis on Mg-chelatase and Fe-chelatase. *Plant Mol.Biol.,* **64:** 733-742.

Hendry, G.A.F. and A.H. Price. 1993. Stress indicators: Chlorophyll and carotenoids *In:* Hendry, G.A.F. and Grimm, J.P. (Eds) Methods in comparative plant ecology. Chapman and Hall, London.

Hidema, J., A. Makino, T. Mae and K. Ojima. 1991. Photosynthetic characteristics of rice leaves aged under different irradiances from full expansion through senescence. *Plant Physiol.,* **97:** 1287-1293.

Hirotsu, N.,A. Makino, A. ushio and T. Mae. 2004. Changes in the thermal dissipation and the electron flow in the water-water cycte in rice grown under conditions of physiologically low temperature. *Plant Cell Physiol.* **45:** 635-644

Hiscox, J.D. and G.F. Israelstam. 1979. A method for the extraction of chlorophyll from leaf tissue without maceration. *Can.J. Bot.* **57:** 1332-1334.

Hussain, F., K.F. Bronson, Y. Singh, B. Singh and S. Peng. 2000. Use of chlorophyll meter sufficiency indices for nitrogen management of irrigated rice in Asia. *Agron.J.,* **92:** 876-879.

Inskeep, W.P. and P.R. Bloom. 1985. Extinction coefficients of chlorophyll*a* and *b* in N, N-dimethylformamide and 80% acetone. *Plant Physiol.* **77:** 483-485.

Ishizaki, T., C. Ohsumi, K. Totsuka and D. Igarashi. 2007. Analysis of glutamate homeostasis by over expression of Fd-GOGAT gene in *Arabidopsis thaliana. Amino Acids,* **38:** 943-950.

Islam, M.S., M.S.U. Bhuiya, S. Rahman and M.M. Hussain, 2009. Evaluation of SPAD and LCC based nitrogen management in rice (*Oryza sativa* L). *Bangladesh J. Agril. Res* **34:** 661-672.

Jinwen, L., Y. Jingping, F. Pinpin, S. Junlan, L. Dongsheng, G. Changsui and C. Wenyue. 2009. Responses of rice leaf thickness, SPAD readings and chlorophylls a/b ratio to different nitrogen supply rates in paddy field. *Field Crop Res.,* **114:** 426-432.

Keller, Y., F. Bouvier, A.D. Harlingue and B. Camata. 1998. Metabolic compartmentation of plastid premyllipid biosynthesis: Evidence for the involvement of a multifunctional CHLP. *Eur. J.Biochem.,* **251:** 413-417.

Kim, S.J., S.C. Lee, S.K. Hong, K.An, G. An and S.R. Kim. 2009. Ectopic expression of a cold responsive OsAsr 1 cDNA gives enhanced cold tolerance in transgenic rice plants. *Mol. Cell.* **27:** 449-458

Kissen, R., P. Winge, D.H. Tran, T.S. Jorstad, T.R. Storseth, T. Christensen and A.M. Bones. 2010. Transcriptional profiling of an Fd-GOGAT/GLU1 mutant in Arabidoposis thaliana reveals a multiple stress response and extensive reprogramming of the transcriptome. *BMC Genomics,* **11:** 190.

Kumar, A.M. and D. Soll. 2000. Antisense HEMAIRNA expression inhibits heme and chlorophyll biosynthesis in Arabidopsis. *Plant Physiol.,* **122:** 49-56.

Kuhl, M., M. Chen and P.J. Ralph. 2005. A niche for cyanobacteria containing Chlorophyll. *Nature,* **433:** 820.

Lazar, D. 1999. Chlorophyll*a* fluorescence induction. *Biochem. Biophys.Acta,* **1412:** 1-28.

Lee, S.C., J.Y. Kim, S.H. Kim, S.J. Kim, K. Lee and S.X Han, 2004a. Trapping and characterization of cold responsive genes from T-DNA tagging lines in rice. Plant Sci. **166:** 69-79

Lee, S.C., K.W. Huh, K. An, G.An and S.R. Kim. 2004b. Ectopic expression of a cold inducible transcription factor, CBF1/DREB1b in transgenic rice (oryze sativa L). *Mol. Cells* **18:** 107-114

Li, F.M., T.C. Ai, S.B. Zhou, X.J. Nie and F.Liu. 2004. Influence of slow release nitrogen fertilizers on lowland rice yield and nitrogen use efficiency. *Chin.J.Soil Sci.,* **35:** 311-315.

Lichtentaler, H.K., J. Schwende, A. Disch and M. Rohmer. 1997. Biosynthesis of isoprenoids in higher plant chloroplast, proceeds via a mevalonate independent pathway. *FEBS Lett.* **400:** 271-274.

Lichtentaler, H.K. and A.R. Wellbum. 1983. Determination of total carotenoids and chlorophylls a and b of leaf extracts in different solvents. *Biochem.Soc.Trans.* **11:** 591-592.

Liu, D.L., J. Nie and J. Xiao. 2002. Study on 15N labeled rice controlled release nitrogen fertilizers in increasing nitrogen utilization efficiency. *Acta. Laser Biol. Sin.,* **11:** 87-92.

Long, J.R., G.H. Ma, J. Zhou and C.F. Song. 2007. Effects of slow release nitrogen fertilizer on the growth and development and nitrogen use efficiency of super hybrid rice Yliangyon 1. *Hybrid Rice,* **22:** 48-51.

Luo, L.F., S.X. Zheng, Y.L. Liao, J. Nie and Y.W. Xiang. 2007. Effect of controlled release nitrogen fertilizer on protein quality of brown rice and key enzyme activity involved in nitrogen metabolism in hybrid rice. *Chin.J. Rice Sci.,* **21:** 403-410.

Malkin, R. and K. Nyogi. 2000. Photosynthesis. *In:* Buchanan, B.B., Gruissem, N. and Jones, R. (Eds) Biochemistry and Molecular Biology of Plants. Am.Soc. Plant Physiol., Rockville, MD. pp.575-577.

Marini, R.P. 1986. Do net gas exchange rates of green and red peach leaves differ? *Hort. Sci.,* **21:** 118-120.

McClung, C.R. 2006. Plant circadian rhythms. *Plant Cell,* **18:** 792-803.

McKinney, G. 1941. Absorption of light by chlorophyll solutions. *J.Biol.Chem.,* **140:** 315-332.

Miah, M.N.M. and Z.U. Ahmed. 2002. Comparative efficiency of the chlorophyll meter technique, urea super granule and prilled urea for hybrid rice in Bangladesh. *In:* Hybrid Rice in Bangladesh: Progress and future strategies. pp.43-50.

Miah, M.N. H., T. Yoshida and Y. Yomamoto. 1997. Effect of nitrogen application during ripening period on photosynthesis and dry matter production and its impact on yield and yield components of semi dwarfindica rice varieties under water culture conditions. *Soil Sci. and Plant Nutr.* **43:** 205-217.

Moran, R. and D. Porath. 1980. Chlorophyll determinations in intact tissue using N.N -dimethyl formamide. *Plant Physiol.,* **65:** 478-479.

Muller, B. and J. Sheen. 2007. Arabidopsis cytokinin signaling pathway. Sci. STKE:Cm5.

Nagata, N., R. Tanaka, S. Satoh and A. Tanaka. 2005. Identification of a vinyl reductase gene for chlorophyll synthesis in Arabidopsis thaliana and implications for the evolution of Prochloroccus species. *Plant Cell,* **17:** 233-240.

Nozue, K., M.F. Covington, P.D. Duek, S. Lorrain, C. Faukhauser, S.L. Harmer and J.N. Maloof. 2007. Rhythmic growth explained by coincidence between internal and external cues. *Nature,* **448:** 358-361.

Paddock, T.N., M.E. Mason, D.F. Lima and G.A. Armstrong. 2010. Arabidopsis photochlorophyllide oxidoreductase A (PORA) restores bulk chlorophyll synthesis and normal development to a porBporC double mutant. *Plant Mol. Biol.,* **72:** 445-457.

Park, M.R., E.A. Cho, S. Rehman and S.J. Yun. 2010. Expression of a sesame geranyl geranyl reductase cDNA is induced by light but repressed by abscisic acid and ethylene. *Pak.J.Bot.,* **42:** 1815-1825.

Peng. S., P.V. Garcia, R.C. laza, A.L. Sanico, R.M. Visperas and K.G. Cassman 1996. Increased N-use efficiency using a chlorophyll meter on high yielding irrigated rice. *Field Crop Res.* **47:** 249-252.

Peng, M., Y.M. Bi, T. Zhu and S.J. Rothstein. 2007. Genome wide analysis of Arabidopsis responsive transcriptome to nitrogen limitation and its regulation by the ubiquitin ligase gene NLA. *Plant Mol. Biol.,* **65:** 775-797.

Pepo, P. 2005. Effect of crop year, genetic and agrotechnical factors on dry matter production and accumulation in winter wheat production. *Cereal Res. Commu.,* **33:** 29-32.

Peter, E. and B. Grimm. 2009. GUN4 is required for post translational control of plant tetrapyrrole biosynthesis. *Mol.Plant,* **2:** 1198-1210.

Peterson, T.A., T.M. Blackmer, D.D. Francis and J.S. Scheppers. 1993. Using chlorophyllmeter to improve N management. A Webguide in Soil Resource Management, University of Nebraska, Lincoln, USA.

Potel, F., M.H. Valadier, S. Ferrasio-Mery, O. Grandjean, H. Morin, L.Gaufichon, S. Boutet-Mercey, J. Lothier, S.T. Rothstein, N. Hirose and A. suzuki. 2009. Assimilation of excess ammonium into amino acids and nitrogen translocation in Arabidopsis thaliana–Roles of glutamate synthases and carbamoyl phosphate synthetase in leaves. *FEBS J.* **276:** 4061-4076.

Prioul, J.L., J. Brangeon and A. Reyss. 1980. Interaction between external and internal conditions in the development of photosynthetic features in a grass leaf I. *Plant Physiol.,* **66:** 762-769.

Ralph, P.J., R. Gademann and N.S. Dennison. 1998. In situ seagrass photosynthesis measured using a submersible pulse-amplitude modulated fluorometer. *Marine Biol.* **132:** 367-373.

Rohmer, M., M. Knani, P. Simonin, B. Sutter and H. Sahm. 1993. Isoprenoid biosynthesis in bacteria: A novel pathway for the early steps leading to isopentenyl diphosphate. *Biochem J.* **295:** 517-524.

Saad, B.R., D. Fabre, D.Mieulet, D.Meynard, M.Dingkuhn and A. Al-Doss. 2012. Expression of the Aeluropus littoralis AlSAP gene in rice confers broad tolerance to abiotic stresses through maintenance of photosynthesis. *Plant Cell Environ.* **35:** 636-643

Saijo, Y.,S. Hata, J.Kyozuka, K. Shimamoto and K. Izui. 2000 over expression of a single Ca^{2+}-dependent protein kinase confers both cold and salt/drought tolerance on rice plants. *Plant J.* **23:** 319-327.

Sakakibara, H., K. Takaei and N. Hirose. 2006. Interactions between nitrogen and cytokinin in regulation of metabolism and development. *Trends Plant Sci.,* **11,** 440-448.

Saric, M. and V. Kovacevic. 1981. Sortna specificnost mineralue ishrane psenice. *In:* Fiziologija psenice (J.Belic Ed) Posebno izdanje SANU, knjiga 53, Beograd. pp. 61-77.

Savidge, B., J.D. Weiss, Y.A. wong, M.W. Lassner, T.A. Mitsky, C.K. Shewmaker, D. Post Beittenmiller and H.E. Valentin. 2002. Isolation and characterization of homogentisate phytyl transferase genes from synechocystis sp. PCC 6803 and Arabidopsis. *Plant Physiol.* **129:** 321-332.

Sakuraba, Y., S.H. Lee, Y.S. Kim, O.K. Park, S. Hortensteiner and N.C. Paek. 2013. Delayed degradation of chlorophylls and photosynthetic proteins in Arabidopsis autophagy mutants during stress induced leaf yellowing. *J.Exp. Bot.* Doi: 10.1093/jxb/eru008.

Schaper, H. and E.K. Chacko. 1991. Relation between extractable chlorophyll and portable chlorophyll meter readings in leaves of eight tropical and subtropical trui-tree species. *J. Plant Physiol.* **138:** 674-677.

Scharf, P.C., S.M. Brouder and R.G. Hoeft. 2006. Chlorophyll meter readings can predict nitrogen need and yield response of corn in the northcentral USA. *Agron.J.* **98:** 655-665.

Schultz, G., J. Soll, E. Fiedler and D. Schulze-Siebert. 1985. Synthesis of prenylquinones in chloroplasts. *Physiol.Plant.,* **64:** 123-129.

Shen, A.L., C.Z. Liu, F.S. Zhang, X.R. Huangpu and Z.L. Kun. 1997. Effects of different application rate of NPK on the growth of rice and N fertilizer utilization ratio under water leakage and non-leakage condition. *Chin.J.Rice Sci.,* **11:** 231-237.

Shoaf, T.W. and B.W. Lium. 1976. Improved extraction of chlorophyll a and b from algae using dimethyl sulphoxide. *Limnol. Oceanogr.,* **21:** 926-928.

Smeal, D. and H. Zhang. 1994. Chlorophyllmeter evaluation for nitrogen management in corn. *Commu. Soil Sci.Plant Anal.,* **25:** 1495-1503.

Soll, J., M. Kemmerling and G. Schultz. 1980. Tocopherol and plasto quinone synthesis in spinach chloroplasts sub fractions. *Arch.Biochem. Biophys.,* **200:** 544-550.

Soll, J., G. Schultz, W. Rudiger and J. Benz. 1983. Hydrogenation of geranylgeraniol two pathways exist in spinach chloroplast. *Plant Physiol.* **71:** 849-854.

Soll, J. 1987. Alpha-Tocopherol and plastoquinone synthesis in chloroplast membranes. *Methods Enzymol,* **148:** 383-392.

Stephenson, P.G., C. Fankhauser and M.J. Terry. 2009. PIF 3 is a repressor of chloroplast development. *Proc.Natl.Acad.Sci.USA,* **106:** 7654-7659.

Stirbet, A. and Govindjee. 2011. On the relation between The kautsky effect (chlorophyll a fluorescence induction) and photosystem 11 : basics and applications of the OJIP fluorescence transient. *J. Photochem. Photobid.* B. **104:** 236-257

Takei, K., H. Sakakibara, M. Taniguchi and T. Sugiyama. 2001. Nitrogen dependent accumulation of cytokimin in root and the translocation to leaf: Implication of cytokinin species that induces gene expression of maize response regulator. *Plant Cell Physiol.* **42:** 85-93.

Takei, K., T. Takahashi, T. Sugiyama, T. Yamaya and H. Sakakibara. 2002. Multiple routes communicating nitrogen availability from roots to shoots: A signal transduction pathway mediated by cytokinin. *J.Exp..Bot.,* **53:** 971-977.

Tian L., I.H. Van Stokkum, R.B. Kochorst A.Jangerius, D. Kirilovsky and H. van Amerongen. 2011. Site, rate and mechanism of photoprotective quenching in cyanobacteria. *J.Am. Chem. Soc.* **133:** 18304-18311.

To, J.P. and J.J. Kieber. 2008. Cytokinin signaling: Two components and more. *Trends Plant Sci.,* **13:** 85-92.

Tucker, M. 2004. Primary Nutrients and Plant Growth. *In:* Essential Plant Nutrients (SCRIBD-Ed) North Carolina Department of Agriculture

Tremkel, M.E. 1997. Improving fertilizer use efficiency controlled release and stabilized fertilizers in agriculture. IFA, Paris. pp 99-104.

Wang, J.,C. Zhang. G. Chen, P. Wang. D.Shi and C. Liu. 2006. Responses of photosynthetic functions to low temperature in flag leaves of rice genotypes at the milky stage. Rice Sci. **13:** 113-119

Yang, C.M. and Y.J. Lee. 2001. Seasonal changes of chlorophyll content in field grown rice crops and their relationships with growth. *Proc. Natl.Sci.Counc. Repub.ChinaB,* **25:** 233-238.

Zhang, M., Y.X. Shi, S.X. Yang and Y.C. Yang. 2001. Statusquo of study of controlled release and slow release fertilizers and progress made in this respect. *J.Chem. Fert. Industry,* **28:** 78-93.

Zhu, Z.L. 2000. Loss of fertilizer nitrogen from plant soil system and the strategies and techniques for its reduction. *Soil Environ.Sci.,* **9:** 1-6.

chapter nine

Proteins

Proteins are the polymers of amino acids which constitute the principal structural unit of protoplasmic organelles, enzymes, and other macromolecules of biological importance. It serves as the energy source and source of nitrogen for organisms. Proteins in plants are synthesized within the body cells through a genetically controlled machinery by virtue of nuclear DNA and ribosomes. DNA produces messenger RNA by transcription. The sequence of m-RNA translates specific sequence of amino acids to produce a polypeptide chain which later undergoes post, synthetic processing transformed into primary, secondary and tertiary proteins essential for maintenance of life in a cell. DNA acts as the central dogma of protein synthesis which especially is controlled by genes present in DNA.

In a protein, amino acids link together by polypeptide bond in which the carboxyl group of one joins amino group of the other and so on. Ribosomal enzymes help in forming of peptide bond and acts as the site for protein synthesis. The sequence of twenty different amino acids on a peptide chain is determined by the sequence of triplet bases on the m-RNA. Mechanism of protein synthesis involves following steps.

(I) Transcription

For protein synthesis, three types of RNAs are involved to carry genetic informations from nucleus to the site for protein synthesis. These are messenger RNA (m-RNA), transfer RNA (t-RNA) and ribosomal RNA (r-RNA). Transcription is the process of synthesis of m-RNA from nuclear DNA which takes place in presence of DNA-directed-RNA polymerase enzyme. m-RNA molecule is transcribed from DNA molecule in the form of templet binding. This process of transcription takes place through following steps.

i) First the parent DNA strands uncoil in presence of RNA polymerase enzyme.
ii) Only one strand called sense strand takes part in templet formation.
iii) On the sense strand a new RNA strand is formed as replica in which uracil residues are inserted in the position specified by adenine in place of thymine.

iv) During transcription the new RNA strand base pairs temporarily with the templet DNA strand to form a short length of hybrid strand of DNA-RNA double helix. Then the RNA "Peels off" shortly after its formation.

v) RNA polymerase is a complex enzyme containing five polynucleotide subunits as holoenzyme. During transcription, this holoenzyme binds to a specific site in the DNA called promoters site which consists of a short sequence recognized by the RNA polymerase enzyme.

vi) RNA polymerase correctly positioned at the promoters site of DNA from where the DNA splits and transcription initiates.

vii) The core enzyme signals specific sequence to terminate the RNA elongation.

In higher plants cells (Eukaryotic) there are three RNA polymerase enzymes involved in transcription.

RNA-polymerase I - It causes formation of r-RNA.

RNA-polymerase II - Helps in transcription of m-RNA.

RNA-polymerase III - Transcribe t-RNA.

After transcription, RNA polymerase also helps in post-transcriptional processing of different RNA. In eukaryotic plant cell m-RNA undergoes special post-transcription modification to yield longer and shorter precursors from which t-RNA and r-RNA are transcribed respectively.

It is experimentally found that in prokaryotic plant cell m-RNAs are processed from heterogeneous nuclear RNA initially transcribed from DNA.

Genetic message for arranging amino acids in specific sequence are carried by m-RNA in terms of triplet base codons in polypeptide chain.

(II) Translation

Translation is the central process of protein synthesis by which polypeptide chain is formed with sequential arrangement of amino acid. It takes place in the ribosome which involves hundreds of enzyme complex and RNAs. The process of translation refers to the transformation of codes into sequence of amino acids. It involves following steps.

i) Activation of amino acids

This process takes place in cytosol (cytoplasm). Amino acids which are preserved in cytosol are in inactive state. These are activated at the expense of energy from ATP in presence of aminoacyl RNA synthetase enzyme. Amino acids bound to the enzyme forming a highly reactive amino-acid-adenylate complex with release of pyrophosphate. This reaction is catalyzed by Mg^{++}.

$$AA + ATP + E \sim AMP + PP$$

(Amino acid) (Enzyme) (Active complex)

ii) Attachment of activated amino acid to the t-RNA

The enzyme bound activated amino acid adenylate attach with respective t-RNA molecule. For each amino acid it is a specific t-RNA having anticodon complementary to the codon of m-RNA. Amino acid attach to the CCA end of t-RNA. This reaction is catalysed by aminoacyl t-RNA synthetase enzyme.

iii) Initiation of the peptide chain

For initiation of polypeptide chain, necessary components include m-RNA, methionly-t-RNA complex, 40-S ribosomal segment 605-ribosomal subunit, GTP, Mg^{++} and initiation factors IF-1, IF-2, IF-3 and imitating codon "AUG".

During initiation the messenger RNA bearing the codes for the polypeptide to be made is bound to the 40-S sub unit of ribosome (30 in Bacteria). Followed by this the initiating amino acid methionine (F-Methionine in Bacteria) binds to its t-RNA to form initiation complex. The t-RNA of the initiating amino acid base-pairs with a specific nucleotide triplet or codon on the m-RNA signals the begging of the polypeptide chain.

This process requires guanosine triphosphate (GTP) and is promoted by three specific cytosolic proteins called initiation factors (IF-1, IF-2, IF-3). Francis Crick opined t-RNA functions as adoptor, so that one part of t-RNA molecule can bind a specific amino acid and other end recognizes a short nucleotide sequence in the messenger RNA coding for that amino acid.

At one binding site t-RNA molecule bears corresponding anticodon UAC covalently attach to the complementary codon AUG on the m-RNA at the "P" side of ribosome. It results is the formation of an initiation complex with both subunits of ribosome. But all other incoming aminoacyl t-RNA complex binds at 'A' side of ribosome.

iv) Elongation of polypeptide chain

After formation of imitating complex at 'P' site the 'A' site is free to accept a specific charged t-RNA. Then in presence of enzyme peptide synthetase a co-valent peptide bond is formed between methionine and the second amino acid. This is regulated by some elongation factors. Energy required for binding each aminoacyl-t-RNA and for the movement of the ribosome along the messenger RNA by one codon comes from the hydrolysis of two molecules of GTP for each residue added to the growing polypeptide. After formation of a dipeptide the t-RNA of P-site after donating amino acid is removed from the ribosome. Then both m-RNA and ribosome move in opposite direction a step further so that the third codon of m-RNA occupies the A-site and the amino acid of the A-site shifts to P-site. The first codon of m-RNA whose translation is completed is now out of the ribosome. Now enzyme peptide synthetase helps in further establishment of second peptide bond between second and third amino acid.

In this way, m-RNA is translated one codon after another beginning with 5 end to 3 end. Polypeptide chain keeps on growing causing elongation of chain by addition of amino acid one by one in specific sequence.

v) Chain termination

There are specific termination codons present in the sequence of codon along m-RNA specified by DNA. These codons are UUA, UAG and UGA. These codons assign no amino acids. When one of these codons appears on the m-RNA, a gap in peptide chain appears and the chain is terminated. These codons are called terminator or nonsense codons. After the chain is terminated it is released from ribosome promoted by releasing factors, i.e. Proteins R_1, R_2 and S. Now two sub units of ribosome dissociate to again engage in formation of a new peptide.

vi) Post-translation processing

In order to achieve its native biologically active form the polypeptides undergo folding into its proper three dimensional conformation. Before or after folding, new polypeptides undergo processing by enzymatic action.

These post-translation modifications involve :

1. Modification of amino terminal and carboxyl terminals
2. Loss of signaling sequence by peptidase activity
3. Phosphorylation of hydroxyl amino acids by ATP
4. Carboxylation of some amino residues
5. Methylation of R-group
6. Attachment of carbohydrate side chains
7. Addition of prosthetic group
8. Formation of disulfide cross links between the overlapping folds

By above process a finished product protein is produced and is directed to the destinations in cell.

9.1. Proteomics

Proteomics can be defined as the systematic analysis of proteome, the protein complement of genome (Pandey and Mann, 2000; Patterson and Aebersold, 2003; Phizicky et al. 2003). This technology allows the global analysis of gene products in various tissues and physiological states of cells. With the completion of genome sequencing projects and the development of analytical methods for protein characterization, proteomics has become a major field of functional genomics. The initial objective of proteomics was the large-scale identification of all protein species in a cell or tissue. The applications are currently diversified to analyze various functional aspects of proteins such as post-translational modifications, protein-protein interactions, activities and structures. Here, general technologies of proteomics will be briefly introduced, followed by the overallprogress in plant proteomics. Although 2DE-based proteomics has proven powerful for the global analysis of proteins, it still retains technical problems that need to be solved (Corthals et al. 2000; Gygi et al. 2000). It is costly and a labor- and time-consuming process, limiting high-throughput analysis of protein expression. In addition, the entire protein profiling and quantification are not possible due to the limited loading capacity and incomplete staining methods.

An alternative method to analyze proteins directly by MS, without gel separation, has been developed to overcome these 2DE associated limitations. It is referred to as multidimensional protein identification technology (MudPIT) or liquid chromatography (LC)-MS/MS that couples capillary, high performanceliquid chromatography (HPLC) to MS/MS and allows automated analyses of peptide mixtures that are generated from complex protein samples (Appella et al. 1995; Washburn et al. 2001; Wolters et al. 2001). Furthermore, quantitative proteomics became feasible using an innovative reagent, termed isotope-coded affinity tag (ICAT), in the LC-MS/MS system (Han et al. 2001). In the post-genomic era, proteomics is positioned at the center of the functional genomics to study gene function on a genome-wide scale. The unique feature of proteomics is its feasibility to analyze the changes occurring at the protein level that cannot be predicted from genomic sequence. Proteins undergo post-translational modifications, proteolysis, recycling, multicomplex formation and subcellulartranslocation that are key events to regulate protein functions in cellular processes. Proteomics can eventually reveal all proteins in a cell or tissue at any given time, including those with post-translational changes. However, it also has technical limitations that are continually being researched. With the improved strategies, proteomics will surely provide powerful tools to disclose gene function in plant biology.

Rice (*Oryza sativa* L.) is an important crop in eastern Asia. A vast number of rice cultivars as well as wild species of rice are widely grown, and their genetic and molecular makeup is being actively investigated. Rice is also considered a model plant in monocots because of the relatively small size of its genome. The rice genome conceivably consists of about 430 million base pairs (Sasaki, 1998), and about 30,000 genes can be expressed in rice plant tissue. High resolution two-dimensional PAGE is useful for separating complex protein mixtures (O'Farrell, 1975). Due to its high resolving power, the technique has been used to study alterations in cellular protein expression in response to various stimuli or as a result of differentiation and development (Celis and Bravo, 1984). The latter approach further allows cDNA cloning from the resultant sequence(s).

The Rice Genome Research Project is a joint project of the National Institute of Agro- biological Sciences and the Institute of the Society for Techno-innovation of Agriculture, Forestry and Fisheries. In addition, this resulted in the establishment of some of the basic tools of rice genome analysis. The Rice Genome Research Project was reorganized into a important project in 1998; objectives of the organization are to completely sequence the entire rice genome and to pursue integrated goals in functional genomics, genome informatics, and applied genomics. Two important objectives in rice proteome research are : (1) to determine whether the cDNA encoding particular proteins from the cDNA library constructed from rice can be identified by a computer search of an amino acid sequence homology and (2) to predict the function of the proteins and study the physiological significance of functional proteins in rice.

Sequencing of a protein separated by two-dimensional PAGE became possible with the introduction of protein electro blotting methods that allow the efficient transfer of a sample from the gel matrix onto a support that is suitable for gas-phase sequencing or related techniques (Matsudaria, 1987). Proteins can also be recognized by their amino acid composition, their exact molecular weight as determined by mass spectrometry (MS),[1] or their partial amino acid sequence. In conjunction with automated gel scanning and computer-assisted analysis, two-dimensional PAGE has contributed greatly to the development of a protein data base (Anderson and Anderson, 1977; Garrels and Franz, 1989; Hirano, 1989).

Gel-separated proteins can be identified rapidly by MS, and if genomic information is also available, such analyses permit the systematic identification of the protein complement of a genome (Yates, 1998). In addition, MS is a powerful tool for the analysis of isoforms, secondary modifications of proteins such as glycosylation and phosphorylation, and proteolysis, which only require low amounts (pico moles to atto moles) of proteins (Wilkins et al. 1999). Such systematic analyses of protein populations are summarized by the term proteomics. Thus, proteomics bridges the gap between genomic sequence information and the actual protein population in a specific tissue, cell or cellular compartment.

Concerning the rice plant, some well known studies have dealt with the construction of proteomes from complex origins, such as the leaf, embryo, endosperm, root, stem, shoot and callus proteome (Komatsu et al. 1993; Komatsu et al. 1999; Zhong et al. 1997; Tsugita et al. 1994). Proteomic studies to date have mainly focused on those changes in genome expression that are triggered by environmental factors. Examples of descriptive proteomes include the global comparison of green and etiolated rice shoots (Komatsu et al. 1999 a,b) and an analysis of defense-associated responses in the rice leaf and leaf sheath following a jasmonic acid treatment (Rakwal and Komatsu 2000). One major advantage of the rice two-dimensional PAGE data base, in which most known proteins are recorded, is the wealth of new proteins on which experiments can be conducted at the biochemical and molecular levels. In addition to facilitating the identification of known proteins, these sequences can be used to prepare oligodeoxyribonucleotides, which are essential for cloning the corresponding cDNA. The proteomics was to separate proteins from rice, to determine their relative molecular weights and isoelectric points, and to perform N-terminal and internal amino acid sequence analysis using a protein sequencer and MS.

9.2. Changes in Seeds

Dormancy and germination are complex traits that are controlled by a large number of genes, which are affected by both developmental and environmental factors. Seed dormancy and germination depend on seed structures, especially those surrounding the embryo, and on factors affecting the growth potential of the embryo. The latter may include compounds that are imported from the mother plant and also factors that are produced by the embryo itself, including several plant hormones. Genetic analysis has identified the crucial role of ABA in seed dormancy, as well as the requirement for gas for germination. QTL and mutant analyses are identifying additional genes. Whether these genes with unknown functions are downstream targets of ABA and GA, or whether they affect seed dormancy/germination in an independent way is currently not known. The molecular identification of all these gene will be important, as will the identification of more target genes. Using whole transcriptome and proteome approaches will be the most efficient way to identify target genes.

Seeds are the most important plant storage organ and play a central role in the life cycle of plants. Since little is known about the protein composition of rice (*Oryza sativa*) seeds, analysis had been undertaken using proteomic methods to obtain a reference map of rice seed proteins and identify important molecules. Overall, 480 reproducible protein spots were detected by two-dimensional electrophoresis on pH 4-7 gels and 302 proteins were identified by MALDI-TOF MS and database searches. Together, these proteins represented 252 gene products and were classified into 12 functional categories, most of which were involved in metabolic pathways.

Database searches combined with hydropathy plots and gene ontology analysis showed that most rice seed proteins were hydrophilic and were related to binding, catalytic, cellular or metabolic processes. These results expand our knowledge of the rice proteome and improve our understanding of the cellular biology of rice seeds(Yang et al. 2013).

9.3. *Classification of Protein Functions*

The 302 identified proteins represented the products of 252 different genes and were classified into 12 categories based on their functions (Fig. 9.1) (Bevan et al. 1998). Protein functions were retrieved online as Gene Ontology information. The 12 categories were: Metabolism (1), Disease/defense (2), Cell structure (3), Energy (4), Signal transduction (5), Protein destination and storage (6), Cell growth/division (7), Protein synthesis (8), Transcription (9), Transporters (10), Intracellular traffic (11) and Unknown protein (12). The functional categories were determined according to Bevan et al. (1998). As shown in Fig. 9.1, 75 spots were involved in metabolic processes and were the most abundant category (24.8%). Proteins related to disease/defense were the second most abundant category (16.9%) and unknown proteins were the third most abundant (16.2%).

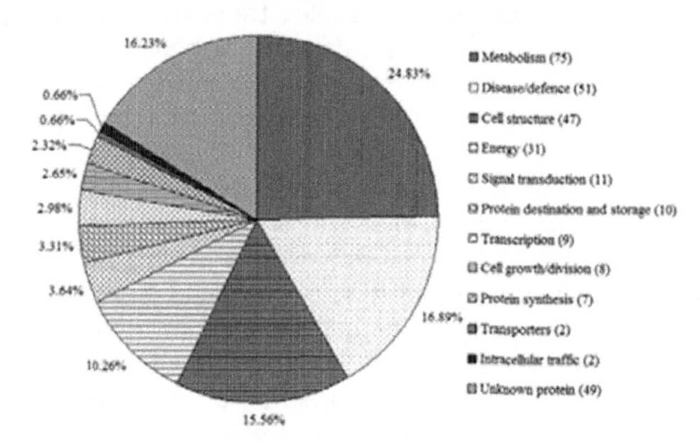

FIGURE 9.1 Fnctional classifications of the idenified proteins. The number of proteins in each category is indicated in parentheses.

9.4. *Bioinformatics Analysis of Identified Proteins*

Proteins with negative GRAVY scores were hydrophilic and those with positive GRAVY scores were hydrophobic. Fig. 9.2 shows that identified proteins with negative GRAVY scores were significantly more abundant than those with positive GRAVY scores. The GRAVY values of most proteins were between -0.6 and 0, indicating that most of them were hydrophilic.

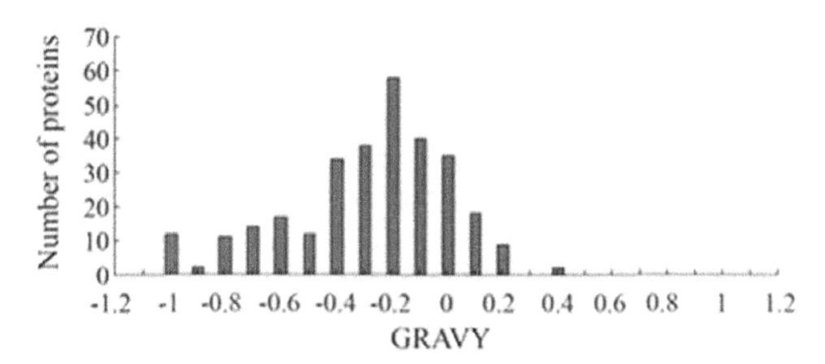

FIGURE 9.2 Hydropathic analysis of all proteins identified by 2-DE. Negative and positive GRAVY values indicate hydrophilic and hydropholbic proteins, respectively.

Figure 9.3 shows the GO analysis of the identified proteins, all of which were classified in terms of cellular component, molecular function, and physiological and biological processes using appropriate software (Gene Ontology Annotation Plot, WEGO). Most of the identified proteins associated with cellular components were involved in cell, cell parts, envelope, macromolecular complex, organelle and organelle parts, while those associated with molecular functions were involved in antioxidant, binding, catalytic, electron carrier, enzyme regulator, nutrient reservoir, transcription regulator and transporter activities. Biological processes involved biological regulation, cellular component organization, cellular process, establishment of localization, localization, metabolic process, multi-organism process, multicellular organismal process, pigmentation, reproduction, reproductive process and response to stimulus (Fig. 9.3).

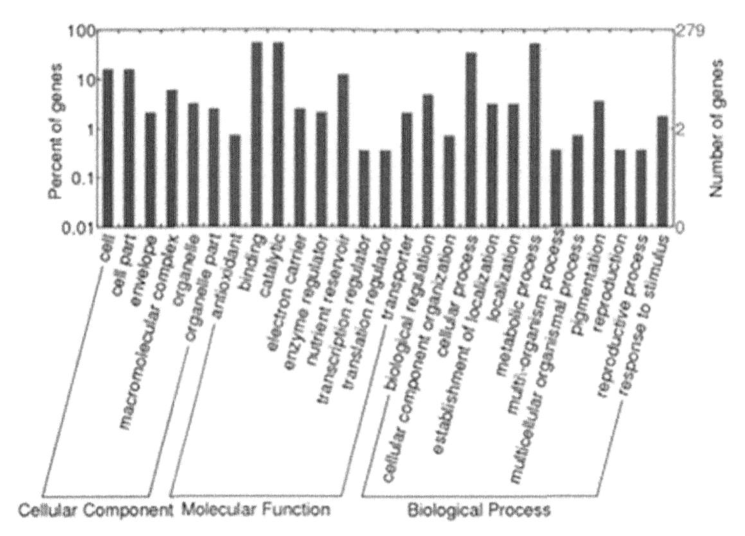

FIGURE 9.3 GO classifications of the identified proteins. All of the proeins were classified ino three main categories and 26 subcategories.

Dry seeds contain translatable, long-lived mRNAs that are stored during seed maturation. Early studies using transcriptional inhibitors supported the view that protein synthesis during the initial phase of germination occurs on long-lived mRNA templates. Rice seeds were treated with the transcriptional inhibitor actinomycin D (Act D), and the embryonic proteins translated from long-lived mRNAs during germination were identified using a proteomic analysis. De novo transcription was not required for germination of rice seeds, since >80% of seeds germinated when transcription was prevented by treatment with Act D. In contrast, germination was completely inhibited in the presence of cycloheximide, an inhibitor of translation. Thus, de novo protein synthesis is necessary for germination of rice seeds. The proteomic analysis revealed that 20 proteins are up-regulated during germination, even after Act D treatment. Many of the up-regulated proteins are involved in carbohydrate metabolism and cytoskeleton formation. These results indicate that some of the germination-specific proteins involved in energy production and maintenance of cell structure in rice seeds are synthesized from long-lived mRNAs. The timing of translation of eight up-regulated proteins was clearly later than that of the other up-regulated proteins under conditions in which transcription was inhibited by Act D, suggesting that translation of long-lived mRNAs in rice seeds is regulated according to the germination phase (Sano et al. 2012).

Although seed germination is a major subject in plant physiological research, there is still a long way to go to elucidate the mechanism of seed germination. Recent (ly, functional genomic strategies have been applied to study the germination of plant seeds. Here, a proteomic analysis of seed germination in rice (*Oryza sativa indica* cv. 9311) – a model monocot–was conducted. Comparison of 2-DE maps showed that there were 148 proteins displayed differently in the germination process of rice seeds. Among the changed proteins, 63 were down-regulated, 69 were up-regulated (including 20 induced proteins). The down-regulated proteins were mainly storage proteins, such as globulin and glutelin, and proteins associated with seed maturation like "early embryogenesis protein" and "late embryogenesis abundant protein", and proteins related to desiccation, such as "abscisic acid-induced protein" and "cold-regulated protein". The degradation of storage proteins mainly happened at the late stage of germination phase II (48 h imbibition), while that of seed maturation and desiccation associated proteins occurred at the early stage of phase II (24 h imbibition). In addition to α-amylase, the up-regulated proteins were mainly those involved in glycolysis such as UDP-glucose dehydrogenase, fructokinase, phosphoglucomutase, and pyruvate decarboxylase. The results reflected the possible biochemical and physiological processes of germination of rice seeds (Yang et al. 2007).

Seed germination begins with water uptake and ends with radicle emergence. A gel-free phosphoproteomic technique was used to investigate the role of protein phosphorylation events in the early stages of rice seed germination. Both seed weight and ATP content increased gradually during the first 24 h following imbibition. Proteomic analysis indicated that carbohydrate metabolism and protein synthesis/degradation-related proteins were predominantly increased and displayed temporal patterns of expression. Analyses of cluster and protein–protein interactions indicated that the regulation of sucrose synthases and alpha-amylases was the central event controlling germination. Phosphoproteomic analysis identified several proteins involved in protein modification and transcriptional regulation that exhibited significantly temporal changes in phosphorylation levels during germination. Cluster analysis indicated that 12 protein modification-related proteins had a peak abundance of phosphoproteins at 12 h after imbibition. These results suggest that the first 12 h following imbibition is a potentially important signal transduction phase for the initiation of rice seed germination. Three core components involved in brassinosteroid signal transduction displayed significant increases in phosphoprotein abundance during the early stages of germination. Brassinolide treatment increased the rice seed germination rate but not the rate of embryonic axis elongation. These findings suggest that brassinosteroid signal transduction likely triggers seed germination (Han et al. 2014)..

Copper is an essential micronutrient for plants. When present at a high concentration in soil, copper is also regarded as a major toxicant to plant cells due to its potential inhibitory effects against many physiological and biochemical processes. The interference of germination-related proteins by heavy metals has not been well documented at the proteomic level. In a study, physiological, biochemical and proteomic changes of germinating rice seeds were investigated under copper stress. Germination rate, shoot elongation, plant biomass and water content were decreased, whereas accumulation of copper and TBARS content in seeds were increased significantly with increasing copper concentrations from 0.2 mM to 1.5 mM followed by germination. The SDS–PAGE showed the preliminary changes in the polypeptides patterns under copper stress. Protein profiles analyzed by two-dimensional electrophoresis (2-DE) revealed that 25 protein spots were differentially expressed in copper-treated samples. Among them, 18 protein spots were up-regulated and sevgen protein spots were down-regulated. These differentially displayed proteins were identified by MALDI-TOF mass spectrometry. The up-regulation of some antioxidant and stress-related proteins such as glyoxalase I, peroxiredoxin, aldose reductase and some regulatory proteins such as DnaK-type molecular chaperone, UlpI protease and receptor-like kinase clearly indicated that excess copper generates oxidative stress that might be disruptive to other important metabolic processes. Moreover, down-regulation of key metabolic enzymes like alpha-amylase or enolase revealed that the inhibition of seed germinations after exposure to excess copper not only affects starvation in water uptake by seeds but also results in failure in the reserve mobilization processes. These results indicate a good correlation between the physiological and biochemical changes in germinating rice seeds exposed to excess copper (Ahsan et al. 2006).

Rice is the staple food for more than fifty percent of the world's population, and is therefore an important crop. However, its production is hindered by several biotic and abiotic stresses. Although rice is the only crop that can germinate even in the complete absence of oxygen (i.e. anoxia), flooding (low oxygen) is one of the major causes of reduced rice production. Rice germination under anoxia is characterized by the elongation of the coleoptile, but leaf growth is hampered. A comparative proteomic approach was used to detect and identify differentially expressed proteins in the anoxic rice coleoptile compared to the aerobic coleoptile. Thirty-one spots were successfully identified by MALDI-TOF MS analysis. The majority of the identified proteins were related to stress responses and redox metabolism. The expression levels of twenty-three proteins and their respective mRNAs were analyzed in a time course experiment.

Abiotic stresses such as drought, salinity, flood and cold vastly affect plant growth and metabolism that ultimately disturbs plant life (Bray et al. 2000; Ahmad and Prasad, 2012a,b). This has a negative impact on global crop production since majority of world's arable lands are exposed to these abiotic stress conditions (Rockström and Falkenmark, 2000). Upto 50–70% decline in major crop productivities have been attributed to abiotic stresses on several occasions (Mittler, 2006). For their survival under these stress conditions, plants respond by modifying several aspects in their metabolic cascade (Dos Reis et al. 2012). These response mechanisms help plants to survive during the stress period as well as to recover following cessation of the stress.

9.5. *Changes in Root*

Abiotic stress responses in plants occur at various organ levels among which the root specific processes are of particular importance. Under normal growth condition, root absorbs water and nutrients from the soil and supplies them throughout the plant body, thereby playing pivotal roles in maintaining cellular homeostasis. However, this balanced system is altered during the stress period when roots are forced to adopt several structural and functional modifications. Examples of these modifications include molecular, cellular and phenotypic changes such as alteration of metabolism and membrane characteristics, hardening of cell wall and reduction of root length (Gowda et al. 2011; Atkinson and Urwin, 2012). These changes are often caused by single or combined effects of several abiotic stress responsive pathways that can be best explored at the global level using high-throughput approaches such as proteomics (Petricka et al. 2012).

Proteomics allow global investigation of structural, functional, abundance and interactions of proteins at a given time point. As a technique proteomics is advantaged over other "omics" tools since proteins are the key players in majority of cellular events. In addition to its capability of complementing transcriptome level changes, proteomics can also detect translational and post-translational regulations, thereby providing new insights into complex biological phenomena such as abiotic stress responses in plant roots (Gygi et al. 1999; Salekdeh et al. 2002). Proteomics studies on root responses against drought, salinity, flood and cold are important with an aim to highlight shared as well as stressor specific protein classes altered due to stress conditions.

Salt stress is one of the major abiotic stresses in agriculture worldwide. A systematic proteomic approach to investigate the salt stress-responsive proteins in rice (*Oryza sativa* L. cv. Nipponbare) was carried. Three-week-old seedlings were treated with 150 mM NaCl for 24, 48 and 72 h. Total proteins of roots were extracted and separated by two-dimensional gel electrophoresis. More than 1100 protein spots were reproducibly detected, including 34 that were up-regulated and 20 down-regulated. Mass spectrometry analysis and database searching helped to identify 12 spots representing 10 different proteins. Three spots were identified as the same protein, enolase. While four of them were previously confirmed as salt stress-responsive proteins, six are novel ones, i.e. UDP-glucose pyrophosphorylase, cytochrome c oxidase subunit 6b-1, glutamine synthetase root isozyme, putative nascent polypeptide associated complex alpha chain, putative splicing factor-like protein and putative actin-binding protein. These proteins are involved in regulation of carbohydrate, nitrogen and energy metabolism, reactive oxygen species scavenging, mRNA and protein processing, and cytoskeleton stability (Yan et al. 2005).

One of the major limitations to crop growth on acid soils is the prevalence of soluble aluminum ions (Al^{3+}). Rice (*Oryza sativa* L.) has been reported to be highly Al tolerant; however, large-scale proteomic data of rice in response to Al^{3+} are still very scanty. An iTRAQ-based quantitative proteomics approach for comparative analysis of the expression profiles of proteins in rice roots in response to Al^{3+} at an early phase reveals that a total of 700 distinct proteins (homologous proteins grouped together) with >95% confidence were identified. Among them, 106 proteins were differentially expressed upon Al^{3+} toxicity in sensitive and tolerant cultivars. Bioinformatics analysis indicated that glycolysis/gluconeogenesis was the most significantly up-regulated biochemical process in response to excess Al^{3+}. The mRNA levels of eight proteins mapped in the glycolysis/gluconeogenesis were further analyzed by qPCR and the expression levels of all the eight genes were higher in tolerant cultivar than in sensitive cultivar, suggesting that these compounds may promote Al tolerance by modulating the production of available energy. The exact roles of these putative tolerance proteins remain to be examined, through (Wang et al. 2014).

Cadmium (Cd) is a non-essential heavy metal that is recognized as a major environmental pollutant. While Cd responses and toxicities in some plant species have been well established, there are few reports about the effects of short-term exposure to Cd on rice, a model monocotyledonous plant,at the proteome level.To investigate the effect of Cd in rice,and monitor the influence of Cd exposure on root and leaf proteomes, root and leaf tissues were separately collected and leaf proteins were fractionated with polyethyleneglycol. Differentially regulated proteins were selected after image analysis and identified using MALDITOFMS. A total of 36 proteins were up or down-regulated following Cd treatment. As expected, total glutathione levels were significantly decreased in Cd-treated roots,and approximately half of the up-regulated proteins in roots were involved in responses to oxidative stress. These results suggested that prompt antioxidative responses might be necessary for the reduction of Cd induced oxidative stress in roots but not in leaves. In addition, RNA gel blot analysis showed that the proteins identified in the proteomic analysis were also differentially regulated at the transcriptional level (Lee et al. 2010).

Aluminum (Al) toxicity is a serious limitation to worldwide crop production. Rice is one of the most Al-tolerant crops and also serves as an important monocot model plant. The study aims to identify Al-responsive proteins in rice, based on evidence that Al resistance is an inducible process. Two Al treatment systems were applied in the study: Al^{3+}-containing simple Ca solution culture and Al^{3+}-containing complete nutrient solution culture. Proteins prepared from rice roots were separated by 2-DE. The 2-DE patterns were compared and the differentially expressed proteins were identified by MS. A total of 17 Al-responsive proteins were identified, with 12 of those being up-regulated and five down-regulated. Among the up-regulated proteins are copper/zinc superoxide dismutase (Cu-Zn SOD), GST, and *S*-adenosylmethionine synthetase 2, which are the consistently known Al-induced enzymes previously detected at the transcriptional level in other plants. More importantly, a number of other identified proteins including cysteine synthase (CS), 1-aminocyclopropane-1-carboxylate oxidase, G protein β subunit-like protein, abscisic acid- and stress-induced protein, putative Avr9/Cf-9 rapidly elicited protein 141, and a 33 kDa secretory protein are novel Al-induced proteins. Most of these proteins are functionally associated with signaling transduction, antioxidation and detoxification. CS, as consistently detected in both Al stress systems, was further validated by Western blot and CS activity assays. Moreover, the metabolic products of CS catalysis, *i.e.* both the total glutathione pool and reduced glutathione, were also significantly increased in response to Al stress. These suggest that antioxidation and detoxification ultimately related to sulfur metabolism, particularly to CS, may play a functional role in Al adaptation for rice (Yang et al. 2007 a).

The root proteome of nitrogen-efficient and nitrogen-inefficient rice cultivars was compared in a study in order to investigate the differential expression of proteins under deficient (1 mM), low (10 mM) and high (25 mM) levels of nitrogen (N). Nitrogen use efficiency (NUE) was assessed by biochemical assays such as N-uptake kinetics and activities of N-assimilation enzymes. Two-dimensional gel electrophoresis and MALDI–TOF–MS analysis resulted in the identification of 504 protein spots (210 and 294 spots in cvs. Rai Sudha and Munga Phool, respectively). A positive correlation was observed between physiological parameters and the concentration of a number of root proteins. Sixty-three spots showed a significant cultivar × N-treatment effect on the level of expression. Functional aspects of eleven spots with major alterations in expression over control were critically analyzed. The data suggest that glutamine synthetase, cysteine proteinase inhibitor-I, porphobilinogen deaminase (fragment) and ferritin were involved in conferring N efficiency to the N-efficient rice cultivars/genotypes. Interestingly, these proteins are involved directly or indirectly in N assimilation. Such studies help us in identifying and understanding the structural or functional protein(s) involved in the response to the level of nitrogen fertilization (Hakeem et al. 2013).

Cold stress is a critical abiotic stress that reduces crop yield and quality. The response of the rice proteome to cold stress has been documented, and differential proteomic analysis has provided valuable information on the mechanisms by which rice adapts to cold stress. A global analysis of the change in protein phosphorylation status in response to cold stress remains to be explored, however. Here, we performed a phosphoproteomic analysis of rice roots following exposure to cold stress using a two-dimensional gelelectrophoresis–based multiplex proteomic approach. Differentially expressed proteins and phosphoproteins were detected and identified by matrix-assisted laser desorption ionization time of flight/time of flight mass spectrometry combined with querying rice protein databases. Nineteen protein gel spots (stained with silver) showed a twofold difference in abundance of protein spots from gels with and without cold stress; these proteins were identified to be involved in redox homeostasis, signal transduction and metabolism. Twelve of the thirteen phosphoprotein gel spots (stained with Pro-Q Diamond) that showed a twofold abundance difference were identified, including the following nine proteins: enolase, glyceraldehyde-3-phosphate dehydrogenase, nucleosidediphosphate kinase, ascorbate peroxidase, adenosine kinase, CPK1 adapter protein 2, ATP synthase subunit alpha, methioninesynthase 1, and tubulin. Phosphorylation site predictors were used to confirm that the identified proteins had putative phosphorylation sites. These results suggest that phosphorylation of some proteins in rice roots is regulated in response to cold stress (Chen et al. 2012).

To gain an enhanced understanding of the mechanism by which gibberellins (GAs) regulate the growth and development of plants, it is necessary to identify proteins regulated by GA. Proteome analysis techniques have been applied as a direct, effective and reliable tool in differential protein expressions. In previous studies, sixteen proteins showed differences in accumulation levels as a result of treatment with GA3, uniconazole, or abscisic acid (ABA), and/or the differences between the GA-deficient semi-dwarf mutant, Tan-ginbozu, and normal cultivars. Among these proteins, aldolase increased in roots treated with GA3, was present at low levels in Tan-ginbozu roots, and decreased in roots treated with uniconazole or ABA. In a root elongation assay, the growth of aldolase-antisense transgenic rice was half of that of vector control transgenic rice. These results indicate that increases in aldolase activity stimulate the glycolytic pathway and may play an important role in the GA-induced growth of roots (Komatsu and Konishi, 2005).

9.6. *Changes in Leaves*

Rice is an important food crop worldwide. Its productivity has been influenced by various abiotic and biotic factors including temperature, drought, salt, microbe, ozone, hormone and glyphosate. The responses of plants to stress are regulated by multiple signaling pathways, and the mechanisms of leaf growth and development in response to stress remain unclear to date. Recently, proteomics studies have provided new evidence for better understanding the mechanisms. The proteins in response to different stress conditions are mainly involved in photosynthesis, signal transduction, transcription, protein synthesis and destination, defense response, cytoskeleton, energy, cell wall and other metabolism. In addition, some stress type-specific proteins have been identified, such as small heat shock proteins under temperature stress, S-like RNase homolog and actin depolymerizing factor under drought stress, ascorbate peroxidase and lipid peroxidation under salt stress, probenazole-inducible protein and rice pathogenesis-related proteins under blast fungus. Many of the proteins including ribulose-1, 5-bisphosphate carboxylase/oxygenase (RuBisCO), molecular chaperones, antioxidases and S-adenosylmethionine synthetase play very important roles in leaves. Thus it is very important to make proteomic characterization of rice leaves in response to various environmental factors.

Of the numerous factors affecting rice yield, how solar radiation is transformed into biomass through rice leaves is the most important. It was essentially analyzed proteomic changes in rice leaves collected from six different developing stages (vegetative to ripening) to study protein expression profiles of rice leaves by running two-dimensional gel electrophoresis. Differential protein expressions, among the six phases, were analyzed by image analysis, which allowed the identification of 49 significantly different gel spots. The spots were further verified by matrix-assisted laser desorption/ionization-time of flight mass spectrometry, in which 89.8% of them were confirmed to be rice proteins. Finally, we confirmed some of the interesting rice proteins by immunoblotting. Three major conclusions can be drawn from these experimental results. (i) Protein expression in rice leaves, at least for high or middle abundance proteins, is attenuated during growth (especially some chloroplast proteins). However, the change is slow and the expression profiles are relatively stable during rice development. (ii) Ribulose-1,5-bisphosphate carboxylase/oxygenase (RuBisCO), a major protein in rice leaves, is expressed at constant levels at different growth stages. Interestingly, a high ratio of degradation of the RuBisCO large subunit was found in all samples. This was confirmed by two approaches, mass spectrometry and immunoblotting. The degraded fragments are similar to other digested products of RuBisCO mediated by free radicals. (iii) The expression of antioxidant proteins such as superoxide dismutase and peroxidase decline at the early ripening stage (Zhao et al. 2005).

Drought is one of the most severe limitations on the productivity of rice. To investigate the response of rice to drought stress, changes in protein expression were analyzed using a proteomic approach in a drought tolerant quantitative trait locus (DT QTL) pyramiding rice line PD86 by way of PEG simulated drought stress. After drought stress for eight days, 23 proteins increased in abundance and the level of five proteins decreased. Twelve of the drought-responsive proteins were identified by mass spectrometry. These proteins are involved in redox metabolism, photosynthesis, cytoskeleton stability, defense, protein metabolism,and signal transduction. Among the identified proteins, peroxiredoxin, ribonuclease and putative chitinase coincided with three QTL regions carried by PD86; in addition, voltage-dependent anion-selective channel protein, ribonuclease and putative chitinase co-localized with DT QTLs on the chromosome. The expression patterns of some of the corresponding genes were further analyzed at the mRNA level using real-time RTPCR. The comprehensive results suggested that the differentially displayed proteins might play a role in redox metabolism, photosynthesis, protein degradation, cytoskeleton organization and programmed cell death in the DT mechanism of rice (Xiong et al. 2010).

To identify the function of differential expression of proteins in different leaves of rice seedlings extracted from 2- to 5-leaf stages, the leaf proteins at the seedling stage of hybrid rice Shanyou 63 were studied by using the approach of plantproteomics, and those proteins were separated with two-dimensional electrophoresis (2-DE) and then analyzed with an imagemaster 2D Elite 5.0. The results showed that the 41 protein spots were detected differential expression, of which 17 new protein spots appeared after the 3-leaf stage, including nine special protein spots, which were only detected at the 3-leaf stage. Thirteen protein spots increased first and then decreased in expression abundance gradually and finally even disappeared. For the other 11 protein spots, three protein spots decreased, but six protein spots were opposite in expression abundance, however, two protein spots expressed in an irregular pattern after the 2-leaf stage. Of the 41 differential leafproteins, 15 protein spots were identified by ESI-Q MS/MS and categorized into four groups of functions. The results indicated that proteins were the carriers of the functions in cells, but were significantly influenced by the changes in cell

function or intercellular environment; hence, the reason that caused the proteomic changes as mentioned earlier might be related to the occurrence of tillers at the rice seedling stage after the 3-leaf stage (Shao et al. 2008).

In an earlier proteomic study it was shown that an actin depolymerization factor(ADF) appeared in vegetative-stage rice leaves in response to drought stress and disappeared on re-watering. This reversible phenomenon has implications for the role of the cytoskeleton in drought responsiveness. Here when extended this study to the reproductive stage, with drought stress starting three days before heading it was focused on three tissues that have a major impact on yield under drought stress: flag leaf, peduncle and anther. In *Oryza sativa* L. cv IR64, the same ADF protein was found to behave in each of these tissues as it did in vegetative-stage leaves, except that ADF continued to accumulate after re-watering in regions of the flag leaf that would not recover from stress. To confirm that a single rice ADF paralogue was responding similarly to drought stress and re-watering in diverse tissues, hence it was conducted *in silico* analysis of the ADF gene family. Thus it detected eleven paralogues spread over seven chromosomes and numbered them *OsADF1-11* in map order. Mass spectrometry of the ADF protein was consistent with drought-induced expression of *OsADF5* on chromosome 3. *OsADF5* is unique among *OsADF* genes in containing an 11-residue insert that expands a small loop that exists between two structures in the protein molecule. Wheat, barley, Italian ryegrass, sugarcane and sorghum have homologues of *OsADF5* but with shorter inserts. The behavior of OsADF5 raises interesting questions about the role of the actin cytoskeleton in drought responsiveness in plants (Liu et al. 2003).

Low temperature is one of the important environmental changes that affect plant growth and agricultural production. To investigate the responses of rice to cold stress, changes in protein expression were analyzed using a proteomic approach. Two-week-old rice seedlings were exposed to 57°C for 48 h, then total crude proteins were extracted from leaf blades, leaf sheaths and roots, separated by 2-DE and stained with

CBB. Of the 250–400 protein spots from each organ, 39 proteins changed in abundance after cold stress, with 19 proteins increasing, and 20 proteins decreasing. In leaf blades, it was difficult to detect the changes in stress-responsive proteins due to the presence of an abundant protein, ribulose bisphosphate carboxylase/oxygenase large subunit (RuBisCO LSU), which accounted for about 50% of the total proteins. To overcome this problem, an antibody-affinity column was prepared to trap RuBisCO LSU, and the remainingproteins in the flow through from the column were subsequently separated using 2-DE. As a result, slight changes in stress responsive proteins were clearly displayed, and four proteins were newly detected after cold stress. From identified proteins, it was concluded that proteins related to energy metabolism were up-regulated, and defense-related proteins were down-regulated in leaf blades, by cold stress. These results suggest that energy production is activated in the chilling environment; furthermore, stress-related proteins are rapidly up-regulated, while defense-related proteins disappear, under long-term cold stress (Hasimoto and Komatsu, 2007).

A study was conducted to investigate salt-stress-related physiological responses and proteomics changes in the leaves of two rice (*Oryza sativa* L.) cultivars. Shoot growth and water content of rice leaves were more severely reduced in Dalseongaengmi-44 than in Dongjin under salt stress. The salt-sensitive Dalseongaengmi-44 exhibited a greater increase in sodium ion accumulation in its leaves than the salt tolerant Dongjin. Comparative analysis of the rice leaf proteins using two-dimensional gel electrophoresis (2-DGE) revealed that a total of 23 proteins were up-regulated under salt stress. Based on matrix-assisted laser desorption ionization-time of flight mass spectrometry and/or electrospray ionization-tandem mass spectrometry analyses, the 23 protein spots were found to represent 16 different proteins. Ten of the identified proteins were previously reported to be salt-responsive proteins, while six–class III peroxidase 29 precursor, beta-1,3-glucanase precursor, OSJNBa0086A10.7 (putative transcription factor), putative chaperon 21 precursor, Rubisco activase small isoform precursor and drought-induced S-like ribonuclease–were novel salt-induced proteins. Under salt stress, fragmentation was increased in several proteins containing the RuBisCO large chain. The results of these physiological and proteomics analyses provide useful information that can lead to a better understanding of the molecular basis of salt-stress responses in rice (Lee et al. 2011).

Three-week old plants of rice (*Oryza sativa* L. cv CT9993 and cv IR62266) developed gradual water stress over 23 days of transpiration without watering, during which period the mid-day leaf water potential declined to approximately –2.4 MPa, compared with approximately –1.0 MPa in well-watered controls. More than 1000 protein spots that were detected in leaf extracts by proteomic analysis showed reproducible abundance within replications. Of these proteins, 42 spots showed a significant change in abundance under stress, with 27 of them exhibiting a different response pattern in the two cultivars. However, only one protein (chloroplast Cu-Zn superoxide dismutase) changed significantly in opposite directions in the two cultivars in response to drought. The most common difference was for proteins to be up-regulated by drought in CT9993 and unaffected in IR62266; or down-regulated by drought in IR62266 and unaffected in CT9993. By 10 days after re-watering, all proteins had returned completely or largely to the abundance of the well-watered control. Mass spectrometry helped to identify 16 of the drought-responsive proteins, including an actin depolymerizing factor, which was one of three proteins detectable under stress in both cultivars but undetectable in well-watered plants or in plants 10 days after re-watering. The most abundant protein up-regulated by drought in CT9993 and IR62266 was identified only after cloning of the corresponding cDNA. It was found to be an S-like RNase homologue but it lacked the two active site histidines required for RNase activity. Four novel drought-responsive mechanisms were revealed by this work: up-regulation of S-like RNase homologue, actin depolymerizing factor and RuBisCO activase, and down-regulation of isoflavone reductase-like protein (Salekdeh et al.2002).

Of the numerous factors affecting rice yield, how solar radiation is transformed into biomass through rice leaves is the most important. Proteomic changes in rice leaves collected from six different developing stages (vegetative to ripening) were analyzed and protein expression profiles of rice leaves was studies by running two-dimensional gel electrophoresis. Differential protein expression among the six phases were analyzed by image analysis, which allowed the identification of 49 significantly different gel spots. The spots were further verified by matrix-assisted laser desorption/ionization-time of flight mass spectrometry, in which 89.8% of them were confirmed to be rice proteins. Finally, we confirmed some of the interesting rice proteins by immune-blotting. Three major conclusions can be drawn from these experimental results. (i) Protein expression in rice leaves, atleast for high or middle abundance proteins, is attenuated during growth (especially some chloroplast proteins). However, the change is slow and the expression profiles are relatively stable during rice development. (ii) Ribulose-1,5-bisphosphate carboxylase/oxygenase (RuBisCO), a major protein in rice leaves, is expressed at constant levels at different growth stages. Interestingly, a high ratio of degradation of the RuBisCO large subunit was found in all samples. This was confirmed by two approaches: mass spectrometry and immune-blotting. The degraded fragments are similar to other digested products of RuBisCO mediated by free radicals. (iii) The expression of antioxidant proteins such as superoxide dismutase and peroxidase decline at the early ripening (Zhao et al. 2005) (Table 9.1).

TABLE 9.1 Identification of growth-dependent proteins in rice leaves

Spot No	Mass (kDa)	pI	Sequence coverage	Matched rate	Protein
14	35.40	5.19	44	13/41	Peroxidase
43	35.01	5.67	48	8/32	Peroxidase
16	31.81	5.23	47	12/84	Putative thioredoxin
23	24.75	5.96	45	10/39	Superoxidedismutase (Mn), mitochondrial precursor
38	16.85	5.56	52	9/45	Superoxide dismutase (Cu,Zn), chloroplast precursor
2	57.52	4.90	35	12/29	RuBisCO-beta (CPN60-beta)
3	57.52	5.10	40	17/29	Chaperonin beta
21	25.67	5.09	59	12/44	Chr06-901 metalic protein
9	37.73	6.20	43	12/37	Putative malate dehydrogenase
11	36.93	5.49	34	15/68	Fructose-bisphosphate aldolase, chloroplast precursor (ALDP)
4, 12, 15, 17, 40	49.65	6.62	52	34/99	RuBisCO large subunit
19, 20, 22, 47	26.40	6.23	26	22/85	RuBisCO large subunit
32, 36	19.15	6.11	30	8/20	RuBisCO small subunit
41	14.97	4.67	50	10/37	RuBisCO small subunit
5	40.41	5.48	59	23/77	CoproporphyringenIII oxidase
28	21.14	4.56	40	7/25	PhotosystemII oxygen evolving complex1
30	20.02	5.43	45	9/40	PhotosystemII oxygen evolving complex2
26	21.59	6.35	75	10/60	Abscisic acid and stress induced protein
34	18.10	5.47	38	6/25	Putative salt induced protein
6	39.94	6.27	44	11/51	Chr03-4135
7	39.93	5.18	54	21/47	Putative plastidic cysteine synthase1

10	37.62	5.13	61	13/44	Putative tyrosine phosphatase
48	37.62	5.13	52	11/44	Putative SHOOT1 protein
13	35.22	6.12	75	17/49	Guanine nucleotide binding beta (GPB-LR)
24	23.07	6.16	33	9/38	Putative ribosome recycling factor, chloroplast precursor
31	19.13	4.61	43	6/22	Ribosomal protein L12
44	39.54	5.88	32	17/22	Structural protein of ribosome
49	28.62	4.50	33	8/21	Hydrogen-transporting two sector ATPase

Unidentified proteins 1, 8, 18, 27, 37

The analysis of stress-responsiveness in plants is an important route to the discovery of genes conferring stress tolerance and their use in breeding programs. Proteomic analysis provides a broad view of plant responses to stress at the level of proteins. In recent years this approach has increased in sensitivity and power as a result of improvements in two-dimensional polyacrylamide gel electrophoresis (2DE), protein detection and quantification, fingerprinting and partial sequencing of proteins by mass spectrometry (MS), bioinformatics, and methods for gene isolation. 2DE provides information on changes in abundance and electrophoretic mobility of proteins, the latter reflecting post-translational modifications such as phosphorylation and free-radical cleavage. It was noted that the technical aspects of proteomics and demonstrated its use in analyzing the response of rice plants to drought and salinity. More than 2000 proteins were detected reproducibly in drought-stressed and well-watered leaves. Out of >1000 proteins that were reliably quantified, 42 proteins changed significantly in abundance and/or position. It was identified several leaf proteins whose abundance increased significantly during drought and declined on re-watering. The three most marked changes were seen with actin depolymerizing factor, a homologue of the S-like ribonucleases and the chloroplastic glutathione-dependent dehydroascorbate reductase. Proteomic comparisons of salt stress-tolerant and stress-sensitive genotypes revealed numerous constitutive and stress-induced differences in root proteins. Among them was caffeoyl-CoA *O*-methyltransferase, an enzyme of lignin biosynthesis. The abundance of ascorbate peroxidase was much higher in salt-tolerant Pokkali than in salt-sensitive IR29 in the absence of stress (Salekdeh et al. 2002).

9.5. *Changes in Tillering*

Tillering in rice (*Oryza sativa* L.) is an important agronomic trait that enhances grain production. A tiller is a specialized grain-bearing branch that is formed on a non-elongated basal internode that grows independently of the mother stem. Transgenic rice over-expressing the transcription factor OsTB1, a homologue of maize TB1 (Teosinte Branched 1), exhibits markedly reduced lateral branching without the propagation of axilary buds being affected. However, the tillering mechanism remains unknown. Therefore, to further understand the mechanism, we applied proteomics methodology to isolate the proteins involved. Using two-dimensional gel electrophoresis and mass spectrometry, our analysis of the basal nodes from two rice cultivars that differ in their numbers of tillers showed that a rice serine proteinase inhibitor, OsSerpin, accumulates in great amounts in high-tillering 'Hwachung' rice. Northern blot analysis revealed that much more OsSerpin transcript is found in 'Hwachung' than in relatively low-tillering 'Hanmaeum', likely because of high levels of transcription. Therefore, data suggest that OsSerpin content determines the extent of lateral branching (Yeu et al. 2007).

9.7. *Changes in Callus Formation*

Callus differentiation is a key developmental process for rice regeneration from cells. To better understand this complex developmental process, it was used a 2-D gel electrophoresis approach to explore the temporal patterns of protein expression at the early stages during rice callus differentiation. This global analysis detected 60 known proteins out of 79 gel spots identified by MS/MS, of which many had been shown to play a role in plant development. Two new proteins were revealed to be associated with the callus differentiation and have been confirmed by Western blot analysis. The results of proteomics experiments were further verified at the mRNA level using microarray and real-time PCR. Comparison of the differentially expressed protein levels with their corresponding mRNA levels at the two callus early differentiation stages showed a good correlation between them, indicating that a substantial proportion of protein changes is a consequence of changed mRNA levels, rather than post-transcriptional effects during callus differentiation, though microarray revealed more expression changes on RNA levels. These findings may contribute to further understanding of the mechanisms that lead to callus differentiation of rice and other plants as well (Yin et al. 2007).

9.8. *Changes in Pollen*

Male reproductive development in rice is very sensitive to various forms of environmental stresses including low temperature. A few days of cold treatment (<20 °C) at the young microspore stage induce severe pollen sterility and thus large grain yield reductions. To investigate this phenomenon, anther proteins at the early stages of microspore development, with or without cold treatment at 12 °C, were extracted, separated by two-dimensional gel electrophoresis, and compared. The cold-sensitive cultivar Doongara and the relatively cold-tolerant cultivar HSC55 were used. The abundance of 37 anther proteins was changed more than 2-fold after 1, 2 and 4 days of cold treatment in cv. Doongara. Among them, one protein was newly induced, 32 protein spots were up-regulated, and four protein spots were down-regulated. Of these 37 protein spots identified, two anther-specific proteins (putative lipid transfer protein and Osg6B) and a calreticulin that were down-regulated and a cystine synthase, a β-6 subunit of the 20 S proteasome, an H protein of the glycine cleavage system, cytochrome *c* oxidase subunit VB, an osmotin protein homologue, a putative 6-phosphogluconolactonase, a putative adenylate kinase, a putative cysteine proteinase inhibitor, ribosomal protein S12E, a caffeoyl-CoA *O*-methyltransferase, and a monodehydroascorbate reductase that were up-regulated. Accumulation of these proteins did not vary greatly after cold treatment in panicles of cv. Doongara or in the anthers of the cv. HSC55. The newly induced protein named *Oryza sativa* cold-induced anther protein (OsCIA) was identified as an unknown protein. The OsCIA protein was detected in panicles, leaves and seedling tissues under normal growth conditions. Quantitative real time RT-PCR analysis of OsCIA mRNA expression showed no significant change between low temperature-treated and untreated plants. A possible regulatory role for the newly induced protein is evident (Imin et al. 2006).

A proteomic analysis to investigate the changing patterns of protein synthesis during pollen development in anthers from rice plants grown under strictly controlled growth conditions was undertaken. Cytological analysis and external growth measurements such as anther length, auricle distances and days before flowering were used to determine pollen developmental stages. This allowed the collection of synchronous anther materials representing six discrete pollen developmental stages. Proteins were extracted from the anther samples and separated by two-dimensional gel electrophoresis to produce proteome maps. The anther proteome maps of different developmental stages were compared and 150 protein spots, which were changed consistently during development, were analyzed by matrix-assisted laser desorption/ionization-time of flight mass spectrometry to produce peptide mass fingerprint (PMF) data. Database searches using these PMF data revealed the identities of 40 of the protein spots analyzed. These 40 proteins represent 33 unique gene products. Four protein spots that could not be identified by PMF analysis were analyzed by N-terminal micro sequencing. Multiple charge-isoforms of vacuolar acid invertase, fructokinase, beta-expansin and profilin were identified. These proteins are closely associated with sugar metabolism, cell elongation and cell expansion, all of which are cell activities that are essential to pollen germination. The existence of multiple isoforms of the same proteins suggests that during the process of pollen development some kind of post-translational modification of these proteins occurs (Kerim et al. 2003).

9.9. Changes in Embryogenesis

The plant embryo is the germination center of the seed. How an embryo forms during seed maturation remains unclear, especially in the case of monocotyledonous plants. Generally, the complex processes of embryogenesis result from the action of a coordinated network of genes. Thus, a large-scale survey of changes in protein abundance during embryogenesis is an effective approach to study the molecular events of embryogenesis. In this study, two-dimensional gel electrophoresis (2DE) was applied to separate rice embryo proteins collected during the three phases of embryogenesis: six days after pollination (DAP), 12 DAP and 18 DAP and then employed matrix-assisted laser desorption-ionization time of flight/time of flight mass spectrometry (MALDI TOF/TOF MS) to identify the phase-dependent differential 2DE spots. A total of 66 spots were discovered to be regulated during embryogenesis, and of these spots, 53 spots were identified. These proteins were further categorized into several functional classes, including storage, embryo development, stress response, glycolysis, and protein metabolism. Intriguingly, the major differential spots originated from three globulins. Thus further examined the possible mechanism underlying the globulins' multiple forms using Western blotting, proteolysis, and blue native gel electrophoresis techniques and found that the multiple forms of globulins were produced as a result of enhanced proteolysis during embryogenesis, indicating that these globulin forms may serve as chaperone proteins participating in the formation of multiple protein complexes during embryogenesis (Zi et al. 2012).

Embryogenesis is the initial step in a plant's life, and the molecular changes that occur during embryonic development are largely unknown. To explore the relevant molecular events, the isobaric tags were used for relative and absolute quantification (iTRAQ) coupled with the shotgun proteomics technique (iTRAQ/Shotgun) to study the proteomic changes of rice embryos during embryogenesis. For the first time, a total of 2165 unique proteins were identified in rice embryos, and the abundances of 867 proteins were actively changed based on the statistical evaluation of the quantitative MS/MS signals. The quantitative data were then confirmed using multiple reactions monitoring (MRM) and were also supported by previous study based on two-dimensional gel electrophoresis (2DE). Using the proteome six days after pollination (DAP) as a reference, cluster analysis of these differential proteins throughout rice embryogenesis revealed that 25% were up-regulated and 75% were down-regulated. Gene Ontology (GO) analysis implicated that most of the up-regulated proteins were functionally categorized as stress responsive, mainly including heat shock-, lipid transfer-, and reactive oxygen species-related proteins. The stress-responsive proteins were thus postulated to play an important role during seed maturation (Zi et al. 2013).

9.10. *Changes in Grain Development*

Rice yield and quality are adversely affected by high temperatures, and these effects are more pronounced at the 'milky stage' of the rice grain ripening phase. Identifying the functional proteins involved in the response of rice to high temperature stress may provide the basis for improving heat tolerance in rice. In the present study, a comparative proteomic analysis of paired, genetically similar heat-tolerant and heat-sensitive rice lines was conducted. Two-dimensional electrophoresis (2-DE) revealed a total of 27 differentially expressed proteins in rice grains, predominantly from the heat-tolerant lines. The protein profiles clearly indicated variations in protein expression between the heat-tolerant and heat-sensitive rice lines. Matrix-assisted laser desorption/ionization time-of-flight/time-of-flight mass spectrometry (MALDI-TOF/TOF MS) analysis revealed that 25 of the 27 differentially displayed proteins were homologous to known functional proteins. These homologous proteins were involved in biosynthesis, energy metabolism, oxidation, heat shock metabolism and the regulation of transcription. Seventeen of the 25 genes encoding the differentially displayed proteins were mapped to rice chromosomes according to the co-segregating conditions between the simple sequence repeat (SSR) markers and the target genes in recombinant inbred lines (RILs). The proteins identified in the present study provide a basis to elucidate further the molecular mechanisms underlying the adaptation of rice to high temperature stress (Liao et al. 2013).

Rice is an important cereal crop and has become a model monocot for research into crop biology. Rice seeds currently feed more than half of the world's population and the demand for rice seeds is rapidly increasing because of the fast-growing world population. However, the molecular mechanisms underlying rice seed development is incompletely understood. Genetic and molecular studies have developed our understanding of substantial proteins related to rice seed development. Recent advancements in proteomics have revolutionized the research on seed development at the single gene or protein level. Proteomic studies in rice seeds have provided the molecular explanation for cellular and metabolic events as well as environmental stress responses that occur during embryo and endosperm development. They have also led to the new identification of a large number of proteins associated with regulating seed development such as those involved in stress tolerance and RNA metabolism. In the future, proteomics, combined with genetic, cytological and molecular tools, will help to elucidate the molecular pathways underlying seed development control and help in the development of valuable and potential strategies for improving yield, quality and stress tolerance in rice and other cereals.

Proteins were extracted from rice grains 10, 20 and 30 days after flowering, as well as from fully mature grains. By merging all of the identified proteins in this study, 4,172 non-redundant proteins were identified with a wide range of molecular weights (from 5.2 kDa to 611 kDa) and *pI* values (from pH 2.9 to pH 12.6). A Genome Ontology category enrichment analysis for the 4,172 proteins revealed that 52 categories were enriched, including the carbohydrate metabolic process, transport, localization, lipid metabolic process and secondary metabolic process. The relative abundances of the 1,784 reproducibly identified proteins were compared to detect 484 differentially expressed proteins during rice grain development. Clustering analysis and Genome Ontology category enrichment analysis revealed that proteins involved in the metabolic process were enriched through all stages of development, suggesting that proteome changes occurred even in the desiccation phase. Interestingly, enrichments of proteins involved in protein folding were detected in the desiccation phase and in fully mature grain (Lee and Koh, 2011).

To characterize global expression trends for proteins involved in specific processes, composite expression profiles were constructed by summing NSpC for each protein in each functional category (according to classification scheme by Bevan et al. (1998). In addition to eight functional categories, it was constructed a composite expression profile of late-embryogenesis-abundant (LEA) proteins. Proteins involved in metabolic processes increased until 20 DAF, and after that, their levels were maintained in fully mature grains with a slight increase at 30 DAF, suggesting that even after the starch had fully accumulated, proteins involved in metabolic processes continued to be present during the desiccation phase. The expression trends of proteins involved in starch biosynthesis and photosynthesis were in accordance with morphological development. Proteins involved in starch biosynthesis increased until 20 DAF, and then decreased slightly at 30 DAF, followed by rapid decrease for fully mature grains, suggesting that starch accumulation was intensive until 20 DAF, and saturated before 30 DAF. Proteins associated with photosynthesis continuously decreased. For the metabolism, starch biosynthesis and photosynthesis, the general trends of composite expression profiles during the grain filling stage were similar to the results previously reported by Xu et al. (2008). Due to dynamic proteomic analyses through eight sequential developmental stages until 20 DAF, Xu et al. (2008) could detect fluctuations of protein expression during grain filling stages for other categories, while those detailed expression patterns could not be revealed in this study because of 10-day sampling interval. However, it could reveal their expression patterns in desiccation phase instead. Proteins involved in glycolysis, TCA-cycle, lipid metabolism and proteolysis increased in fully mature grains. Proteins involved in glycolysis, TCA cycle and lipid metabolism showed similar expression trends and the levels of expression for these proteins increased slightly during grain development, being highest for fully mature grains. The roles of glycolysis and the TCA-cycle, which are closely related and provide energy and carbon skeletons for various primary metabolites, increased in fully mature grain. This was also observed for some of the proteins involved in lipid metabolism that have catalytic activity in fatty acids such as acyl-CoA synthetase. The next growth stage of fully mature grain is germination, during which large amounts of energy and nutrition are required, so remobilization of reserves in the endosperm and increases of these proteins in germinating seeds is critical. Thus, the accumulation of such proteins in mature grain may reflect the fact that a certain level of proteins is required for germination. Proteins involved in proteolysis were also increased in fully mature grains, which also may represent preparation for germination. However, the expression trend of proteolysis fluctuated (Fig. 9.4).

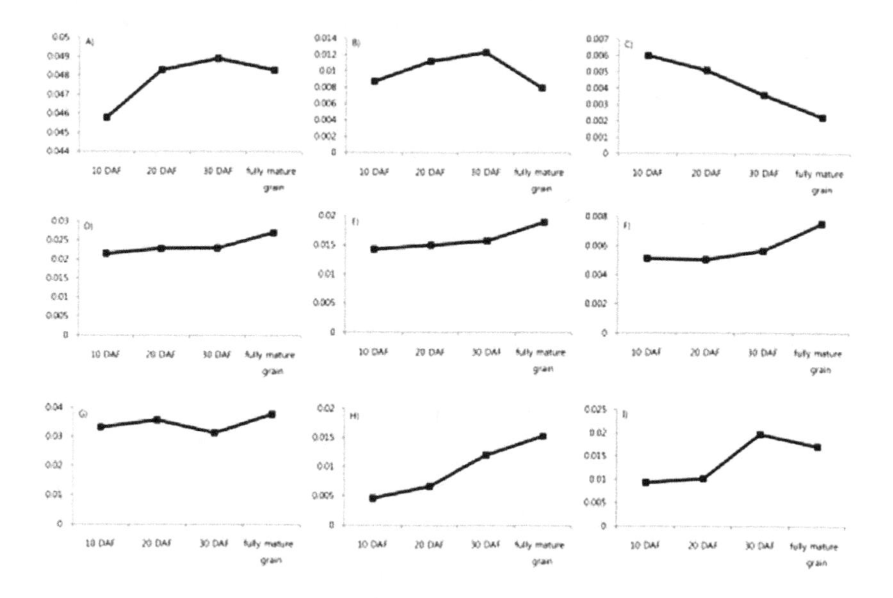

A. Metabolism (102), B. Starch Biosynthesis (7), C. Photosynthesis (11), D. Glycolysis (23), E.TCA-Cycle (17), F.Lipid metabolism (19), G. Proteolysis (100),, H.LEA protein (10), I.Chaperone (20)* No. of non-redundant proteins.

FIGURE 9.4 Proteomics changes during rice grain development.

Cultivars of rice (*Oryza sativa* L.), especially of the type with large spikelets, often fail to reach the yield potential as expected due to the poor grain-filling on the later flowering inferior spikelets (in contrast to the earlier-flowering superior spikelets). The study showed that the size and grain weight of superior spikelets (SS) was greater than those of inferior spikelets (IS), and the carbohydrate supply should not be the major problem for the poor grain-filling because there was adequate amount of sucrose in IS at the initial grain-filling stage. High resolution two-dimensional gel electrophoresis (2-DE) in combination with Coomassie-brilliant blue (CBB) and Pro-Q Diamond phosphoprotein fluorescence stain revealed that 123 proteins in abundance and 43 phosphoproteins generated from phosphorylation were significantly different between SS and IS. These proteins and phosphoproteins were involved in different cellular and metabolic processes with a prominently functional skew toward metabolism and protein synthesis/destination. Expression analyses of the proteins and phosphoproteins associated with different functional categories/subcategories indicated that the starch synthesis, central carbon metabolism, N metabolism and cell growth/division were closely related to the poor grain-filling of IS. Functional and expression pattern studies also suggested that 14-3-3 proteins played important roles in IS poor grain-filling by regulating the activity of starch synthesis enzymes. The proteome and phosphoproteome obtained from this study provided a better understanding of the molecular mechanism of the IS poor grain-filling. They were also expected to be highly useful for improving the grain filling of rice (Zhang et al. 2014).

During seed maturation, the water content of seeds decreases remarkably. Mature seeds can germinate after imbibition since the embryos are protected by mechanism of desiccation tolerance. To better understand the mechanism of desiccation tolerance in seeds, analysis of the fluctuation of stress-related proteins in the desiccation phase of rice seeds by a real-time RT-PCR and gel-based proteomic approach. Based on the changes in water content of developing rice seeds, we defined stages from the beginning of dehydration (10 to 20 days after flowering) and the desiccation phase (20 to 40 days after flowering). The proteomic analysis revealed that late embryogenesis abundant proteins, small heat shock proteins and antioxidative proteins accumulate at the beginning of dehydration and remain at a high level in the desiccation phase, suggesting that these proteins are involved in acquisition of desiccation tolerance. The fluctuation in levels of mRNA encoding some stress-related proteins did not precisely reflect the change in levels of these proteins. Therefore, proteomic analysis, which provides an accurate assessment of changes in protein levels, is a more efficient technique than transcriptomics for inferring the role of stress-related proteins in rice seeds. List of the 88 differentially expressed spots in response to anoxia, detected by computer-assisted 2D gel differential analysis. Of these 88 spots, 50 were upregulated and 38 downregulated (Table 9.2).

TABLE 9.2 Identification of stress-related proteins in mature seeds of rice

Spot No.	Protein name	pI	Mr (kDa)	Score	Accession No.
1	18.1 kDa heat shock protein	5.9	19.6	75	LOC-Os03g15960
2	16.9 kDa heat shock protein	6.3	18.2	85	LOC-Os01g04370
3a	70.0 kDa heat shock protein	4.9	81.5	84	LOC-Os12g14070
3b		4.8	81.0	85	
4	Glyoxalate1	5.6	39.5	72	LOC-Os08g09250
5	Putative stress induced protein sti1	6.4	59.8	102	LOC-Os02g43020
6	Peroxiredoxin	6.6	30.2	152	LOC-Os07g44430
7a	Putative aldose reductase	6.7	39.4	132	LOC-Os05g39680
7b		6.3	33.1	72	
8	Putative embryonic abundant protein, group3	6.3	34.2	79	LOC-Os01g16920
9	Late embryogenesis abundant protein1	6.6	43.8	89	LOC-Os03g20680
10a	Putative embryo specific protein	7.3	44.8	71	LOC-Os03g07180
10b		7.4	44.8	87	
10c		7.5	44.0	70	
11	Putative 20kDa chaperonin	5.1	35.2	87	LOC-Os09g26730
12	Putative chaperonin21 precursor	5.7	34.2	102	LOC-Os06g09679

pI = Experimental isoelectric point; Mr = Experimental molecular size; Score = Probability of the observed match; Accession No. = MSU Rice Genome Annotation.

9.11. *Changes in Heterosis*

Heterosis is a common phenomenon in which the hybrids exhibit superior agronomic performance than either inbred parental lines. Although hybrid rice is one of the most successful apotheoses in crops utilizing heterosis, the molecular mechanisms underlying rice heterosis remain elusive. To gain a better understanding of the molecular mechanisms of rice heterosis, comparative leaf proteomic analysis between a super hybrid rice LYP9 and its parental cultivars 9311 and PA64s at tillering, flowering and grain-filling stages were carried out. A total of 384 differentially expressed proteins (DP) were detected and 297 DP were identified, corresponding to 222 unique proteins. As DP were divided into those between the parents (DP(PP)) and between the hybrid and its parents (DP(HP)), the comparative results demonstrate that proteins in the categories of photosynthesis, glycolysis, and disease/defense were mainly enriched in DP. Moreover, the number of identified DP(HP) involved in photosynthesis, glycolysis, and disease/defense increased at flowering and grain-filling stages as compared to that at the tillering stage. Most of the up-regulated DP(HP) involved in the three categories showed greater expression in LYP9 at flowering and grain-filling stages than at the tillering stage. In addition, CO_2 assimilation rate and apparent quantum yield of photosynthesis also showed a greater increase in LYP9 at flowering and grain-filling stages than at the tillering stage. These results suggest that the proteins involved in photosynthesis, glycolysis and disease/defense as well as their dynamic regulation at different developmental stages may be responsible for heterosis in rice (Zhang et al. 2012).

Heterosis in which the hybrids exhibit superior agronomic performance to both parents such as biomass production, grain yield, and stress tolerance. Although heterosis breeding is one of the main techniques in rice breeding, the molecular mechanisms responsible for this basic biological phenomenon are not well understood. To further get an insight into the molecular mechanisms of rice heterosis, comparative root proteomic analysis between a super-hybrid rice *LYP9* and its parental cultivars *9311* and *PA64S* at seedling stage were performed. Total proteins were extracted and subjected to two-dimensional gel electrophoresis. The scatter plots analysis results showed that the *LYP9*'s expression profiles were more similar to *9311* than *PA64S*. A total of 11 differentially expressed protein spots were detected and identified by matrix-assisted laser desorption/ionization-time of flight/time of flight mass.

9.12. Changes in Biotic Stress

Proteomic data from two-dimensional gel electrophoresis showed that 14 candidate proteins were significantly up- or down-regulated in the *spl5* mutant compared with WT. These proteins are involved in diverse biological processes including pre-mRNA splicing, amino acid metabolism, photosynthesis, glycolysis, reactive oxygen species (ROS) metabolism, and defence responses. Two candidate proteins with a significant up-regulation in *spl5* â€" APX7, a key ROS metabolism enzyme and Chia2a, a pathogenesis-related protein â€" were further analyzed by qPCR and enzyme activity assays. Consistent with the proteomic results, both transcript levels and enzyme activities of APX7 and Chia2a were significantly induced during the course of lesion formation in *spl5* leaves. Many functional proteins involving various metabolisms were likely to be responsible for the lesion formation of *spl5* mutant. Generally, in *spl5*, the up-regulated proteins involve in defence response or PCD, and the down-regulated ones involve in amino acid metabolism and photosynthesis. These results may help to gain new insight into the molecular mechanism underlying *spl5*-induced cell death and disease resistance in plants.

Rice (*Oryza sativa* L.) is one of the most important crops in the world. It is the main staple food of more than half of the world's population. Since rice has a genome that is significantly smaller than those of other cereals, it is an ideal model plant for genetic and molecular studies, particularly among the monocots. Draft sequences of rice genomes have been reported for subspecies *indica* and *japonica*. Furthermore, the complete map-based genome sequences of chromosomes 1 and 4 for cultivar Nipponbare have been reported. The challenge ahead for the plant research community is to identify the functions, post-translational modifications, and the regulation of proteins encoded by the plant's genes. Understanding the biological functions of novel genes is a more difficult proposition than merely obtaining the nucleotide or peptide sequences. This is because the existing information on amino acid sequences of known proteins in the database is derived primarily from genetic and biochemical studies, which are by nature focused and labor intensive; the use of a large number of plant species as experimental systems; and the extensive range of unique plant-produced secondary metabolites.

Rice is not only a very important crop but also a model plant for biological research, because its genome is smaller than those of other cereals, making it suitable for efficient genetic analysis and transformation. The completion of draft genome sequences of *Oryza sativa* L. ssp. *indica* and *Oryza sativa* L. ssp. *japonica* , and ofa complete map-based sequence of chromosome 1 and chromosome 4 of *Oryza sativa* L. cv. Nipponbare, provides a rich resource for understanding the biological processes of rice. Now it is necessary to identify the function, regulation and PTM, if any, of each encoded protein. Gaining an understanding of the biological functions of novel genes is a more ambitious goal than obtaining just their sequences; however, the wealth of information on nucleotide sequences that is being generated through genome projects far outweighs what is currently available on the amino acid sequences of known proteins. Because the analysis of proteins is the most direct approach to defining the function of their associated genes, analysis of the proteome linked to genome sequence information is very useful for functional genomics.

Several studies have dealt with the construction of proteomes for complex samples from rice, such as leaf, embryo, endosperm, root, stem, shoot and callus . The proteomes of Golgi, mitochondria and other subcellular compartments have also been studied. As part of rice proteomics research, a system for direct differential display using 2-DE has been developed for the identification of rice proteins that vary in expression under different physiological conditions and among different tissues. This system readily visualizes changes in the protein profile, directly and rapidly selects proteins from the gels that show altered expression, and then analyzes their structure by comparison with entries in the rice proteome database, or by MS and Edman sequencing. The recently constructed rice proteome database website (http://gene64.dna.affrc.go.jp/RPD/) provides extensive information on the progress of rice proteome research. Proteomics analysis of various tissues and organelles has revealed diverse functional categories of proteins. Although many ubiquitous proteins have been identified that share similar functions in different tissues and organelles, most of these proteins are specific to either tissues or subcellular compartments. These results highlight the diversity of proteomes within the rice plant, and hence the urgent need to analyze additional tissues and subcellular organelles to gain a comprehensive understanding of the proteins encoded by the rice genome.

A major advantage of the rice proteome database, in which known proteins are recorded along with where and when they are expressed, is the wealth of newly identified proteins on which further experiments can be conducted at the biochemical and molecular levels. Rice proteomics studies to date have focused mainly on changes in genome expression that are triggered by environmental factors. A complete functional understanding of the proteome requires, however, full characterization of the PTMs of proteins and the complex networks of protein-protein interactions. The aim of proteomics studies is a more systematic and comprehensive survey of the proteomes of rice; specifically, to separate proteins extracted from rice, to determine the *N*-terminal and internal amino acid sequences, and to construct a rice proteome database. In addition to facilitating the identification of known proteins, these amino acid sequences can be used to prepare oligodeoxyribonucleotides, which are essential for cloning the corresponding cDNA, and initial attempts towards determining the physiological significance of some proteins identified from rice. Finally, approaches to proteomics analysis, such as phosphoproteome analysis and protein-protein interaction analysis, are also important.

References

Ahmad, P. and M.N.V. Prasad. 2012a. Abiotic stress responses in plants: Metabolism, productivity and sustainability. Springer, New York doi: 10.1007/978-1-4614-0634-1

Ahmad, P. and M.N.V. Prasad. 2012b. Environmental adaptations and stress tolerance in plants in the era of climate change. Springer, New York doi: 10.100T/978-1-4614-0815-4

Ahsan, N., D.G. Lee, S.H. Lee, K.Y. Kang, J.J. Lee and P.J. Kim. 2007. Excess copper induced physiological and proteomic changes in germinating rice seeds. *Chemosphere*, **67:** 1182-1193.

Anderson, L. and N.G. Anderson. 1977. High resolution two dimensional electrophoresis of human plasma proteins. *Prac.Natl.Acad.Sci.USA*, **74:** 5421-5425.

Appella, E., E.A. Padlan and D.F. Hunt. 1995. Analysis of the structure of naturally processed peptides bound by class I and class II major histocompatibility complex molecules. *EXS*, **73:** 105-119.

Atkinson, N.J. and P.E. Urwin. 2012. The interaction of plant biotic and abiotic stresses: from genes to the field. *J.Exp.Bot.* **63:** 3523-3544.

Bevan, M., I. Bancraft, E. Bent, K. Love, H. Goodman, C. Dean, R. Bergkamp, W. Dirkse, M. Van Staveren and W. Stiekma. 1998. Analysis of 1.9 Mb of contiguous sequence from chromosome 4 of Arabidopsis thaliana. *Nature*, **391:** 485-488.

Bray, E.A., J. Bailey-Sevres and E. Weretilnyk. 2000. *In:* Cruissem, W., Buchannan, B. and Jones, R.(eds) Biochemistry and molecular biology of plants. Rockville, MD. pp.1158-1203.

Celis, J.E. and R. Bravo. 1984. Two dimensional gel electrophoresis of proteins– Methods and applications. Academic Press, New York.

Chen, J., L. Tian, H. Xu, D. Tian, Y. Luo, C. Ren, L. Yang and J. Shi. 2012. Cold induced changes of protein and phosphoprotein expression patterns from rice roots as revealed by multiplex proteomic analysis. *Plant Omics J.*, **5:** 194-199.

Corthals, G.L., V.C. Wasinger, D.F. Hochstrasser and J.C. Sanchez. 2000. The dynamic range of protein expression: A challenge for proteomic research. *Electrophoresis*, **21:** 1104-1115.

Dos Reis, S.P., A.M. Lima and C.R.B. De Souza. 2012. Recent molecular advances on downstream plant responses to abiotic stress. *Int.J.Mol.Sci.*, **13:** 8628-8647.

Garrels, J.I and B.R. Franz. 1989. Transformation sensitive and growth related changes of protein synthesis in REF52 cells. *J.Biol.chem.* **264:** 5229-5321.

Ghosh, D. and J. Xu. 2014. Abiotic stress responses in plant roots: A proteomics perspective. *Front.Plant Sci.* **5:** 6 doi:10.3389/fpls.2014.00006

Gowda, V.R.P., A. Henry, A. Yamauchi, H.E. Shashidhar and R. Serraj. 2011. Root biology and genetic improvement for drought avoidance in rice. *Field Crops Res.*, **122:** 1-13.

Gygi, S.P., Y. Rochon, B.R. Franza and R. Aebersold. 1999. Correlation between protein and mRNA abundance in yeast. *Mol.Cell.Biol.*, **19:** 1720-1730.

Gygi, S.P., G.L. Corthals, Y. Zhang, Y. Rochon and R. Aebersold. 2000. Evaluation of two dimensional gel electrophoresis based proteomic analysis technology. *Proc. Natl. Acad. Sci. USA*, **97:** 9390-9395.

Hakeem, K.R., B.A. Mir Mohd., I. Qureshi, A. Ahmad and M. Iqbal 2013. Physiological studies and proteomic analysis for differentially expressed proteins and their possible role in the root of N-efficient rice (*Oryza sativa L.*) *Mol.Breeding*, **32:** 785-798.

Han, C., P. Yang, K. Sakata and S. Komatsu. 2014. Quantitive proteomics reveals the role of protein phosphorylation in rice embryos during early stages of germination. *J.Proteome Res.*, **13:** 1766-1782.

Han, D.K., J. Eng, H. Zhou and R. Aebersold 2001. Quantative profiling of differentiation-induced microsomal proteins using isotope-coded affinity tags and mass spectrometry. *Nat.Biotechnol.*, **19:** 946-951.

Hashimoto, M. and S. Komatsu. 2007. Proteomic analysis of rice seedlings during cold stress. *Proteomics,* **7:** 1293-1302.

Hirano, H.1989. Microsequence analysis of winged bean seed proteins electro-blotted from two dimensional gel. *J.ProteinChem.,* **8:** 115-130.

Imin, N., T. Kerim, J.J. Weinman and B.G. Rolfe. 2006. Low temperature treatment at the young microspore stage induces protein changes in rice anthers. *Mol. Cell. Proteomics,* **5:** 274-292.

Kerim, T., N. Imin, J.J. Weinman and B.G. Rolfe. 2003. Proteome analysis of male gametophyte development in rice anthers. *Proteomics,* **3:** 738-751.

Komatsu, S., H. Kajiwara and Hirano. 1993. A rice protein library: A data file of rice proteins separated by two dimensional electrophoresis. *Theor. Appl. Genet.* **86:** 953-962.

Komatsu, S., A. Muhammad and R. Rakwal. 1999a. Separation and characterization of proteins from green and etiolated shoots of rice: Towards a rice proteome. *Electrophoresis,* **20:** 630-636.

Komatsu, S., R. Rakwal and Z. Li. 1999b. Separation and characterization of proteins in rice suspension cultured cells. *Plant Cell, Tissue and Organ Culture,* **55:** 183-192.

Komatsu, S. and H. Konishi. 2005. Proteome analysis of rice root proteins regulated by gibberellin. *Genomics Proteomics Bioinformatics* **3:** 132-142.

Lee, D.G., K.W. Park, J.Y. An, Y.G. Sohn, J.K. Ha, H.Y. Kim, D.W. Bae, K.H. Lee, N.J. Kang, B.H. Lee, K.Y. Kang and J.J. Lee. 2011. Proteomics analysis of salt-induced leaf proteins in two rice germplasms with different salt sensitivity. *Can. J. Plant Sci.* **91:** 337-349.

Lee, K.T., D.W. Bae, S.H. Kim, H.J. Han, X. Lin, H.C. Park, C.O. Lim, S.Y. Lee and W.S. Chung. 2010. Comparative proteomic analysis of the short term responses of rice roots and leaves to cadmium. *J. Plant Physiol.,* **167:** 161-168.

Lee, J. and H.J. Koh. 2011. A label-free quantitive shotgun proteomics analysis of rice grain development. *Proteome Sci.,* **9:** 61-71.

Liao, J. L., H.W. Zhou, H.Y. Zhang, P.A. Zhong and Y.J. Huang. 2013. Comparative proteomic analysis of differentially expressed proteins in the early milky stage of rice grains during high temperature stress. *J. Exp. Bot.* doi: 10.1093/jxb/ert435.

Liu, J.M., Raveendran, R. Mushtaq, X. Ji, X. Yang, R. Bruskiewich, S. Katiyar, S. Cheng, R. Lafitte and J. Bennett. 2003. Proteomic analysis of drought responsiveness in rice: OSADF5. *In:* Tuberosa. R. Phillips and Gale, M. (eds) Proceedings of the International Congress "In the Wake of double helix: From the green revolution to the gene revolution". Bologna, Italy.

Mastsudaria, P.J. 1987. Sequence from picomole quantities of proteins electro-blotted onto polyvinylidene difluoride membranes. *J. Biol. Chem.,* **262:** 10035-10038.

Mittler, R. 2006. A biotic stress, the field environment and stress combination. *Trends in Plant Sci.* **11:** 15-19.

O'Farrell, P.H. 1975. High resolution two dimensional gel electrophoresis of proteins. *J. Biol. Chem.,* **250:** 4007-4021.

Pandey, A. and M. Mann. 2000. Proteomics to study genes and genomes. *Nature,* **405:** 837-846.

Patterson, S.D. and R.H. Aebersold. 2003. Proteomics: The first decade and beyond. *Nat. Genet. Suppl.* **33:** 311-323.

Petricka, J.J., M.M. Schaner, M. Megraw, N.W. Breakfield, J.W. Thompson, S. Georgiev, E.J. Soderblom, U. Ohler, M.A. Mosela, U. Grossnikilans and P.N. Benfey. 2012. The protein expression landscape of the Arabidopsis root. *Proc. Natl. Acad. Sci. USA,* **109:** 6811-6818.

Phizicky, E., P.I. Basliaeus, H. Zhu, M. Snyder and S. Fields. 2003. Protein analysis on a proteomic scale. *Nature,* **422:** 208-215.

Rakwal, R. and S. Komatsu. 2000. Role of jasmonate in the rice self defense mechanism using proteome analysis. *Electrophoresis.* **21:** 2492-2500.

Rockstorm, J. and M. Falkenmark. 2000. Semiarid crop production from a hydrological perspective: Gap between potential and actual yields. *Crit.Rev.Plant Sci.,* **19:** 319-346.

Salekdeh, G.H., J. Siopongco, L.J. Wade, B. Ghareyazie and J. Bennett. 2002. Proteomic analysis of rice leaves during drought stress and recovery. *Proteomics,* **2:** 1131-1145.

Salekdeh, G.H., J. Siopongco, L.J. Wade, B. Ghareyazie and J. Bennett. 2002. A proteomic approach to analyzing drought and salt responsiveness in rice. *Field Crops Res.,* **76:** 199-219. Doi:10.1016/S0378-4290(02)00040-0.

Sano, N., H. Permana, R. Kumada, Y. Shinozaki, T. Tanabata, T. Yamoda, T. Hirasawa and M. Kanekatsu. 2012. Proteomic analysis of embryonic proteins synthesized from long lived mRNAs during germination of rice seeds. *Plant Cell Physiol.,* **53:** 637-698.

Sasaki, T. 1998. The rice genome project in Japan. *Proc.Natl. Acad.Sci.USA,* **95:** 2027-2029.

Shao, C.H., G.R. Lin, J.Y. Wang, C.F. Yue and W.X. Lin. 2008. Differential proteomic analysis of leaf development at rice (*Oryza sativa*) seedling stage. *Agri.Sci.in China,* **7:** 1153-1160.

Tsugita, A., T. Kawakami, Y. Uchiyama, M. Kamo, N. Miyatake and Y. Noze. 1994. Separation and characterization of rice proteins. *Electrophoresis,* **15:** 708-720.

Wang, Z.Q., X.Y. Xu, Q.Q. Gong, C. Xie, W. Fan, J.L. Yang, Q.S. Lin and S.J. Zhengi. 2014. Root proteome of rice studied by iTRAQ provides integrated insight into aluminium stress tolerance mechanisms in plants. *J. Proteomics,* **98:** 189-205.

Washburn, M.P., D. Wolters and J.R. Yates 3rd. 2001. Large scale analysis of the yeast proteome by multidimensional protein identification technology. *Nat. Biotechnol.,* **19:** 242-247.

Wilkins, M.R., E. Gasteiger, A.A. Gooley, B.R. Herbert, M.P. Molloy, P.A. Binz, K. Ou, J.C. Sanchez, A. Bairoch, K.L.Willams and D.F.J. Hochstrasser. 1999. High-throughput mass spectrometric discovery of protein post-translational modifications. *Mol.Biol.,* **289:** 645-657.

Wolters, D.A., M.P. Washburn and J.R. Yates 3rd. 2001. An automated multidimensional protein identification technology for shotgun proteomics. *Anal.Chem.,* **73:** 5683-5690.

Xiang, X., S. Ning and D. Wei. 2013. Proteomic profiling of rice roots from a super-hybrid rice cultivar and its parental lines. *Plant Omics J.,* **6:** 318-324.

Xiong, J.H., B.Y. Fu, H.X. Xu and Y.S. Li. 2010. Proteomic analysis of PEG-simulated drought stress responsive proteins of rice leaves using a pyramiding rice line at the seedling stage. *Bot.Studies,* **51:** 137-145.

Xu, D.Q., J. Huang, S.Q. Guo, X. Yang, Y.M. Bao, H.J. Tang and H.S. Zhang. 2008. Over expression of a TFIIIA-type zinc finger protein gene ZFP252 enhances drought and salt tolerance in rice (*Oryza sativa* L.). *FEBS Letter,* **582:** 1037-1043.

Yan, S., Z. Tang, W. Su and W. Sun. 2005. Proteomic analysis of salt stress responsive proteins in rice root. *Proteomics,* **5:** 235-244.

Yang, Q., Y. Wang, J. Zhang, W. Shi, C. Qian and X. Peng. 2007a. Identification of aluminium-responsive proteins in rice roots by a proteomic approach: Cysteine synthase as a key player in Al response. *Proteomics,* **7:** 737-749.

Yang, Y., L. Dai, H. Xia, K. Zhu, H. Liu and K. Chen. 2013. Protein profile of rice (*Oryza sativa*) seeds. *Genet. Mol.Biol.*, **36:** 87-92.

Yang, P., X. Li, X. Wang, H. Chen, F. Chen and S. Shen. 2007. Proteomic analysis of rice (*Oryza sativa*) seeds during germination. *Proteomics,* **7:** 3358-3368.

Yates, J.R. 3rd. 1998. Mass spectrometry and the age of the proteome. *J. Mass Spectrom.,* **33:** 1-19.

Yeu, S.Y., B.S. Park, W.G. Sang, Y.D. Choi, M.C. Kim, J. T. Song, N.C. Paek, H.J. Koh and H. Seo. 2007. The serine proteinase inhibitor Osserpin is a potent tillering regulator in rice. *J. Plant Biol.,* **50:** 600-604.

Yin, L., Y. Tao, K. Zhao, J. Shao, X. Li, G. Liu, S. Liu and L. Zhu. 2007. Proteomic and transcriptomic analysis of rice mature seed-derived callus differentiation. *Proteomics,* **7:** 755-768.

Zhao, C., J. Wang, M. Cao, K. Zhao, J. Shao, T. Lei, J. Yin, G.G. Hill, N. Xu and S. Liu. 2005. Proteomic changes in rice leaves during development of field grown rice plants. *Proteomics,* **5:** 961-972.

Zhang, C., Y. Yin, A. Zhang, Q. Lu, X. Wen, Z. Zhu, L. Zhang and C. Lu. 2012. Comparative proteomic study reveals dynamic proteome changes between super hybrid and its parents at different developmental stages. *J. Plant Physiol.,* **169:** 387-398.

Zhang, Z., H. Zhao, J. Tang, Z. Li, D. Chen and W. Lin. 2014. A proteomics study on molecular mechanism of poor grain-filling of rice (*Oryza sativa* L.) inferior spikelets. *PLOS One* doi:10.1371/journal pone0089140

Zhong, B., H. Karibe, S. Komatsu, H. Ichimura, Y. Nagamura, T. Sasaki and H. Hirano. 1997. Screening of rice genes from a cDNA catalog based on the sequence data file of proteins separated by two-dimensional electrophoresis. *Breeding Sci.* **47:** 245-251.

Zi, J., J. Zhang, Q. Wang, L. Lin, W. Tong, X. Bai, J. Zhao, Z. Chen, X. Fu and S. Liu. 2012. Proteomics study of rice embryogenesis: Discovery of the embryogenesis dependent globulins. *Electrophoresis,* **33:** 1129-1138.

Zi, J., J. Zhang, Q. Wang, B. Zhou, J. Zhong, C. Zhang, X. Qin, B. Wen, S. Zhang, X. Fu and S. Liu. 2013. Stress responsive proteins are actively regulated during rice (*Oryza sativa*) embryogenesis as indicated by quantitative proteomics analysis. *PLOS One* doi:101371/journal pone 074229

chapter ten

Nucleic Acids

Nucleic acids are polymeric complex and each unit contains a phosphate nucleoside. In the polymeric chain different purine and pyrimidine may be attached to the sugar base. Purine and pyrimidine are the cyclic nitrogenous compounds. The purine and pyrimidine attached to sugar base are known as nucleosides. When nucleosides are bonded with phosphate it is called nucleotides. An account of purine and pyrimidine associated with nucleic acids is given hereunder.

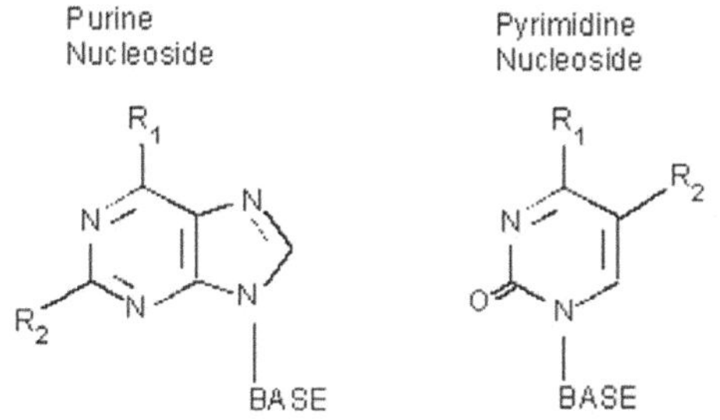

Purine Nucleoside

	R_1	R_2
Adenosine	NH_2	H
Guanosine	=O	NH_2

Pyrimidine Nucleoside

	R_1	R_2
Cytidine	NH_2	H
Thymidine	=O	CH_3
Uracil	=O	H

FIGURE 10.1 The polymeric structure of DNA may be described in terms of monomeric units of increasing complexity. Condensation polymerization of these leads to the DNA formulation outlined above. Finally, a 5′-monophosphate ester, called a *nucleotide* may be drawn as a single monomer unit. Since a monophosphate ester of this kind is a strong acid (pK_a of 1.0), it will be fully ionized at the usual physiological pH (ca.7.4). Isomeric 3′-monophospate nucleotides are also known, and both isomers are found in cells. They may be obtained by selective hydrolysis of DNA through the action of nuclease enzymes. Anhydride-like di- and tri-phosphate nucleotides have been identified as important energy carriers in biochemical reactions, the most common being ATP (adenosine 5′-triphosphate).

The high molecular weight nucleic acid, DNA, is found chiefly in the nuclei of complex cells, known as *eukaryotic cells,* or in the nucleoid regions of *prokaryotic cells,* such as bacteria. It is often associated with proteins that help to pack it in a usable fashion. In contrast, a lower molecular weight but much more abundant nucleic acid, RNA, is distributed throughout the cell, most commonly in small numerous organelles called *ribosomes*. Three kinds of RNA are identified, the largest subgroup (85 to 90%) being ribosomal RNA, rRNA, the major component of ribosomes, together with proteins. The size of rRNA molecules varies, but is generally less than a thousandth the size of DNA. The other forms of RNA are messenger RNA, mRNA, and transfer RNA, tRNA. Both have a more transient existence and are smaller than rRNA. However, there are small RNAs.

All these RNAs have similar constitutions, and differ from DNA in two important respects. As shown in diagram (Fig. 10.2), the sugar component of RNA is ribose, and the pyrimidine base uracil replaces the thymine base of DNA. The RNAs play a vital role in the transfer of information (transcription) from the DNA library to the protein factories called ribosomes, and in the interpretation of that information (translation) for the synthesis of specific polypeptides.

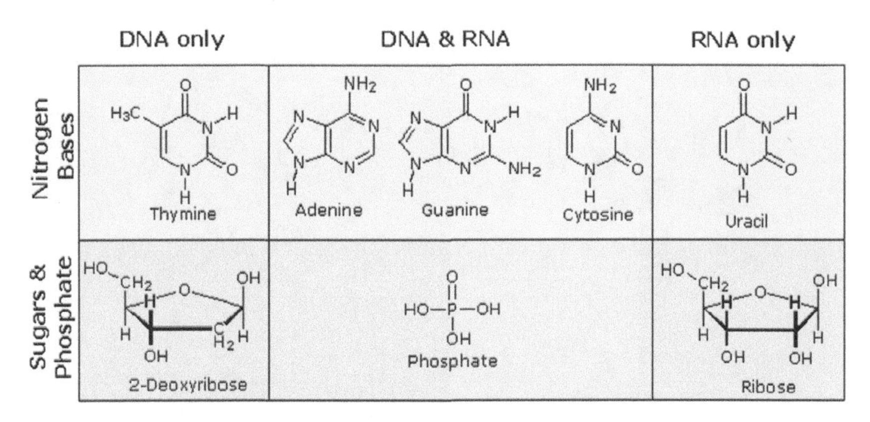

FIGURE 10.2 Components of nucleic acids.

10.1. Secondary Structure of DNA

In the early 1950s the primary structure of DNA was well established, but a firm understanding of its secondary structure was lacking. Indeed, the situation was similar to that occupied by the proteins a decade earlier, before the alpha helix and pleated sheet structures were proposed by Linus Pauling. Many researchers grappled with this problem, and it was generally conceded that the molar equivalences of base pairs (A & T and C & G) discovered by Chargaff would be an important factor. Rosalind Franklin, working at King's College, London, obtained X-ray diffraction evidence that suggested a long helical structure of uniform thickness. Francis Crick and James Watson, at Cambridge University, considered hydrogen bonded base pairing interactions, and arrived at a double stranded helical model that satisfied most of the known facts, and has been confirmed by subsequent findings.

10.2. *Base Pairing*

Careful examination of the purine and pyrimidine base components of the nucleotides reveals that three of them could exist as hydroxy pyrimidine or purine tautomers, having an aromatic heterocyclic ring. Despite the added stabilization of an aromatic ring these compounds prefer to adopt amide-like structures, with the more stable tautomer.

The biosynthesis of nucleic acids is, in brief, similar as described in the chapter PROTEINS already.

10.3. *Rice Nucleic Acids*

It has been noted that *Oryza* species has a wide variation in nuclear DNA content as mentioned by different scientists. The following table is given in this context (Table 10.1).

TABLE 10.1 Nuclear DNA content of rice species as determined by flow cytometry

Genome scientific name	Group	Nuclear DNA content	
		pg/2C	Mbp/1C
Oryza glaberrinma	AgAg	0.76	366
Oryza longistaminata	A1A1	0.78	376
Oryza sativa (indica)	AA	0.87	419
Oryza sativa (japonica)	AA	0.86	415
Oryzasativa (javanica)	AA	0.88	424
Oryza officinalis	CC	1.14	550
Oryza elchingeri	CC	1.17	564
Oryza atistraietisis	EE	1.99	960
Oryza ridleyi	Unknown	1.31	632
Oryza latifolia	CCDD	2.32	1124
Oryza minuta	BBCC	2.33	1124

A comparative study of yield and purity of total RNA in major cereals showed the distinctive differences (Table 10.2).

TABLE 10.2 Yield and purity of total RNA prepared by the new protocol and standard Phenol-SDS method evaluated by UV light absorption spectra and ratios of A260/A280 and A260/A230

Seed Species	A_{260}/A_{280}[a]		A_{260}/A_{230}[a]		RNA Yield[b] (μg/100 mg FW[c])	
	Phenol-SDS	New Protocol	Phenol-SDS	New Protocol	Phenol-SDS	New Protocol
Wheat	1.54 ± 0.25	1.98 ± 0.02	0.34 ± 0.10	1.85 ± 0.03	18.3 ± 3.2	48.5 ± 4.1
Rice	1.69 ± 0.10	1.95 ± 0.02	0.77 ± 0.09	2.01± 0.02	25.7 ± 2.5	50.3 ± 3.2
Maize	1.57 ± 0.14	1.98 ± 0.03	0.90± 0.05	1.89 ± 0.03	28.5 ± 1.9	51.5 ± 2.8

SDS : Sodium dodecyl sulfate; FW : Fresh weight.
[a] Ten independent RNA extraction replicas of each species that were measured for analysis values of A260/A280 or A260/A230 are mean ± SD.
[b] RNA yields using the phenol-SDS protocol were significantly different from yields with the new protocol at the 95% confidence level

One of the critical aspects of our DNA isolation protocol is the incubation of seed/grain before isolation of DNA. It was observed that grinding of seed/grain without incubation in the buffer results in distinct DNA degradation while grinding the seed/grain, after 30-45 min incubation always gave good quality and quantity DNA. Incubation of seeds in the buffer softens the hard tissue due to imbibition, which helps in smooth and easy grinding. In the present study, it was observed that DNA isolation procedure is highly amenable for DNA isolation from half seed also. Hence, the DNA can be used for non-destructive analysis of segregating progeny since the selected remnant half seeds containing the embryo part can be germinated later. From each dehusked seed/grain of rice, it was obtained 1.8-2.0 μg of DNA. The same protocol has also been tested and found to be effective for isolation of DNA from leaf with slight modification. Fresh leaf tissue can be ground directly in extraction buffer using a

TABLE 10.3 DNA yield obtained (in micro gram) from different DNA isolation protocols.

Modified CTAB protocol	Genotype	Tissue used	Protocol of Pal et al. 2001	Protocol of Chunwongse et al. 1993	Nucleon phytopure kit
1.8	Swarna	Single seed	1.2	1.5	1.7
1.5	Pusa Basmati 1	Milled grain	1.3	0.9	1.3
1.7	KRH2	Single seed	1.2	1.5	1.5
2.0	IR58025A	Single seed	1.4	1.6	1.7

spot test plate as per the procedure of Zheng et al. (1995). From 3 cm leaf piece, 4-5 μg of DNA can be isolated. In a modest laboratory, a team of 2-3 personnel could handle DNA extraction from about 800-1000 seed/grain of rice per day. The isolated DNA was highly intact, devoid of shearing and comparable to those isolated using the protocols of Pal et al. (2001), Chunwongse et al. (1993) and Nucleon phytopure kit (Table 10.3).

This new and rapid protocol of DNA isolation from single seed/grain and leaf tissues is fast, consistent and inexpensive. Besides, this method does not involve the use of phenol, which is hazardous. We recommend this method for use in marker-based seed/grain purity assays and also for rapid genotyping in marker-assisted breeding programs.

10.4. Seed Germination

According to Mukherjee et al. (1971), in rice germination the degradation of phytate continues gradually with time,and finally the concentrations of DNA and RNA Pi increased,indicating enhanced cell division and net protein synthesis.

The RNA level in the grain also decreased during germination faster than the starch level, coinciding with the increase in RNase level (Table 10.4). The drop in RNA was significant after four days germination in light and after three days in the dark. The increase in RNase level was also significant after four days germination in light and after three days germination in the dark. Its level was still high after seven days germination (Palmiano and Juliano, 1972). Matsushita (1958) similarly found a progressive decrease in rice endosperm RNA during germination.

Dry mature seeds contain a large number of mRNA species. Cotton was the first plant found to store RNA in a mature dry seed (Dure and Waters, 1965); after the 1990s, stored RNA was found to be universal in the mature dry seeds of plant species (Kimura and Nambara, 2010; Nakabayshi et al. 2005). Stored mRNA in seeds reflects gene expression

TABLE 10.4 Phosphorus fractions in rice seeds during germination

Germination (h)	P fraction	P fraction	P fraction	P fraction	P fraction
	Phytate (mgP/ gdw)	Lipid (mgP/ gdw)	Inorganic (mgP/ gdw)	Ester (mgP/ gdw)	RNA + DNA (mgP/ gdw)
0	2.67	0.41	0.24	0.08	0.06
24	1.48	1.39	0.64	0.10	0.05
48	1.06	1.54	0.89	0.11	0.08
72	0.60	1.71	0.86	0.12	0.12

patterns during seed germination. Affymetrix arrays can provide a comprehensive description and real-time changes at the whole-transcriptome level during seed germination and have been used to investigate the biological processes of seed germination in many plants such as *Arabidopsis*, barley, rice and maize.

Small RNAs (sRNAs) are common and effective modulators of gene expression in eukaryotic organisms. To characterize the sRNAs expressed during rice seed development, massively parallel signature sequencing (MPSS) was performed, resulting in the obtainment of 797 399 22-nt sequence signatures, of which 111 161 are distinct ones. Analysis on the distributions of sRNAs on chromosomes showed that most sRNAs originate from interspersed repeats that mainly consist of transposable elements, suggesting the major function of sRNAs in rice seeds is transposon silencing. Through integrative analysis, 26 novel miRNAs and 12 miRNA candidates were identified. Further analysis on the expression profiles of the known and novel miRNAs through hybridizing the generated chips revealed that most miRNAs were expressed preferentially in one or two rice tissues. Detailed comparison of the expression patterns of miRNAs and corresponding target genes revealed the negative correlation between them, while few of them are positively correlated. In addition, differential accumulations of miRNAs and corresponding miRNA*s suggest the functions of miRNA*s other than being passenger strands of mature miRNAs, and in regulating the miRNA functions.

Many hypotheses have been proposed regarding causes of seed ageing such as loss of vigour and viability in terms of germination due to many physiological changes in deteriorated seeds. However the exact molecular changes involved in seed deterioration have not been very well studied in order to support the above hypothesis. DNA and RNA orchestrate gene activity implicated in life and cell death processes (Bushell et al. 2004; Hoeberichts and Woltering, 2003). It is therefore surprising that the integrity of nucleic acids during seed ageing has received relatively little attention in the recent literature. Changes in nucleic acid content (Brockelhurst and Fraser, 1980; Sen and Osborne, 1974, 1977; Thompson et al. 1987) and more recently, DNA fragmentation (Kranner et al. 2006; Osborne, 2000), have been reported in seeds in relation to maturation and germination rate. Single and double strand breaks of DNA accumulate in ageing seeds (Tuteja et al. 2001) and DNA fragmentation was correlated with seed death induced by drying in 'recalcitrant', i.e. desiccation intolerant, seeds (Faria et al. 2005; Kranner et al. 2006).

The molecular lesions are associated with loss of viability (Osborne et al. 1974). Integrity of macromolecules such as DNA may not be maintained during seed ageing and cleavage of DNA molecules to lower molecular weight fragments occurs without loss of DNA, by virtue of their larger size, susceptible to shear. Prolonged ageing up to 12 days further showed fragmentation of DNA. As seeds lost viability during ageing, DNA was gradually degraded into inter nucleosomal fragments, resulting in 'DNA laddering'. Nucleolytic events such as DNA laddering are known characteristics of programmed cell death (PCD) (Iise et al. 2011). Seed deterioration at the DNA level pointed out that chromosomal aberration, point mutations, and decreased activity of DNA repair enzymes are some of the major events occurring during the process of aging in seeds (Murata et al. 1981).

10.5. *Changes in Seedlings*

RNA contents were reduced in rice leaves after infection by *Pyricularia oryzea*, Decrease was greater with race IE-2 than with IC-22. No clear cut differences were found between healthy, resistant and susceptible plants. But differences between susceptible and healthy plants after inoculation were significant. A significant increase in DNA contents was observed after inoculation with either race of *P. oryzae* (Padhi and Chakraborti, 1984).

The effect of an industrial effluent on cellular macromolecular composition and indophenol photochemical activity of isolated chloroplasts of rice (*Oryza sativa* L. CV. Mushoori) seedlings have been investigated. Each experiment consisted of two parts. One was the effect of various concentrations and the other was the time-dependent changes induced by the undiluted effluent. The total pigments, proteins and nucleic acids of rice seedlings declined with an increase in effluent concentration and the time of incubation. The loss in contents of macromolecules like deoxyribonucleic acid (DNA), ribonucleic acid (RNA), and protein was relatively more marked in the root than in the shoot. RNA and chlorophyll (Chl) contents of the seedlings were found most susceptible to effluent stress. Loss in Hill reaction activity measured as photoreduction of 2,6-dichlorophenol indophenol (DCPIP) of isolated chloroplasts could be correlated in a general way with the loss of pigments, proteins and nucleic acids. Diphenyl carbazide- and Mn^{2+}-induced restoration of loss in DCPIP photoreduction suggests that the damage of oxygen evolving systems is the initial site of action of the effluent (Behera and Misra, 1983).

In rice, the molecular mechanism for the regulation of potassium starvation responses has not been investigated in detail. Here, a combined physiological and whole genome transcriptomic study of rice seedlings exposed to a brief period of potassium deficiency then replenished with potassium. Results reveal that the expressions of a diverse set of genes annotated with many distinct functions were altered under potassium deprivation. Findings highlight altered expression patterns of potassium-responsive genes majorly involved in metabolic processes, stress responses, signaling pathways, transcriptional regulation, and transport of multiple molecules including K^+. Interestingly, several genes responsive to low-potassium conditions show a reversal in expression upon resupply of potassium. The results of this study indicate that potassium deprivation leads to activation of multiple genes and gene networks, which may be acting in concert to sense the external potassium and mediate uptake, distribution and ultimately adaptation to low potassium conditions. The interplay of both upregulated and downregulated genes globally in response to potassium deprivation determines how plants cope with the stress of nutrient deficiency at different physiological as well as developmental stages of plants (Shankar et al. 2013).

Nitrogen is an essential mineral nutrient required for plant growth and development. Insufficient nitrogen (N) supply triggers extensive physiological and biochemical changes in plants. In this study, we useing Affymetrix GeneChip rice genome arrays to analyse the dynamics of rice transcriptome under N starvation. N starvation induced or suppressed transcription of 3518 genes, representing 10.88% of the genome. These changes, mostly transient, affected various cellular metabolic pathways, including stress response, primary and secondary metabolism, molecular transport, regulatory process and organismal development. 462 or 13.1% transcripts for N starvation expressed similarly in root and shoot. Comparative analysis between rice and *Arabidopsis* identified 73 orthologous groups that responded to N starvation and demonstrated the existence of conserved N stress coupling mechanism among plants. Additional analysis of transcription profiles of microRNAs revealed differential expression of miR399 and miR530 under N starvation, suggesting their potential roles in plant nutrient homeostasis(Cai et al. 2012).

The endoplasmic reticulum (ER) stress response is widely known to function in eukaryotes to maintain the homeostasis of the ER when unfolded or misfolded proteins are overloaded in the ER. To understand the molecular mechanisms of the ER stress response in rice (*Oryza sativa* L.), it was previously analyzed the expression profile of stably transformed rice in which an ER stress sensor/transducer *OsIRE1* was knocked-down, using the combination of preliminary microarray and quantitative RT-PCR. In this study, to obtain more detailed expression profiles of genes involved in the initial stages of the ER stress response in rice, analysis of RNA sequencing of wild-type and transgenic rice plants produced by homologous recombination in which endogenous genomic *OsIRE1* was replaced by missense alleles defective in ribonuclease activity (Wakasa et al. 2014).

At least 38,076 transcripts were investigated by RNA sequencing, 380 of which responded to ER stress at a statistically significant level (195 were upregulated and 185 were downregulated). Further, 17 genes from the set of 380 ER stress-responsive genes were identified that were not included in the probe set of the currently available microarray chip in rice. Notably, three of these 17 genes were non-annotated genes, even in the latest version of the Rice Annotation Project Data Base (RAP-DB, version IRGSP-1.0). Therefore, RNA sequencing-mediated expression profiling provided valuable information about the ER stress response in rice plants and led to the discovery of new genes related to ER stress.

10.6. *Stress Responses*

The decrease in growth of rice caused by NaCl salinity is accompanied by a decrease in the contents of DNA, RNA and protein in the embryo axis. As was in the case of seedling growth the levels of polyamines (putrescine, spermidine and spermine) and the activity of agmatine deiminase (EC 3.5.3.12), an enzyme involved in the biosynthesis of polyamines, in endosperm and embryo axis also were markedly lowered by salt. These biochemical changes induced by NaCl are observed with respect to growth and development of rice seedlings(Prakash et al. 1988).

Mutations in DNA lead to substitutions of one nucleic acid with another nucleic acid. An example is a replacement of "ATG" with "AGG" in a string of three nucleic acids. Strings of three nucleic acids code for a 'codon' that is then translated into a specific amino acid. An example of a codon is "ATG" coding for the amino acid "methionine." Strings of amino acids make up proteins, and proteins are what orchestrate processes that make living things *'alive!'* Some nucleotide sequence replacements result in changes in amino acid sequences. These amino acid sequence changes sometimes change the protein structure enough that typical functioning of the protein and/or its pathway is disrupted or changed. In a study the population-level processes where mutations first occur in order to understand how such genetic variation leads to long-term diversity of protein sequences and functions. Combining the work on genetic variation with trait variation and environmental variation helps us understand how all three types of variation (genetic, trait and environment) result in adaptation and evolution over time. It was noted primarily on population-level processes because adaptation and evolution are the result of changes in frequencies and are not individual-level processes and pathway.

A T-DNA insertional mutant OsTEF1 of rice gives 60-80% reduced tillering, retarded growth of seminal roots, and sensitivity to salt stress compared to wild type Basmati 370. The insertion occurred in a gene encoding a transcription elongation factor homologous to yeast elf1, on chromosome 2 of rice. Detailed transcriptomic profiling of OsTEF1 revealed that mutation in the transcription elongation factor differentially regulates the expression of more than 100 genes with known function and finely regulates tillering process in rice by inducing the expression of cytochrome P450. Along with different transcription factors, several stress associated genes were also affected due to a single insertion. In silico analysis of the TEF1 protein showed high conservation among different organisms. This transcription elongation factor predicted to interact with other proteins that directly or indirectly positively regulate tillering in rice (Paul et al. 2012).

The highly conserved plant microRNA, miR156, is an essential regulator for plant development. In Arabidopsis (*Arabidopsis thaliana*), miR156 modulates phase changing through its temporal expression in the shoot. In contrast to the gradual decrease over time in the shoot (or whole plant), we found that the miR156 level in rice (*Oryza sativa*) gradually increased from young leaf to old leaf after the juvenile stage. However, the miR156-targeted rice SQUAMOSA-promoter binding-like (SPL) transcription factors were either dominantly expressed in young leaves or not changed over the time of leaf growth. A comparison of the transcriptomes of early-emerged old leaves and later-emerged young leaves from wild-type and miR156 overexpression (miR156-OE) rice lines found that expression levels of 3,008 genes were affected in miR156-OE leaves. Analysis of temporal expression changes of these genes suggested that miR156 regulates gene expression in a leaf age-dependent manner, and miR156-OE attenuated the temporal changes of 2,660 genes. Interestingly, seven conserved plant microRNAs also showed temporal changes from young to old leaves, and miR156-OE also attenuated the temporal changes of six microRNAs. Consistent with global gene expression changes, miR156-OE plants resulted in dramatic changes including precocious leaf maturation and rapid leaf/tiller initiation. Results indicate that another gradient of miR156 is present over time, a gradual increase during leaf growth, in addition to the gradual decrease during shoot growth. Gradually increased miR156 expression in the leaf might be essential for regulating the temporal expression of genes involved in leaf development (Xie et al. 2012).

Nitrogen is essential for plants. The synthesis of cellular proteins, amino acids, nucleic acids, purine and pyrimidine nucleotide are dependent upon N. It is the most abundant mineral element in plant tissues which is derived from the soil. However, excess N may cause significant biochemical changes in plants and may lead to nutritional imbalances (Mills and Jones 1979). Generally it is needed in most rice soils. Rice plants require a large amount of N at the early and mid tillering stages to maximize the number of panicles (De Datta 1981).

10.7. Vegetative Growth

Tillering in rice (*Oryza sativa*L.) is an important agronomic trait for grain production, and also a model system for the study of branching in monocotyledonous plants. Rice tiller is a specialized grain-bearing branch that is formed on the unelongated basal internode and grows independently of the mother stem (culm) by means of its own adventitious roots. Rice tillering occurs in a two-stage process: the formation of an axillary bud at each leaf axil and its subsequent outgrowth. Although the morphology and histology and some mutants of rice tillering have been well described, the molecular mechanism of rice tillering remains to be elucidated. Here it is reported the isolation and characterization of MONOCULM 1 (MOC1), a gene that is important in the control of rice tillering. The moc1 mutant plants have only a main culm without any tillers owing to a defect in the formation of tiller buds. MOC1 encodes a putative GRAS family nuclear protein that is expressed mainly in the axillary buds and functions to initiate axillary buds and to promote their outgrowth (Xu et al. 2012).

Tiller initiation and panicle development are important agronomical traits for grain production in *Oryza sativa* L. (rice), but their regulatory mechanisms are not yet fully understood. In this study, T-DNA mutant and RNAi transgenic approaches were used to functionally characterize a unique rice gene, *LAGGING GROWTH AND DEVELOPMENT 1* (*LGD1*). The *lgd1* mutant showed slow growth, reduced tiller number and plant height, altered panicle architecture and reduced grain yield. The fewer unelongated internodes and cells in *lgd1* led to respective reductions in tiller number and to semi-dwarfism. Several independent *LGD1*-RNAi lines exhibited defective phenotypes similar to those observed in *lgd1*. Interestingly, *LGD1* encodes multiple transcripts with different transcription start sites (TSSs), which were validated by RNA ligase-mediated rapid amplification of 5' and 3' cDNA ends (RLM-RACE). Additionally, GUS assays and a luciferase promoter assay confirmed the promoter activities of *LGD1.1* and *LGD1.5*. *LGD1* encoding a von Willebrand factor type A (vWA) domain containing protein is a single gene in rice that is seemingly specific to grasses. GFP-tagged LGD1 isoforms were predominantly detected in the nucleus, and weakly in the cytoplasm. *In vitro* northwestern analysis showed the RNA-binding activity of the recombinant C-terminal LGD1 protein. Our results demonstrated that LGD1 pleiotropically regulated rice vegetative growth and development through both the distinct spatiotemporal expression patterns of its multiple transcripts and RNA binding activity. Hence, the study of *LGD1* will strengthen our understanding of the molecular basis of the multiple transcripts, and their corresponding polypeptides with RNA binding activity, that regulate pleiotropic effects in rice (Thangasamy et al. 2012).

10.8. Changes in Reproductive Development

Molecular events that follow the establishment of floral organ primordia, ultimately culminate into development of male (pollen) and female (embryo sac) gametophytes in specialized sex organs known as the androecium and gynoecium. The gametes thus formed undergo fertilization and develop into seeds. Understanding the underlying gene regulatory networks that control the development of reproductive floral organs, and the male and female gametophytes therein, involves :

- identification of the genetic components involved,
- their classification into pair wise protein-protein and protein-DNA interactomes, followed by
- construction of biologically realistic gene regulatory networks.

To unravel these networks to understand developmental mechanisms in terms of mechanistic models and, thus, it will pave the way for translating genetic interactions into phenotypic traits. In this regard, it is necessary to carry out whole genome microarray-based transcriptome analysis of more than twenty tissues/stages of rice vegetative and reproductive development, which has helped in the identification of several co-expressed groups of genes. These groups either show similar up-regulation profiles or express in a tissue or developmental stage specific manner and, thus, forming putative interactomes. The transcriptomic analysis has been refined to include subtractive logic in order to shortlist genes that express specifically in individual tissues/stages of development for validation of function and/or promoter activities. For gene function validation, RNAi/miRNA based silencing and ectopic expression strategies in transgenic rice and/or Arabidopsis are being followed. Moreover, promoter activities are being determined by driving expression of GUS and/or GFP reporter gene in transgenic systems (Communication from Sanjoy Kapoor).

Rice Mutant Database (RMD, http://rmd.ncpgr.cn) is an archive for collecting, managing and searching information of the T-DNA insertion mutants generated by an enhancer trap system. We have generated ~129 000 rice mutant (enhancer trap) lines that are now being gathered in the database. Information collected in RMD includes mutant phenotypes, reporter-gene expression patterns, flanking sequences of T-DNA insertional sites, seed availability and others, and can be searched by respective ID, keyword, nucleotide sequence or protein sequence on the website. This database is both a mutant collection for identifying novel genes and regulatory elements and a pattern line collection for ectopic expression of target gene in specific tissue or at specific growth stage (Zhang et al. 2006).

MicroRNAs (miRNAs), present widely in eukaryotes, are a class of endogenous, non-coding small RNAs (20–24 nt) that regulate gene expression by targeted protein-coding gene mRNA sequence-specific cleavage, translation repression or DNA methylation at the post-transcriptional level, often resulting in gene silencing (Voinnet, 2009; Wu et al. 2010; Jones-Rhoades et al. 2006). miRNAs genes are transcribed by RNA polymerase II into long, specific, hairpin-structure primary transcripts (pri-miRNAs). miRNAs are generated from pri-miRNAs by Dicer-like1 (DCL1) cleavage with the help of the RNA-binding protein DAWDLE (DDL), the C_2H_2-zinc finger protein SERRATE (SE), the double -stranded RNA-binding protein HYPONASTIC LEAVES1 (HYL1), and other factors. Mature miRNAs are methylated by the S-adenosyl methionine-dependent methyltransferase Hua Enhancer 1 (*HEN1*), and are incorporated into the *ARGONAUTE* (*AGO*) proteins to form the RNA-induced silencing complexes (RISC) that are involved in gene silencing (Miyoshi et al. 2010; Chen, 2005). miRNAs have an important function in diverse biological and metabolic processes in plants, including hormonal regulation, defense responses, tissue development, phase transition, flowering, and adaptation to a variety of biotic and abiotic stresses (Eldem et al. 2013;Chuck et al. 2009).

Studies in *Arabidopsis* (Yamaguchi et al. 2009), rice (Xie et al. 2012; Zhu et al. 2009), *Ipomoea nil* (Glazinska et al. 2009), and the early-flowering mutant of trifoliate orange have shown that miRNAs, such as miR156 and miR172, regulate the expression of developmental factors involved in flowering. MiR156, one of the most highly conserved plant microRNAs, is part of an intrinsic pathway for controlling the transition from vegetative growth to flowering in plants. The regulation of this pathway is based on changes in the miR156 content, miR156 declines from the vegetative stage to the reproductive stage, but the levels of its targets, *SPL* (Squamosa Promoter Binding Protein Like) transcriptional factors, increase during the same period. However, *SPLs* control the transition between the juvenile and flowering stages by regulating the expression of a class of MADS box genes, which induced flowering (Wang et al. 2009; Wu et al. 2009; Poethig, 2009). On the other hand, miR172 controls flowering time and floral organogenesis by regulating expression of the transcription factor gene *APETALA2* (*AP2*) and other *AP2*-like genes. Like *AP2* mutants, overexpressing of miR172 plants flower earlier and produce abnormal floral organs. The roles of miR172 and *AP2*-like genes are important.

Rice (*Oryza sativa*) is an excellent model monocot with a known genome sequence for studying embryogenesis. Reports on the transcriptome profiling analysis of rice developing embryos using RNA-Seq as an attempt to gain insight into the molecular and cellular events associated with rice embryogenesis. RNA-Seq analysis generated 17,755,890 sequence reads aligned with 27,190 genes, which provided abundant data for the analysis of rice embryogenesis. A total of 23,971, 23,732 and 23,592 genes were identified from embryos at three developmental stages (3–5, 7 and 14 DAP), while an analysis between stages allowed the identification of a subset of stage-specific genes. The number of genes expressed stage-specifically was 1,131, 1,443 and 1,223, respectively. In addition, investigation on transcriptomic changes during rice embryogenesis based on our RNA-Seq data was done. A total of 1,011 differentially expressed genes (DEGs) ($\log_2$Ratio ≥1, FDR ≤0.001) were identified; thus, the transcriptome of the developing rice embryos changed considerably. A total of 672 genes with significant changes in expression were detected between 3–5 and 7 DAP; 504 DEGs were identified between 7 and 14 DAP. A large number of genes related to metabolism, transcriptional regulation, nucleic acid replication/

processing, and signal transduction were expressed predominantly in the early and middle stages of embryogenesis. Protein biosynthesis-related genes accumulated predominantly in embryos at the middle stage. Genes for starch/sucrose metabolism and protein modification were highly expressed in the middle and late stages of embryogenesis. In addition, we found that many transcription factor families may play important roles at different developmental stages, not only in embryo initiation but also in other developmental processes. These results will expand our understanding of the complex molecular and cellular events in rice embryogenesis and provide a foundation for future studies on embryo development in rice and other cereal crops (Xu et al. 2012a).

Spatial and temporal changes in the distribution of mRNA sequences during anther and pollen development in rice (*Oryza sativa*) were investigated by in situ hybridization with [3H]polyuridylic acid ([3H]poly(U)) and a cloned rice histone gene probe. Annealing of sections with [3H] poly(U) showed that poly(A)-containing RNA (poly(A)+RNA) was uniformly distributed in the cells of the anther primordium. During the formation of the archesporial initial, the primary parietal cell, the primary sporogenous cell and tapetum, there was no differential accumulation of poly(A)+RNA in their progenitor cells. Preparatory to meiosis, there was a sharp decrease in poly(A)+RNA concentration in the epidermis and middle layer of the anther wall, although the label persisted in the endothecium, tapetum and microsporocytes. Poly(A)+RNA concentration decreased in these cell types during meiosis and attained very low levels in the disintegrating tapetum and the persistent endothecium of the post-meiotic anther. Pollen development was characterized by the absence of [3H]poly(U) binding sites in the uninucleate microspores and by their presence in the vegetative and generative cells of the bicellular pollen grain. In anther sections hybridized with [3H]histone probe, gene expression was only detected in the endothecium of the premeiotic anther and in the bicellular pollen grains (Raghavan, 1989).

Rice pollen and seed development are directly related to grain yield. To further improve rice yield, it is important to functionally annotate the genes controlling pollen/seed development and to use them for rice breeding. A genome-wide expression analysis was first carried out with an emphasis on genes being involved in rice pollen and seed development. Based on the transcript profiling, we have identified and functionally classified 82 highly expressed pollen-specific, 12 developing seed-specific and 19 germinating seed-specific genes. We then presented the utilization of the maize transposon *Dissociation* (*Ds*) insertion lines for functional genomics of rice pollen and seed development and as alternative germplasm resources for rice breeding. A two-element *Activator/Dissociation* (*Ac/Ds*) gene trap tagging system was established that generated around 20,000 *Ds* insertion lines. These were subjected to these lines for screens to obtain high and low yield *Ds* insertion lines. Some interesting lines have been obtained with higher yield or male sterility. Flanking Sequence Tags (FSTs) analyses showed that these *Ds*-tagged genes encoded various proteins including transcription factors, transport proteins, unknown functional proteins and so on. They exhibited diversified expression patterns. Our results suggested that rice could be improved not only by introducing foreign genes but also by knocking out its endogenous genes. This finding might provide a new way for rice breeder to further improve rice varieties (Jiang and Ramachandran, 2011).

Gene expression throughout the reproductive process in rice (*Oryza sativa*) beginning with primordia development through pollination/fertilization to zygote formation was analyzed. Twenty five stages/organs of rice reproductive development including early microsporogenesis stages were analyzed with 57,381 probe sets, that identified around 26,000 expressed probe sets in each stage. Fine dissection of 25 reproductive stages/organs combined with detailed microarray profiling revealed dramatic, coordinated and finely tuned changes in gene expression. A decrease in expressed genes in the pollen maturation process was observed in a similar way with Arabidopsis and maize. An almost equal number of ab initio predicted genes and cloned genes which appeared or disappeared coordinated with developmental stage progression. A large number of organ-/stage-specific genes were identified; notably 2,593 probe sets for developing anther, including 932 probe sets corresponding to ab initio predicted genes. Analysis of cell cycle-related genes revealed that several cyclin-dependent kinases (CDKs), cyclins and components of SCF E3 ubiquitin ligase complexes were expressed specifically in reproductive organs. Cell wall biosynthesis or degradation protein genes and transcription factor genes expressed specifically in reproductive stages were also newly identified. Rice genes homologous to reproduction-related genes in other plants showed expression profiles both consistent and inconsistent with their predicted functions. The rice reproductive expression atlas is likely to be the most extensive and most comprehensive data set available, indispensable for unraveling functions of many specific genes in plant reproductive processes that have not yet been thoroughly analyzed (Fujita et al. 2010).

By using two line hybrid rice Liangyoupeijiu and Peiliangyou 500, the RNA and protein contents in ovaries of superior and inferior grains were analyzed. At heading, total RNA contents per unit weight in superior grain ovaries were higher than those in inferior grain ovaries, and protein content contrarily. On the fifth day after heading, total RNA contents per unit weight in superior grains ovaries were still higher, the difference of the mRNA was not significant, and protein contents per unit weight were lower compared with those in inferior grain ovaries. Whether superior grains or inferior grains, the contents of total RNA, mRNA and protein in a single grain increased from heading to the fifth day after heading, and the contents in superior grains were higher than those in inferior grains at every stage. Two protein bands were found in superior grains but none in inferior grains at heading, and only one in inferior grains at the fifth day after heading. It suggested that the gene related to grains filling in superior grains expressed earlier and more briskly than that in inferior grains (Huang and Zou 2004)

10.9. Changes in Abiotic Stress

There were 5,284 genes detected to be differentially expressed under drought stress. Most of these genes were tissue- or stage-specific regulated by drought. The tissue-specific down-regulated genes showed distinct function categories as photosynthesis-related genes prevalent in leaf, and the genes involved in cell membrane biogenesis and cell wall modification over-presented in root and young panicle. In a drought environment, several genes, such as *GA2ox, SAP15,* and *Chitinase III,* were regulated in a reciprocal way in two tissues at the same development stage. A total of 261 transcription factor genes were detected to be differentially regulated by drought stress. Most of them were also regulated in a tissue- or stage-specific manner. A *cis*-element containing special CGCG box was identified to over-present in the upstream of 55 common induced genes, and it may be very important for rice plants responding to drought environment.

Genome-wide gene expression profiling revealed that most of the drought differentially expressed genes (DEGs) were under temporal and spatial regulation, suggesting a crosstalk between various development cues and environmental stimuli.

Understanding the molecular mechanisms that underline plant responses to drought stress is challenging due to the complex interplay of numerous different genes. Here, network-based gene clustering to uncover the relationships between drought-responsive genes from large microarray datasets. 2,607 rice genes were identified that showed significant changes in gene expression under drought stress; 1,392 genes were highly intercorrelated to form 15 gene modules. These drought-responsive gene modules are biologically plausible, with enrichments for genes in common functional categories, stress response changes, tissue-specific expression and transcription factor binding sites. We observed that a gene module (referred to as module 4) consisting of 134 genes was significantly associated with drought response in both drought-tolerant and drought-sensitive rice varieties. This module is enriched for genes involved in controlling the response of the plant to water and embryonic development, including a heat shock transcription factor as the key regulator in the expression of ABRE-containing genes. These results suggest that module 4 is highly conserved in the ABA-mediated drought response pathway in different rice varieties. Moreover, our study showed that many hub genes clustered in rice chromosomes had significant associations with QTLs for drought stress tolerance. The relationship between hub gene clusters and drought tolerance QTLs may provide a key to understand the genetic basis of drought tolerance in rice (Zhang et al. 2012).

To identify cold-, drought-, high-salinity-, and/or abscisic acid (ABA)-inducible genes in rice (*Oryza sativa*), a rice cDNA microarray was prepared including about 1,700 independent cDNAs derived from cDNA libraries prepared from drought-, cold-, and high-salinity-treated rice plants. It was confirmed stress-inducible expression of the candidate genes selected by microarray analysis using RNA gel-blot analysis and finally identified a total of 73 genes as stress inducible including 58 novel unreported genes in rice. Among them, 36, 62, 57 and 43 genes were induced by cold, drought, high salinity and ABA, respectively. This observed a strong association in the expression of stress-responsive genes and found 15 genes that responded to all four treatments. Venn diagram analysis revealed greater cross talk between signalling pathways for drought, ABA and high-salinity stresses than between signalling pathways for cold and ABA stresses or cold and high-salinity stresses in rice. The rice genome database search enabled us not only to identify possible known cis-acting elements in the promoter regions of several stress-inducible genes but also to expect the existence of novel cis-acting elements involved in stress-responsive gene expression in rice stress-inducible promoters. Comparative analysis of Arabidopsis and rice showed that among the 73 stress-inducible rice genes, 51 already have been reported in Arabidopsis with similar function or gene name.Transcriptome analysis revealed novel stress inducible genes,suggesting some differences between Arabidopsis and rice in their response to stress (Rabbani et al. 2003).

The quantity of nucleic acids is reduced sharply with prolonged salinity treatment andRNA synthesis appeared to be more sensitive to salt stress than the synthesis of DNA. The activity of RNase in sensitive genotype is much higher than that of the tolerant one. There is a positive relationship between the amount of protein and RNA, while a negative correlation was found between the activity of RNase and the level of RNA. The protein to RNA ratio is similar in both tolerant and sensitive genotypes, indicating similar effectiveness of their RNA in protein production. The ratio of RNA to DNA reflecting the activity of DNA is higher in salt tolerant genotype.

Suppression of RNase activity and decrease in RNA content of endosperms were observed under salt treatment. The suppression of RNase activity was about 2 and 4 folds with 0.15 M and 0.30 M NaCl respectively. Salinity inhibited the rate of disappearance of protein from endosperm. Increased protease activity was noticed up to 48 h followed by sharp decrease during latter phases.

High hydrostatic pressure (HHP) is an extreme thermal-physical stress affecting multiple cellular activities. Recently, it was found that HHP treatment caused various physiological changes in rice. To investigation of the molecular mechanisms of plant response to HHP, we constructed forward and reverse subtracted cDNA libraries of rice seeds treated with 75 MPa hydrostatic pressure for 12h by suppression subtractive hybridization in combination with mirror orientation selection. Of 97 clones isolated through microarray dot-blot and sequenced, 45 were unique genes. Among these 45 unique cDNAs, 29 clones showed significant sequence similarity to known genes, 12 were homologous to genes with unknown function, and the remaining four clones did not match any known sequences. Most of the genes with known function were involved in metabolism, defense response, transcriptional regulation, transportation regulation, and signal transduction (Liu et al. 2008).

References

Behera, B.K. and B.N. Misra. 1983. Analysis of the effect of industrial effluent on pigments, proteins, nucleic acids, and the 2,6 dichlorophenol indophenols Hill reaction of rice seedlings. *Environ. Res.* **31:** 381-389.

Bennett, M.D. and J.B. Smith.1991. Nuclear DNA amounts in angiosperms. *Philo. Trans. Royal Soc. London* **B 334:** 309-345.

Brockelhurst, P.A. and R.S.S. Fraser. 1980. Ribosomal RNA integrity and rate of seed germination. *Planta,* **148:** 417-421.

Bushell, M., M. Stoneley, P. Sarnow and A.E. Willis. 2004. Translation inhibition during the induction of apoptosis: RNA or protein degradation? *Biochem. Soc. Trans.* **32:** 606-610.

Cai, H., Y. Lu, W. Xie, T. Zhu and X. Lian. 2012. Transcriptome response to nitrogen starvation in rice. *J. Biosci.* **37:** 731-747.

Chen, X. 2005. Micro RNA biogenesis and function in plants. *FEBS Letters,* **579:** 5923-5931.

Chuck, G., H. Candela and S. Hake. 2009. Big impacts by small RNAs in plant development. *Curr.Opin. Plant Biol.* **12:** 81-86.

Chunwongse, J., G.B. Martin and S.D. Tanksley. 1993. Pre-germination genotypic screening using PCR amplification of half seeds. *Theor. Appl. Genet.* **86:** 694-698.

De Datta, S.K. 1981. Principles and practices of rice production. John Wiley & Sons, New York.

Dure, L. and L. Waters. 1965. Long lived messenger RNA: Evidence from cotton seed germination. *Science,* **147:** 410-412.

Eldem, V., S. Okay and T. Unver. 2013. Plant micro RNAs: New players in functional genomics. *Turk. J. Agri. Forest* **37:** 1-21.

Faria, J.M., J. Buitink, A.A.M. Van Lammeren and H.W.M. Hilhorst. 2005. Changes in DNA and microtubules during loss and re-establishment of desiccation tolerance in germinating Medicago trumneatula seeds. *J. Exp. Bot.* **56:** 2119-2130.

Fujita, M., Y. Horiuchi, Y. Ueda, Y. Mixuta, T. Kubo, K. Yano, S. Yamaki, K. Tsuda, T. Nagata, M. Niihama, Kato, S. Kikuchi, K. Hamada, T. Mochizuki, T. Ishimizu, H. Twai, N. Tsutsumi and N. Kurata. 2010. Rice expression atlas in reproductive development. *Plant Cell Physiol.* **51:** 2060-2081.

Glazinska, P., A. Zienkiewicz, W. Wojciechowski and J. Kopcewicz. 2009. The putative miR172 target gene in APETALA-2 like is involved in the photoperiodic flower induction of Ipomoea nil. *J. Plant Physiol.* **166:** 1801-1813.

Hoeberichts, F.A. and E.J. Woltering. 2003. Multiple mediators of plant programmed cell death: Interplay of conserved cell death mechanisms and plant specific regulators. *Bioassays,* **25:** 47-57.

Huang, S.M. and Y.B. Zou. 2004. Difference of nucleic acid and protein metabolism in superior and inferior grains of rice. *Chin. J. Rice Sci.* **18:** 374-376.

Iise, K., C. Hongyug and W. Hugh. 2011. Inter-nucleosomal DNA fragmentation and loss of RNA integrity during seed ageing. *Plant Growth Regul.,* 63-72.

Jiang, S.Y. and S. Ramachandran. 2011. Functional genomics of rice pollen and seed development by genome-wide transcript profiling and Ds insertion mutagenesis. *Int.J. Biol. Sci.* **7:** 28-40.

Jinyu. Z., T. Risheng, Y. Timei and W. Guangnan. 1986. Dynamic changes of nucleic acids during rice development in relation to organogenesis. *Acta Agron. Sin.* **12:** 171-176.

Jones-Rhoades, M.W., D.P. Bartel and B. Bartel. 2006. Micro RNAs and their regulatory roles in plants. *Annu. Rev. Plant Biol.* **57:** 19-53.

Kimura, M. and E. Nambara. 2010. Stored and neosynthesized mRNA in Arabidopsis seeds: Effects of cycloheximide and controlled deterioration treatment on the resumption of transcription during inhibition. *Plant Mol. Biol.* **73:** 119-129.

Kranner, I., S. Birtic, K.M. Anderson and H.W. Pritchard. 2006. Glutatthione half cell reduction potential: A universal stress marker and modulator of programmed celldeath? *Free Radical Bio. Med.* **40:** 2155-2165.

Liu, X., M. Zhang, J. Duan and K. Wu. 2008. Gene expression analysis of germinating rice seeds responding to high hydrostatic pressure. *J. Plant physiol.* **165:** 1855-1864.

Mills, H.A. and J.B. Jones Jr. 1979. Nutrient deficiencies and toxicities in plants : Nitrogen. *J. Plant Nutr.* **1:** 101-122.

Matsushita, S. 1958. Studies on the nucleic acids in plants. III. Changes of the nucleic acid contents during germination stage of the rice plant. *Mem. Res. Inst. Food Sci. Kyoto Univ.* **14:** 30-32.

Miyoshi, K., T. Miyoshi and H. Siomi 2010. Many ways to generate micro RNA like small RNAs: Non-canonical pathways for micro RNA production. *Mol. Genet. Genom.,* **284:** 95-103.

Mukherji, S., B. Dey, A.K. Paul and S.M. Sircar. 1971. Changes in phosphorus fractions and phytase activity of rice seeds during germination. *Physiol.Plant.,* **25:** 94-97.

Murata, M., E.E. Roos and T. Tsuchiya. 1981. Chromosome damage induced artificial seed ageing in barley. I. Germinability and frequency of aberrant anaphases at first miotosis. *Can J. Genet. Cytol.* **23:** 267-280.

Nakabayashi, K., M. Okamoto, T. Koshiba, Y. Kamiya and E. Nambara. 2005. Genome-wide profiling of stored mRNA in Arabidopsis thaliana seed germination: Epigenetic and genetic regulation of transcription in seed. *Plant J.* **41:** 697-709.

Osborne, D.J., B.E. Roberts, P.I. Payne and S. Sen. 1974. Protein synthesis and viability in rye embryos. In : Mechanisms of regulation of plant growth. *Plant Physiol.* **12:** 805-812.

Osborne, D.J. 2000. Hazards of a germinating seed: Available water and the maintenance of genomic integrity. *Israel J. Plant Sci.* **48:** 173-179.

Padhi, B. and N.K. Chakraborti. 1984. Changes in nucleic acid contents in rice plants inoculated with Pyricularia oryzae. *J. Phytopath.* **109:** 372-375.

Pal, S., S. Jain and R.K. Jain. 2001. DNA isolation from milled rice samples for PCR based molecular marker analysis. *RGN*, **18:** 94.

Palmiano, E.P. and B.O. Juliano. 1972. Biochemical changes in the rice grain during germination. *Plant Physiol.* **49:** 751-756.

Paul, P., A. Awasthi, A.K. Rai, S.K. Gupta, R. Prasad and T.R. Sharma. 2012. Reduced tillering in Basmati rice T-DNA insertional mutant OSTEF1 associates with differential expression of stress related genes and transcription factors. *Func. Integ. Genomics*, **12:** 291-304.

Poethig, R.S. 2009. Small RNAs and developmental timing in plants. *Curr. Opin. Genet. Dev.* **19:** 374-378.

Prakash, L., M. Dutt and G. Pathapasenan. 1988. NaCl alters contents of nucleic acids, protein, polyamines and activity of agmatine deiminase during germination and seedling growth of rice (*Oryza sativa* L). *Aust. J. Plant Physiol.* **15:** 769-776.

Rabbani, M.A., K. Maruyama, H. Abe, M.A. Khan, K. Katsura, Y. Ito, K. Yoshiwara, M. Seki, K. Shinozaki and K. Yamaguchi–Shinozaki. 2003. Monitoring expression profiles of rice genes under cold, drought and high salinity stress and abscisic acid application using cDNA microarray and RNA gel blot analysis. *Plant Physiol.* **133:** 1755-1767.

Raghavan, V. 1989. mRNAs and a cloned histone gene are differentially expressed during anther and pollen development in rice (*Oryza sativa* L). *J. Cell Sci.* **92:** 217-229.

Rajendrakumar, P., K. Sujatha, K.S. Rao, P. Nataraj Kumar, B.C. Viraktamath, S.M. Balachandran, A.K. Biswal and R.M. Sundaram. 2006. A protocol for isolation of DNA suitable for rapid seed and grain purity assessments in rice. *Rice Genet. Newsletter,* **23:** 92.

Shankar, A., A. Singh, P. Kanwar, A.K. Srivastava, A. Pandey, P. Suprasanna, S. Kapoor and G.K. Pandey. 2013. Gene expression analysis of rice seedling under potassium deprivation reveals major changes in metabolism and signaling components. *PLoS ONE* doi: 10.1371/ journal.Pone. 0070321.

Sen, S. and D.J. Osborne. 1974. Germination of rye embryos following hydration dehydration treatments — enhancement of protein and RNA synthesis and earlier induction of DNA replication. *J. Exp. Bot.* **25:** 1010-1019.

Sen, S. and D.J. Osborne. 1977. Decline in ribonucleic acid and protein synthesis with loss of viability during the early hours of inbibition of rye (*S. cereale L.*) embryos. *Biochem. J.* **166:** 33-38

Sun, L.M., X.Y. Ai, W.Y. Li, W.W. Guo, X.X. Deng, C.G. Hu and J.Z. Zhang. 2012. Identification and comparative profiling of miRNAs in an early flowering mutant of trifoliate orange and its wild type by genome wide deep sequencing. *PLoS ONE* doi: 10.1371/ journal.Pone. 0043760.

Thangasamy, S., P.W. Chen, M.H. Lai, J.Chen and G.Y. Jauh. 2012. Rice LDG1 containing RNA binding activity affects growth and development through alternative promoters. *Plant J.* **71:** 288-302.

Thompson, S., J.A. Bryant and P.A. Brockelhurst. 1987. Changes in levels and integrity of ribosomal RNA during seed maturation and germination in carrot (*Daucus carota* L) *J.Exp. Bot.*, 38: 1343-1350.

Tuteja, N., M.B. Singh, M.K. Misra, P.L. Bhalla and R. Tuteja. 2001. Molecular mechanisms of DNA damage and repair: Progress in Plants. *Crit. Rev. Biochem. Mol. Biol.* **36:** 337-397.

Voinnet, O. 2009. Origin, biogenesis and activity of plant micro RNAs. *Cell*, **136:** 669-687.

Wakasa, Y., Y. Oono, T. Yazawa, S. Hayshi, K. Ozawa, H. Handa, T. Matsumoto and F. Takaiwa. 2014. RNA sequencing-medicated transcriptome analysis of rice plants in endoplasmic reticulum stress conditions. *BMC Plant Biol.* **14:** 101 doi: 10.1186/1471-2229-14-101.

Wang, J.W., B. Czech and Weigel. 2009. MiR156 regulated SPL transcription factors define an endogenous flowering pathway in Arabidopsis thaliana. *Cell*, **38:** 738-749.

Wu, L., H. Zhou, Q. Zhang, J. Zhang, F. Ni, C. Liu and Y. Qi. 2010. DNA methylation mediated by a micro RNA pathway. *Mol. Cell*, **38:** 465-475.

Wu, G., M.Y. Park, S.R. Conway, J.W. Wang, D. Weigel and R.S. Poethig. 2009. The sequential action of miR 156 and miR 172 regulates developmental timing in Arabidopsis. *Cell*, **138:** 750-759.

Xie, K., J. Shen, X. Hou, J. Yau, X. Li, J. Xiao and L. Xiong. 2012. Gradual increase of miR156 regulates temporal expression changes of numerous genes during leaf development in rice. *Plant Physiol.* **158:** 1382-1394.

Xu, C., Y. Wang, Y. Yu, J. Duan, Z. Liao, G. Xiong, X. Meng, G. Liu, Q. Qian and J. Li. 2012. Degradation of MONOCULM1 by APC/C TAD1 regulates rice tillering. *Nature comms.* 3 doi: 10.1038/ncomms 1743.

Xu, H., Y. Gao and J. Wang. 2012a. Transcriptomic analysis of rice (Oryza sativa) developing embryos using the RNA seq technique. *PLoS ONE* doi : 10.1371/journal. Pone. 0030646.

Yamaguchi, A., M.F. Wu, L. Yang, G. Wu, R.S. Poethig and D. Wagner. 2009. The micro RNA regulated SBP-box transcription factor SPL3 is a direct upstream activator of LEAFY, FRUIT FULL and APETALA1. *Dev. Cell.* **17:** 268-278.

Zhang, J., C. Li, C. Wu, L. Xiong, G. Chen, Q. Zhang and S. Wang. 2006. RMD: A rice mutant database for functional analysis of the rice genome. *Nucleic Acids Res.* (Database issue) D745-D748.

Zhang, L., S. Yu, K. Zuo, L. Luo and K. Tang. 2012. Identification of gene modules associated with drought response in rice by network based analysis. *PLoS ONE* doi :10.1371/journal pone.0033748.

Zheng, K., N. Huang, J. Bennett and G.S. Khush. 1995. PCR-based marker assisted selection in rice breeding. IRRI Discussion Paper Series No.12, IRRI, Manila, Philippines.

Zhiwu, L. and H.N. Trick. 2005. Rapid method for high quality RNA isolation from seed endosperm containing high levels of starch. *Biotechniques*, **38:** 872-876.

Zhu, Q.H., M.M. Upadhyaya, F.Gubler and C.A. Helliwell. 2009. Over expression of miR172 causes loss of spikelet determinacy and organ abnormalities in rice. *BMC Plant Biol.*, **9:** 149 doi : 10.1186/1471-2229-9-149.

Index

An environmentally friendly book printed and bound in England by www.printondemand-worldwide.com

This book is made of chain-of-custody materials; FSC materials for the cover and PEFC materials for the text pages.